AF744926

Hartley Library
UNIVERSITY OF
Southampton
One Week Loan

University of Southampton
WITHDRAWN

NanoScience and Technology

Springer
Berlin
Heidelberg
New York
Barcelona
Hong Kong
London
Milan
Paris
Singapore
Tokyo

Physics and Astronomy ONLINE LIBRARY
http://www.springer.de/phys/

NanoScience and Technology

Series Editors: P. Avouris K. von Klitzing R. Wiesendanger

The series NanoScience and Technology is focused on the fascinating nano-world, mesoscopic physics, analysis with atomic resolution, nano and quantum-effect devices, nanomechanics and atomic-scale processes. All the basic aspects and technology-oriented developments in this emerging discipline are covered by comprehensive and timely books. The series constitutes a survey of the relevant special topics, which are presented by leading experts in the field. These books will appeal to researchers, engineers, and advanced students.

Sliding Friction
Physical Principles and Applications
By B. N. J. Persson

Scanning Probe Microscopy
Analytical Methods
Editor: R. Wiesendanger

Mesoscopic Physics and Electronics
Editors: T. Ando, Y. Arakawa, K. Furuya, S. Komiyama,
and H. Nakashima

Biological Micro- and Nanotribology
Nature's Solutions
By M. Scherge and S. N. Gorb

Series homepage – http://www.springer.de/phys/books/nst/

Matthias Scherge Stanislav N. Gorb

Biological Micro- and Nanotribology

Nature's Solutions

With 164 Figures

Springer

Dr. Matthias Scherge
Head of Basic and Tribology Research
IAVF Antriebstechnik AG
Im Schlehert 32
76187 Karlsruhe, Germany

Dr. Stanislav S. Gorb
Biologische Mikrotribologie Gruppe
MPI für Entwicklungsbiologie
Spemannstrasse 35
72076 Tübingen, Germany

Series Editors:
Professor Dr. Phaedon Avouris
IBM Research Division, Nanometer Scale Science & Technology
Thomas J. Watson Research Center, P.O. Box 218
Yorktown Heights, NY 10598, USA

Professor Dr., Dres. h. c. Klaus von Klitzing
Max-Planck-Institut für Festkörperforschung, Heisenbergstrasse 1
70569 Stuttgart, Germany

Professor Dr. Roland Wiesendanger
Institut für Angewandte Physik, Universität Hamburg, Jungiusstrasse 11
20355 Hamburg, Germany

ISSN 1434-4904
ISBN 3-540-41188-7 Springer-Verlag Berlin Heidelberg New York

Library of Congress Cataloging-in-Publication Data.

Scherge, Matthias 1962– . Biological micro- and nanotribology / Matthias Scherge and Stanislav N. Gorb. p.cm. – (Nanoscience and technology). Includes bibliographical references and index. ISBN 3-540-41188-7 (alk. paper). 1. Tribology. 2. Nanotechnology. I. Gorb, Stanislav N., 1965– . II. Title. III. Series. TJ1075.S348 2001 621.8'9–dc21 00-066117.

This work is subject to copyright. All rights are reserved, whether the whole or part of the material is concerned, specifically the rights of translation, reprinting, reuse of illustrations, recitation, broadcasting, reproduction on microfilm or in any other way, and storage in data banks. Duplication of this publication or parts thereof is permitted only under the provisions of the German Copyright Law of September 9, 1965, in its current version, and permission for use must always be obtained from Springer-Verlag. Violations are liable for prosecution under the German Copyright Law.

Springer-Verlag Berlin Heidelberg New York
a member of BertelsmannSpringer Science+Business Media GmbH

http://www.springer.de

© Springer-Verlag Berlin Heidelberg 2001
Printed in Germany

The use of general descriptive names, registered names, trademarks, etc. in this publication does not imply, even in the absence of a specific statement, that such names are exempt from the relevant protective laws and regulations and therefore free for general use.

Typesetting: Camera-ready by the authors using a Springer TEXmacropackage
Cover design: *design & production*, Heidelberg

Printed on acid-free paper 57/3111/tr - 5 4 3 2 1

UNIVERSITY OF SOUTHAMPTON LIBRARY

TJ 1045 SCH
00534178

Foreword

Ever since the genesis of life, and throughout the course its further evolution, Nature has constantly been called upon to act as an engineer in solving technical problems. Organisms have evolved a variety of well-defined shapes and structures. Although often intricate and fragile, they can nonetheless deal with extreme mechanical loads. Some organisms live attached to a substrate; others can also move, fly, swim and dive. These abilities and many more are based on a variety of ingenious structural solutions. Understanding these is of major scientific interest, since it can give insights into the workings of Nature in evolutionary processes. Beyond that, we can discover the detailed chemical and physical properties of the materials which have evolved, can learn about their use as structural elements and their biological role and function. This knowledge is also highly relevant for technical applications by humans.

Many of the greatest challenges for today's engineering science involve *miniaturization*. Insects and other small living creatures have solved many of the same problems during their evolution. Zoologists and morphologists have collected an immense amount of information about the structure of such living micromechanical systems. We have now reached a sophistication beyond the pure descriptive level. Today, advances in physics and chemistry enable us to measure the adhesion, friction, stress and wear of biological structures on the micro- and nanonewton scale. Furthermore, the chemical composition and properties of natural adhesives and lubricants are accessible to chemical analysis. And not least, computer-aided reconstructions can be exploited to mimic biomechanical systems in the workshop of the engineer.

The present book by Scherge and Gorb is devoted to biomechanical systems with frictional surfaces or adhesive secretions to attach parts of the body to each other, or to attach the organism to a substrate. Such systems have fascinated microscopists for centuries, but only recently has the use of novel methods enabled us to really understand the functioning of at least some of these systems. This book is an excellent example of interdisciplinary science in action, drawing as it does on approaches from physics, engineering, tribology, structural biology, and materials science. The text is not confined to the broad variety of natural devices adapted for attachment: In order to show the different principles of morphology, ultrastructure and biomechanics of friction systems, it also presents several examples of experiments and outlines the general rules governing the interrelationship between the design of biological structures, their properties, and their functions. This account is based on knowledge accumulated in long-term studies of the principles of microtribology *(Scherge)* and ultrastructural and experimental studies of insect attachment devices *(Gorb)*. It provides a first-class introduction to the principles of biological micro- and nanotribology. Together with the data on ultrastructure, material properties, and attachment–detachment performances of biological frictional systems, the book is an excellent starting point for engineers and physicists working with biological systems and for biologists starting to study friction and adhesion. The information presented will undoubtedly also be useful in areas of high-technology, such as micromechanics and the science of composite materials.

Max Planck Institut für
Entwicklungsbiologie, September 2000

U. Schwarz

Preface

This book is intended for scientists and engineers who want to become familiar with biological micro- and nanotribology as a new interdisciplinary field of research combining methods and knowledge of physics, chemistry, mechanics and biology. The aim of biological micro- and nanotribology is to gather information about friction, adhesion and wear of biological systems and to apply this new knowledge to the design of Micro–Electro–Mechanical Systems, the development of monolayer lubricants, the development of new adhesives, or the construction of artificial joints.

The book is divided into five main sections, where the first part is titled **Basics and Physical Tools**, the second part **Biological Friction Systems**, the third part **Test Equipment**, and the fourth part **Case Studies**. The last part contains the **Appendix**.

The first part deals with basic subjects such as microfriction and adhesion theory and measurement, and the applicability and usage of contact models, as well as lubrication. Especially emphasized is the action of thin liquid films and their influence on adhesion and friction. A detailed study is devoted to water, since this is the most common adsorbate film. The tribological properties are explained using a silicon ball sliding on a flat silicon sample. This system is treated as a model and reference system to provide a platform on which the data obtained from the biological tests can be interpreted.

The second part deals with explicit studies of biological friction and adhesion phenomena, where the fundamental problems are discussed. An overview on biological friction and adhesion systems is given. From studies of biological macrotribology (human joints, etc.), the field of biological microtribology will be defined. As they appear in the text, the biological terms are explained. For more information the reader can make use of the glossary at the end of the book.

In the third part technical equipment is introduced. A detailed description is devoted to novel test equipment to take account of recent major advances in measurement and characterization. Modern instruments are introduced, including laser interferometer- and fiber-optics-based microfriction and adhesion testers and scanning probe microscopes. This part also focuses on biological preparation techniques applied to maximize the outcome of the experiments. In addition, supporting technical analysis methods like microscopy

(SEM, TEM), chemical analysis (for example photoelectron spectroscopy) and thin film characterization are discussed.

Topics discussed in the fourth part of the book are special case studies performed with the great green bush cricket (*Tettigonia viridissima*). This part combines the knowledge of Parts I, II and III to analyze friction, indentation and adhesion between the attachment pads of the cricket and various kinds of substrate. Attention is paid also to the problems of sample aging and stability of the test parameters.

Biologists, chemists, physicists and tribologists and many other applied scientists will find this book a useful addition to their libraries. Moreover, this book is intended as an introduction to the higher levels of micromechanical analysis and is well suited for, for example, graduate students who want to become active in interdisciplinary research. The majority of the sections close with an itemized summary of the most important results.

We wish to express our gratitude to I.U. Ahmed, J. Berger, S. Granick, H.J. Haefke, J. Halbritter, Y. Jiao, V. Kastner, J.A. Schaefer, M. Schoen, U. Schwarz and H. Schwarz for their fruitful discussion and help. Additional thanks go to A. Lahee and C. Ascheron of Springer-Verlag as well as to G. Hartung, G. Hungenbach, X. Li, A. Opitz and R. Nevchoupa for improving the text. We would also like to thank K. von Klitzing and R. Wiesendanger for giving us the opportunity to publish in the "NanoScience and Technology" series.

Special thanks go to our families for their love and support.

Karlsruhe *M. Scherge*
Tübingen *S. N. Gorb*
January 2001

Contents

Part IV. Case Studies

Part V. Appendix

Part I

Basics and Physical Tools

1. Introduction

Micro- and nanotribology – considered as the mechanical interaction of moving bodies – is the science of friction, adhesion, lubrication and wear on the length scale of micrometers to nanometers and the force scale of millinewtons (mN) to nanonewtons (nN). This rather young field of (micro- and nano-) science was boosted by the advent of new analysis techniques, for example the Atomic Force Microscope (AFM), allowing insight into fundamental processes. In the microrange the data storage industry used tribology to solve problems concerning the head–disk–interface in magnetic disk drives. Due to miniaturization of the mechanical parts the forces at the interface decreased dramatically and have reached the nanorange for the case of IBM's millepede, an AFM-based storage device [1]. More and more conventional mechanical devices, for instance motors [2], are becoming miniaturized. Former mesoscale applications like accelerometers in cars are being replaced by silicon-based mechanical microsystems (see [3] for an overview).

Biological insect microsystems and technical microsystems have many things in common. First of all, the mechanical interaction occurs on identical length and force scales [4,5]. In both types of system surface properties, for example wettability, microstructure or surface chemistry, have a strong impact on the performance of the system. In a few cases biological design examples were used for technical systems, as shown in Fig. 1.1 [6]. Here, the technical microsystem adopts the friction-induced motion of a worm. Internal microdrives accomplish a harmonica-like motion of the "techno-insect".

Whereas the mechanical properties of technical microsystems with respect to friction and adhesion are mainly determined by the physico-chemical state of the topmost atomic layers of the solid, biological microsystems are also strongly influenced by contributions of the bulk due to their elastic and viscoelastic constitution.

The other main difference between biological and technical microsystems is their performance. While insect micromechanics always operates in a lubricated state, no reliable concepts exist for technical microsystems. For example, the joints of the cricket, shown in Fig. 1.2, are lubricated. In contrast, the application of a liquid to a micromechanical gear system would prevent the whole system from operating, since capillary action induces forces higher

Fig. 1.1. Perhaps the world's smallest artificial worm. With a length of about 1.6 cm and a diameter of 2 mm, the worm contracts periodically and uses friction to move. The worm could be used to investigate the interior of arteries (courtesy of D. Riemer, photograph provided by V. Steger).

than the forces generated by the Micro-Electro-Mechanical System (MEMS). Yet, without lubrication MEMS are prone to high friction and wear.

Another important field of research is the prevention of adhesion between microparts requiring laborious surface treatments in MEMS technology [7]. Whereas adhesion is perfectly controlled by an insect to accomplish the attachment to smooth surfaces, micromanipulation and microassembly [8] (for instance with micrograppers) are severely obstructed by the water-induced adhesion [9].

Insect systems repel water by design, thus eliminating unwanted adhesion and friction. However, when necessary – for example for attachment – adhesion and friction are mediated by hydro-carbon liquids produced by the insect. In addition, insect mechanics uses deformable mechanical components to achieve intimate mechanical contact with the substrate. Many detachment processes are triggered by a quick change of the mechanical form. In contrast, technical microsystems mainly rely on stiff and undeformable mechanical components. So far, evolutionary optimized biological systems are scarcely adapted to technical applications. This is mainly caused by insufficient comprehension of microbiological mechanisms.

Therefore, overcoming micro- and nanotribological problems will determine whether MEMS will change our daily life as real-world applications or remain as models in the laboratories of microtechnology [10,11]. In the text the authors use the term *microsystem* for both biological and technical applications. Molecular systems are not the topic of this book; for more information see [12].

Fig. 1.2. *Tettigonia viridissima.* This insect is used in Part IV for friction, adhesion and indentation experiments.

The mechanisms explained in this book occur exclusively on the microscale and it is the aim of this book to explain the tribological properties on this scale [13]. However, in order to understand the fundamental properties on the microtribological scale it is necessary to deal also with nanoscale properties. The distinction between micro and nano is arbitrary, since the phenomena cover both length and force scales. For the purpose of systematization the authors adapt the definitions given in [14].

Friction is considered

- atomic, if the scan size is of the order of 1 nm $\times$ 1 nm,
- nano, if the scan size is tens of 1 nm times tens of 1 nm,
- micro, if the scan size ranges from 1 μm $\times$ 1 μm and upwards.

Indentation and adhesion are considered

- atomic, if the force is of the order of 1 nN,
- nano, if the force ranges over tens of nN,
- micro, if the force ranges from 1 μN to hundreds of μN.

Where applicable, the findings in the micro- and nanorange are compared with results obtained in macroscale experiments. This is accomplished in a condensed manner and the reader is referred to the literature, given in the text.

2. Physical Principles of Micro- and Nanotribology

In this chapter the physical principles are described mainly by presenting results of experiments performed with a model system comprising a silicon ball sliding on a flat silicon sample (referred to as the SILICON model system). Friction, lubrication and wear problems are discussed to obtain a comprehensive picture of tribology and adhesion. In addition, mechanical properties such as roughness, hardness, elasticity, viscoelasticity and contact theory are also introduced.

2.1 Fundamental Definitions

The fundamental definitions will be discussed using the SILICON model system. This setup is treated as the reference system [15–24]. In each experiment

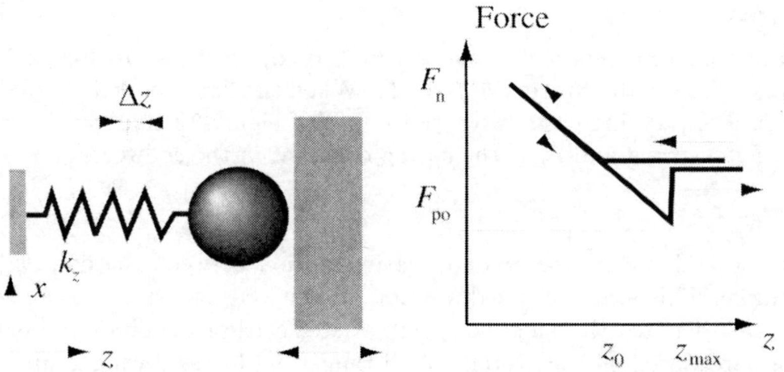

Fig. 2.1. Basic setup for adhesion experiments, left-hand side. A flat sample is moved towards a ball supported by a spring of known stiffness (k_z). In this figure only the contribution in the z-direction is shown. The right-hand part of the figure shows a force–distance curve to measure indentation and adhesion (pull-off force F_{po}). At z_0 the two samples make contact. Any further compression of the spring increases the normal force. This initial part of the curve describes the indentation. Upon retraction the normal force decreases. The extension of the curve to negative force values depends on the adhesive properties of the materials.

the ball is attached to a spring to determine the force arising either from adhesion or friction. For decoupled measurements of normal and tangential forces a double-leaf spring is used.

Figure 2.1 shows the basic setup for adhesion measurement. When the flat sample is moved in the $-z$-direction, the spring is compressed and a normal force F_n is generated according to

$$F_n = -k_z \Delta z \ , \tag{2.1}$$

where Δz is the difference of the absolute z-values in compressed and initial positions of the spring and k_z is the spring constant in the z-direction. In macrotribological experiments instead of using a spring the normal force is commonly applied by weights acting on a lever. However, to generate a normal force of 1 μN a piece with a mass of 0.1 μg would have to be attached. This is very difficult, since such a tiny piece is hard to handle. Furthermore, the mechanics that guides the lever would absorb the potential energy by friction and inertia.

When the motion of the flat sample is reversed, the spring expands until the initial position is reached. Provided that there is adhesion between ball and flat the spring expands further. The contact finally breaks at z_{max}, when the force of the spring equals the attracting force. This value is used to determine the adhesion force from (2.1). In the literature this force is also called the pull-off force [25], since it combines several influences contributing to adhesion. F_{po} acts in the same direction as the normal force

$$F_{po} = -k_z(z_{max} - z_0) \ . \tag{2.2}$$

For friction experiments a different setup is used, as shown in Fig. 2.2. Equation (2.3) defines the tangential force F_t. When the flat sample is moved in the x-direction, F_t increases with the opposite sign. Δx manifests the expansion of the spring and k_x is the spring constant in the x-direction

$$F_t = -k_x \Delta x \ . \tag{2.3}$$

Between $x = 0$ and $x = x_s$ there is no relative motion between the ball and the flat sample. This state is called stiction. When the ball starts to slide at $x = x_s$, F_{fs} denotes the singular point of static friction. The dynamic friction F_{fd} is regarded as the average of all tangential forces during sliding. Figure 2.2 shows an ideal tangential force curve. In reality the part of the curve succeeding x_s is noisy or contains friction-induced instabilities

$$F_{fs} = -k_x x_s \ , \tag{2.4}$$

$$F_{fd} = -\frac{1}{N} k_x \sum_{i=1}^{N} x_i = -\frac{1}{N} \sum_{i=1}^{N} F_{t_i} \ . \tag{2.5}$$

Here N is the number of samples taken between x_s and x_{max}. Below x_s the ball and the flat sample stick and move at the same velocity. After exceeding

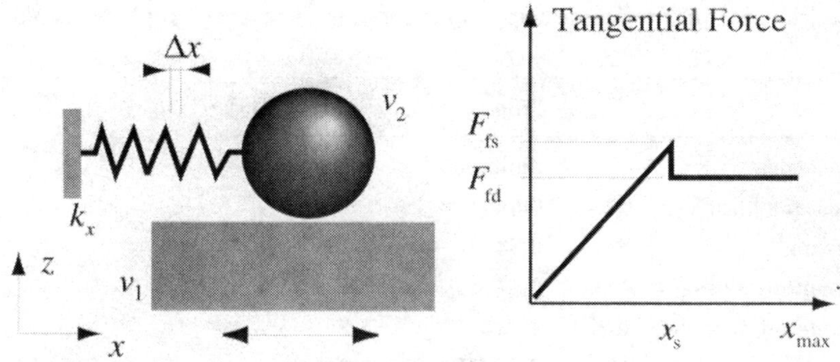

Fig. 2.2. Basic setup for friction experiments. The tangential force is the product of the spring deflection in the x-direction and the stiffness k_x. The normal force is exerted by the spring using its stiffness in the vertical direction. The friction force can basically be divided into a static F_{fs} and a dynamic F_{fd} contribution. v_1 is the velocity of the oscillating flat sample. v_2 is the friction-induced velocity of the ball.

the static friction force the velocity of the ball becomes zero for the case of fluctuation-free sliding. The relative sliding velocity v_{rel} is equal to v_1. For the case of fluctuations, v_{rel} ranges between zero and v_1 ($v_{rel} = v_1 - v_2$).

Equations (2.4) and (2.5) when divided by the normal force give the static and dynamic coefficient of friction, μ_s and μ_d, respectively. However, throughout the book the coefficient of friction is not used as a descriptive parameter. On the macroscale μ can be assigned a general value, since the subtle influences of single contact points (asperities) are averaged due to large dimensions and high normal forces. However, on the microscale μ hides intrinsic properties or is not applicable in the case of fluctuations.

2.2 Model Materials

As shown in the previous section the SILICON model system was introduced as the reference system for the biological tests. This section focuses on the selected materials and their physical properties. For the ball, a 1 mm sphere of polished silicon was used (Ball Semiconductor Inc.). The balls are made from material originally used as substrate material for integrated chip manufacturing, thus having identical mechanical properties like regular wafer silicon used as the flat counter sample. Since silicon is always covered by an oxide, Table 2.1 shows the physical properties of silicon and its oxide (silica). Bare silicon exposed to air at room temperature (relative humidity, r.h.=42%) develops an oxide film of about 0.5 nm after one day [26].

For both materials an abundance of experimental data exists covering the whole range from mechanical to electronic properties, for example hardness [27] or lattice vibrations [28,29]. Silicon has been used for many years as the

Table 2.1. Physical properties of silicon and silica used in the SILICON model system

Property	Silicon	Silica (SiO_2)
crystal structure	diamond	amorphous
lattice constant [nm]	0.543	–
density [g/cm^3]	2.33	2.2
average roughness [nm]	0.17	0.2
thermal conductivity [W/(m·K)]	1.412	1.2
thermal expansion [$\times 10^{-6}$/K]	2.0 (RT)	0.9
electrical resistivity [Ω·cm]	0.1	10^{11} – 10^{12}
Young's modulus [GPa]	150	73
Poisson ratio	0.28	0.17 [a]
mineral hardness (Mohs)	6.5	7
Hamaker constant [J]	2.6×10^{-19} [b]	8.3×10^{-21} [c]

[a] S. Spinner, J. Am. Ceram. Soc., **45**(1962)394
[b] H. Krupp, Adv. Colloid Interface Sci., **1**(1967)111
[c] for two silicon oxide surfaces in water

base material for integrated circuits. Thus, not only the material itself but also many procedures to modify the surface of the material are well known [30–32]. With a ball diameter of 1 mm a contact area of about 90 μm^2 is formed with the oxide-covered silicon sample (the calculation of the contact area is shown in Sect. 2.3.4). This leads to a contact pressure of 5 MPa for a normal force of 500 μN. Since this pressure is about 14,000 times lower than the moduli of elasticity, the mechanical degradation of the solid surface can be neglected.

The properties of the SILICON model system can be summarized as follows:

- The setup can be used to measure friction, adhesion and indentation.
- A simple ball on flat sample setup allows the decoupled determination of normal and tangential forces due to the use of a double-leaf spring.
- Both ball and flat samples are commercially available products with approximately identical physical properties.
- Silicon is one of the most extensively investigated materials. Not only is the material well known but also the procedures to prepare and structure the Si surface are established standards.
- Due to moderate contact pressure mechanical degradation of the material can be neglected.

2.3 Mechanical Properties

2.3.1 Roughness

To characterize a surface in terms of roughness several widely accepted parameters are defined. Figure 2.3 shows a roughness profile of a surface. The average roughness R_a represents a ratio of area above and below an imaginary line of length l that is unity. Since absolute values are taken, the R_a horizon is above this line:

$$R_a = \frac{1}{l} \int_{x=0}^{x=l} |z(x)| \mathrm{d}x \, . \tag{2.6}$$

The RMS (root-mean-square) roughness is the square root of the standard deviation of R_a:

$$RMS = \sqrt{\frac{1}{l} \int_{x=0}^{x=l} [\bar{z} - z(x)]^2 \mathrm{d}x} \, , \tag{2.7}$$

where $\bar{z}$ is the mean value of all $z(x)$. Very often, especially in scanning probe microscopy, the peak-to-valley distance is also given:

$$z_{pv} = |z_{max} - z_{min}| \, . \tag{2.8}$$

A great portion of work has been devoted to the modeling of roughness and numerical methods [33]. There are basically two approaches to this problem [34–36]. The first approach deals with a stochastic distribution of asperities. The height distribution is described by the standard deviation and skewness. The spatial distribution can be described by means of a power spectrum reflecting the spectral variation of roughness features. The spatial frequency corresponds to an inverse length. Therefore, when a surface is characterized by repeating patterns (for example a surface-textured material), these features are represented by distinct peaks in the power spectrum. If the surface

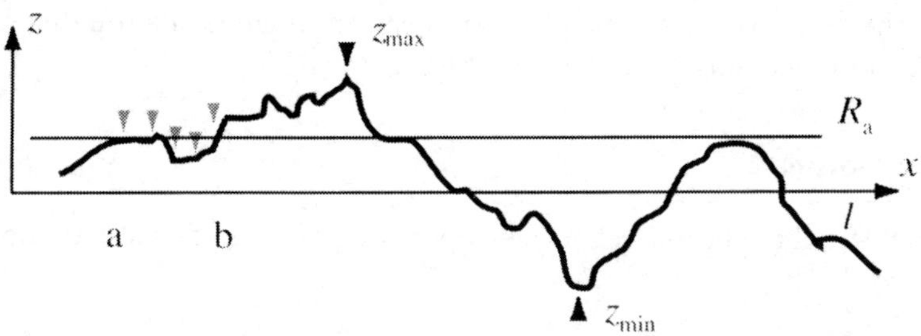

Fig. 2.3. Roughness measurement and definitions. The roughness values depend on the sampling interval between the points (a) and (b). In order to obtain proper values, the size of the scanning probe has to be appropriate to the size of the features that are to be scanned in order to minimize errors.

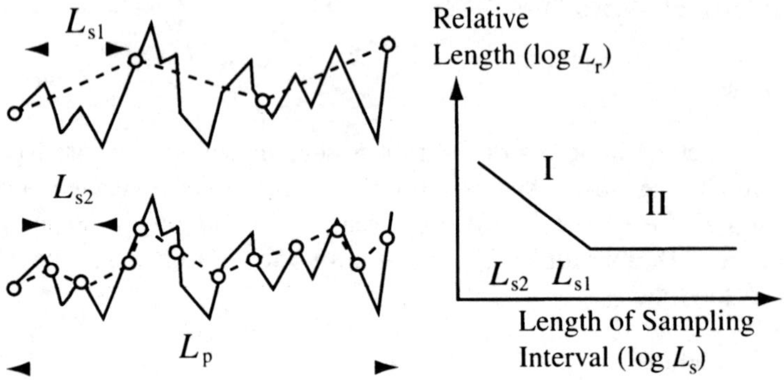

Fig. 2.4. Fractal analysis of a roughness profile. According to the length of the sampling interval the profile is approximated with different sampling errors.

is composed of completely random structures, then the spectral density only shows a noisy background.

The second approach, the fractal approach [37], attempts to describe a surface with scale-invariant and self-similar basic features. It assigns to a surface a fractal dimension with values between two for an area and three for a volume. A strong fact that supports this kind of approach is that measured values of roughness depend on the size of the probe that is used to scan the profile [38]. The result are either low- or high-frequency cut-offs. This is shown in Fig. 2.4. A surface profile is sampled in different sampling intervals, L_{s1} and L_{s2}. The relative length can be calculated by:

$$L_r = \frac{nL_s}{L_p} , \tag{2.9}$$

where n is the number of samples and L_p is the projected length. When the relative length is plotted versus the length of the sampling interval, two distinct ranges (I and II) can be detected. The slope of the curve in range I is, thereby, equal to the fractal dimension. The point where range I and II meet denotes the cross-over from rough to smooth. In range II the roughness does not depend on the length of the sampling interval.

2.3.2 Elastic Bodies

Hooke's law reflects the relation between stress σ and strain ε of elastic bodies in form of:

$$\sigma = \varepsilon E , \tag{2.10}$$

where E is the modulus of elasticity (Young's modulus), which has the dimension of a pressure (Pa). The stress is calculated by dividing the normal force by the geometric contact area A_g that it acts on:

$$\sigma = \frac{F_n}{A_g} \ . \tag{2.11}$$

Note that A_g is always larger than the real contact area A_c (see Fig. 2.7). This means that the stress in the actual contact area can be much larger. The strain is expressed as the dimensionless ratio of the difference in length Δl of a strained body and its initial length l_0:

$$\varepsilon = \frac{\Delta l}{l_0} \ . \tag{2.12}$$

2.3.3 Viscoelastic Bodies

The behavior of a material is called viscoelastic if it combines flexible deformations according to Hooke's law with Newtonian viscous flow. The elastic deformation applies to solid bodies within a certain range only. Basically, a deformation can be called inelastic, if it cannot be described by Hooke's law. Such deformations can be caused, for example by changes of temperature or by exceeding of the yield stress when mechanical load results in irreversible modifications of the crystalline structure. Many materials do not return to their initial state after termination of the acting force or temperature. Moreover, repeated deformation processes can lead to material fatigue. Most, if not all biological tissues possess viscoelastic behavior [39].

Viscoelastic deformation is described commonly by spring-dashpot models as shown in Fig. 2.5. Depending upon the arrangement a different mode results in response to the acting force. The simplest of these models is the Maxwell model, which consists of a spring and an absorber filled with oil (viscosity η). In the model the strain

$$\varepsilon = \varepsilon_{el}(t) + \varepsilon_{vis}(t) = \text{const.} \ , \tag{2.13}$$

which consists of a time-dependent (t) elastic and viscous contribution, is kept constant. From the Hookean behavior of the spring

$$\varepsilon_{el}(t) = \frac{\sigma(t)}{E} \ , \tag{2.14}$$

and the Newtonian behavior of the dashpot

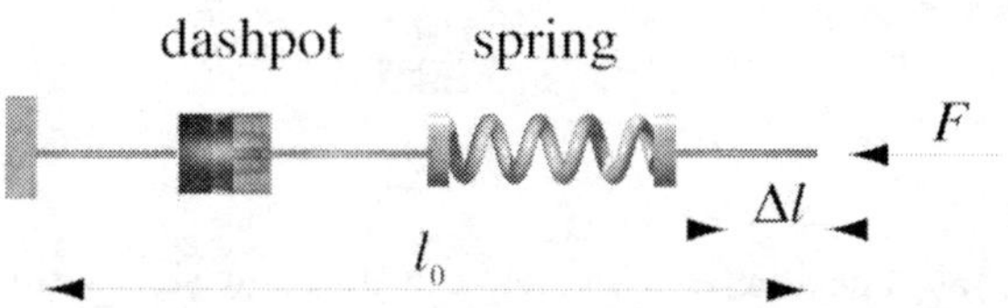

Fig. 2.5. Basic viscoelastic model elements. The dashpot is a shock absorber filled with a viscous liquid. When the force F acts, the whole structure is compressed by the length Δl.

$$\varepsilon_{\mathrm{vis}}(t) = \frac{1}{\eta}\int \sigma(t)\mathrm{d}t \;, \tag{2.15}$$

one obtains the following equation after differentiation ($\mathrm{d}\varepsilon/\mathrm{d}t$):

$$\frac{\mathrm{d}\sigma(t)}{\mathrm{d}t} + \frac{E}{\eta}\sigma(t) = 0 \;, \tag{2.16}$$

with the solution:

$$\sigma(t) = \sigma_0 \mathrm{e}^{\frac{-Et}{\eta}} = \sigma_0 \mathrm{e}^{\frac{-t}{t'}} \;, \tag{2.17}$$

where $t' = \eta/E$ is called the relaxation time and σ_0 is the initial stress. In analogy to the behavior of an ideal spring, where an applied force induces a compression, the behavior of ideal liquids is described according to the principles of hydrodynamics. This means that an applied force is proportional to the deformation rate, however independent of the deformation.

Figure 2.6 summarizes three fundamental models describing the viscoelastic behavior of materials. In the Maxwell model, Fig. 2.6a, a force acts simultaneously on spring and dashpot, where the spring responds immediately, while the dashpot slowly follows leading to linearly increasing strain. During relaxation – when the strain is kept constant – the stress decreases exponentially. In the Kelvin–Voigt model, Fig. 2.6b, the acting force is distributed

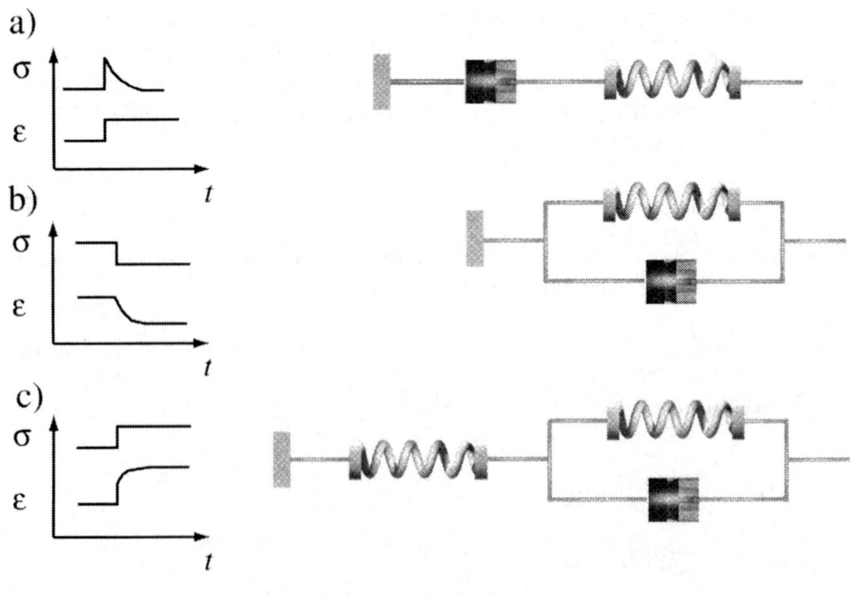

Fig. 2.6. Viscoelastic models. (**a**) The Maxwell model: the diagram shows the stress relaxation to applied strain. (**b**) The Kelvin–Voigt model: strain relaxation as response after stopping the stress. (**c**) The Thompson model: strain response to applied stress.

over both elements. A force that is suddenly applied leads to a delayed response of the system. When the force ceases to act the system returns to its initial state. The standard model that describes the behavior of a viscoelastic body is the model of Thomson Fig. 2.6c, which consists of a Kelvin–Voigt element connected in series with an additional spring. This model combines the dynamic characteristics of the two other models. Upon loading, the serial spring causes instant elasticity followed by creep. During relaxation the system shows asymptotic behavior.

These simple models, while illustrating stress relaxation might not be sufficient to describe complex biological materials. In order to describe those systems the three models need to be combined. It is obvious that the stress depends not only on deformation and temperature but also on their history:

$$\sigma(t) = f(\text{history of deformation, history of temperature}) \,. \tag{2.18}$$

This is the general constitutive equation for a viscoelastic material [40].

Viscoelasticity is characterized by the following features:

- Relaxation: a material, which was subjected to an instantaneous deformation and is kept in this state, shows a decrease of the internal stress.
- Creep: in a material, which remains under constant stress, a progressive deformation can be detected.
- Hysteresis: a material, which is stressed and relieved continuously, possesses a modulus of elasticity that differs for the loading and unloading phase.

2.3.4 Contact Mechanics

If two bodies contact each other in a point or a line, then the action of the compressive forces results in deformation. As a consequence, the real area of contact changes as shown in Fig. 2.7. This, of course, has a strong impact on the adhesion and friction. Contact mechanics represents a sophisticated synthesis of elasticity theory, fracture mechanics and surface science. The mechanical contact between solid bodies can be described quite well with continuum models, although they neglect atomic details. Most of the existing contact models are based on the assumption of the contact of ideally smooth spheres with radii r_1 and r_2. The radius of curvature R is expressed by

$$\frac{1}{R} = \frac{1}{r_1} + \frac{1}{r_2} \,. \tag{2.19}$$

When one of the radii goes to infinity, then the contact of sphere and flat sample can be calculated. The deformation (flattening or indentation) is considered as being small in relation to the dimensions of the sphere. Recent models also incorporate the effect of roughness as well as the action of attractive forces inside the contact and in the vicinity of the contact radius.

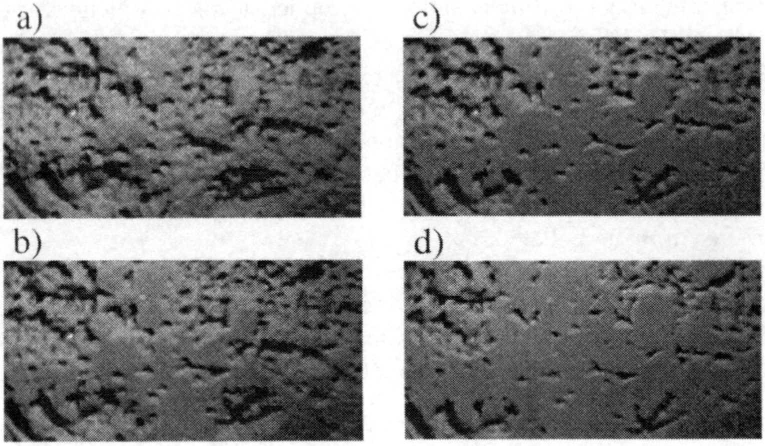

Fig. 2.7. Contact area as a function of increasing normal force [from (**a**) to (**d**)] demonstrating that the real contact area (when compared with the geometric contact area) is only a fraction of the geometric contact area. Contact models treat the asperities as spheres. A real surface can be simulated by a finite number of asperity contacts.

A comprehensive treatment of contact mechanics can be found in [36]. Contact mechanics with respect to tribology is treated in [41]. Departing from mechanical contacts without adhesion, contact models considering adhesive contributions and extended models are presented. The contact problem in this book is confined to our SILICON model system. However, on this base an adaptation to biological systems is possible.

The Hertz Model

The theory of Hertz [42] is based on Boussinesq equations and assumes that the material is homogeneous and isotropic and that Hooke's law (see Sect. 2.3.2) applies. This theory considers deformations of both spheres, so that a finite contact area is formed. For the case of the SILICON model system ($r_2 \to \infty$) the contact area A_c is given by:

$$A_\mathrm{c} = \pi \left(\frac{RF_\mathrm{n}}{K} \right)^{\frac{2}{3}} . \qquad (2.20)$$

The radius of curvature reduces to the radius of the ball, $R = r_1$, and K is the effective elastic modulus according to

$$\frac{1}{K} = \frac{3}{4} \left(\frac{1-\nu_1^2}{E_1} + \frac{1-\nu_2^2}{E_2} \right) , \qquad (2.21)$$

where ν_i are the Poisson ratios and E_i are the Young's moduli of the SILICON model system. The ball is flattened by the distance δ

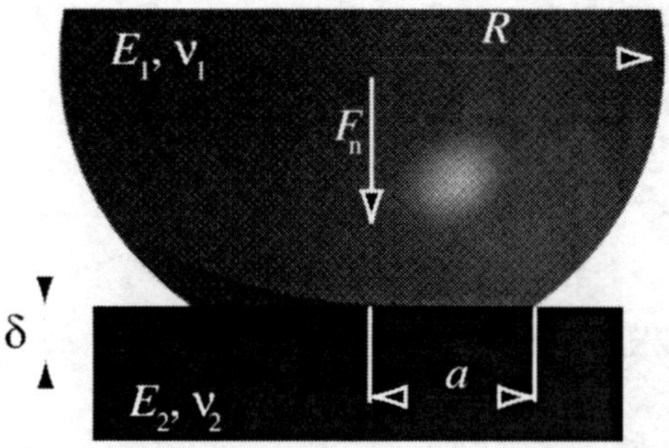

Fig. 2.8. Hertz model to calculate the common contact area of a ball and a flat sample under compression. Both ball and flat are considered ideally smooth. The indentation δ has to be small compared with the diameter of the ball.

$$\delta = \frac{a^2}{R} , \tag{2.22}$$

with a the radius of the contact area; see Fig. 2.8. Without applied normal force this radius disappears and the ball returns its initial form, no hysteresis effects occur. The theory can be applied to dry surfaces where adhesion is neglected. As will be shown in the following sections, capillarity is the dominating contribution to friction in the microrange. Therefore, besides the exact knowledge of the deformation the area of the solid that is immersed in and wetted by the liquid must also be considered in order to determine the resulting contact area correctly. This is of great importance, since in the microrange the capillary force equals or is larger than the applied normal force. As an extension of the Hertz model the generalized Hertz model particularly incorporates strong capillary forces. The model considers adhesion by means of an additional constant. This is the force that occurs at the breaking of the capillary bridge. The model has to be solved numerically and the reader is referred to the literature [43].

The Johnson–Kendall–Roberts Model

As an improvement of the Hertz model the theory of Johnson, Kendall and Roberts includes adhesive contributions in the contact area [36,44]. In contrast to the Hertz model the contact area increases due to interfacial energy minimization (see Fig. 2.9):

$$A_{\mathrm{c}} = \pi \left[\frac{R}{K} \left(F_{\mathrm{n}} + 6\pi\gamma R + \sqrt{12\pi\gamma R F_{\mathrm{n}} + (6\pi\gamma R)^2} \right) \right]^{\frac{2}{3}} , \tag{2.23}$$

with γ the interfacial energy, representing the action of adhesion. When no normal force is applied (F_{n}=0) a finite value of the contact radius a_0 remains and is given by:

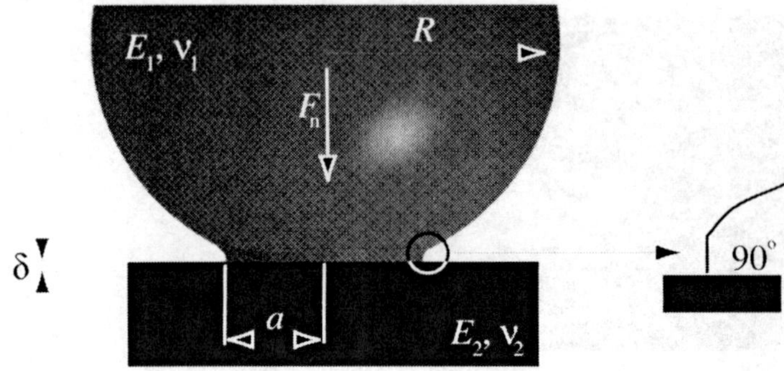

Fig. 2.9. JKR model to calculate the contact area of the ball and the flat with adhesive properties. The action of adhesion changes the shape of the contact region leading to the problem of infinite stress at the edge of the contact area at $x = a$.

$$a_0 = \left(\frac{12\pi\gamma R^2}{K}\right)^{\frac{1}{3}} . \tag{2.24}$$

The indentation is equal to

$$\delta = \frac{a^2}{R} - \sqrt{\frac{16\pi\gamma a}{3K}} . \tag{2.25}$$

Even in the absence of a normal force, the ball sticks to the surface. Furthermore, there are solutions of (2.23) for negative normal forces. When the contact breaks, the negative normal force equals the pull-off force, which can be calculated from

$$F_{\text{po}}^{\text{JKR}} = -3\pi\gamma R . \tag{2.26}$$

Thus, the JKR model incorporates also hysteresis effects. A deficiency of the JKR theory is that at the edge of the contact area it predicts an infinitely high stress [45]. This is caused by the fact that the short-range attractive action outside of the contact area is zero; see Fig. 2.10.

The Derjaguin–Muller–Toporov Model

The theory developed by Derjaguin, Muller and Toporov (DMT) [46–48] considers also adhesion with long-range forces. This means that also outside of the contact area attraction exists. The interaction is approximated by Lennard–Jones potentials. The model assumes that the surfaces in the contact deform due to attractive forces. This deformation is calculated by taking the Hertzian approach. The stress distribution in the contact is shown in Fig. 2.10. The pull-off force between the ball and the flat sample of the SILICON model system can be described by

$$F_{\text{po}}^{\text{DMT}} = -4\pi\gamma R . \tag{2.27}$$

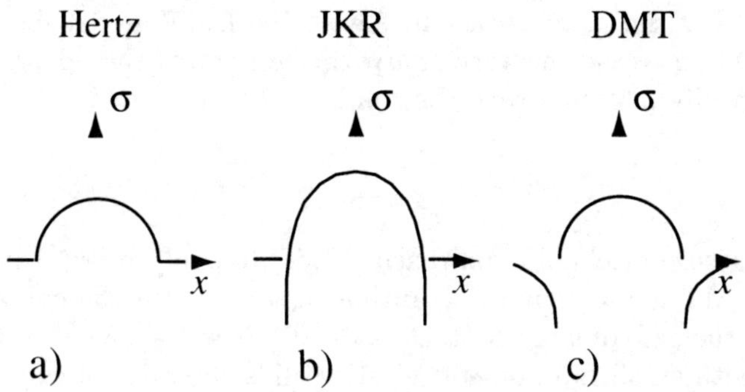

Fig. 2.10. Stress distribution in the contact region and outside according to the Hertz (**a**), JKR (**b**) and DMT (**c**) models. The shape of the stress curve in the negative range is characterized by the attractive force. In JKR theory attraction is incorporated only in the direct contact area as a delta function. The DMT theory considers adhesion also as a long-range contribution.

Equation (2.27) applies only to the case of an undeformable flat with equal adhesive (van der Waals) properties for both solids. The DMT model does not consider hysteresis effects. For the case of no external normal force the radius of the remaining contact area is determined by

$$a_0 = \left(\frac{4\pi\gamma R^2}{2K} \right)^{\frac{1}{3}} . \qquad (2.28)$$

Contrary to the models discussed so far, the DMT model requires numeric solutions with a computer. This disadvantage is caused by the fact that the surface profile is assumed as in the Hertz model. Despite attractive forces no deformations take place before contact. Thus, the spherical shape remains unchanged. When pulling back the sticking ball the solids will separate when the contact area is zero.

After much discussion in the literature [49–54] many scientists agreed on a dimensionless number β introduced by Derjaguin [48] to decide whether the chosen experimental setup should be described by the JKR or the DMT model:

$$\beta = \frac{64}{3\pi} \left[\frac{\gamma^2 R}{\pi K^2 s^3} \right]^{1/3} , \qquad (2.29)$$

where s is the separation between two solids with the strongest attraction. When β is large (compared with unity) the JKR theory is valid. Applying (2.29) to our SILICON model system (1 mm hydrophilic Si ball versus hydrophilic Si(100) flat), β is equal to 15.6, with γ=72 mN/m for a water-covered surface, $s \approx 0.3$ nm [45] and K=50 GPa. Since Si is hydrophilic (i.e., oxidized), the values for silica were used; see Table 2.1. This shows that the JKR model applies, since the radius and surface energy are large. The ball, however, is not perfectly hard because of the water coverage, which is subject to compression. For the nanocontact in AFMs the DMT model should be applied [55,56]. In this case the very sharp tip penetrates the adsorbed water film and physically interacts with the solid.

Extended Models

Maugis applied a fracture mechanics approach [57,58] to analyze the transition from JKR to DMT mode. The same author also extended the contact mechanics model to the case of large contact radii [59]. It was shown that for spherical particles with small radii on soft elastic solids the ratio of contact radius over particle radius can be so large that the parabolic approximation used in the JKR model does not hold. By means of exact sphere profiles the theory was extended.

Other geometries like cones or paraboloids were considered by Sneddon [60]. Departing from the Boussinesq equation, simple equations are given for arbitrary indenters.

For biological samples the appropriate models can be summarized as follows:

- All contact models are valid for ideally smooth surfaces, which do not apply to biological surfaces. However, since those surfaces can be extremely flexible and soft an intimate contact can be established.
- Completely elastic solids showing no adhesion can be treated with the Hertz model.
- When capillarity is involved, the generalized Hertz model can be applied. In order to minimize the computational effort, the boundary conditions of the actual experimental situation should be considered carefully.
- The JKR model is applicable for large soft bodies with high surface energies [61].
- Small and hard bodies with low surface energy should be treated with the DMT model [62].
- While the JKR model accounts for short-range attraction forces, the DMT model covers long-range attraction.
- Data obtained with scanning probe techniques can be analyzed also by contact theory. However, the geometry as well as the mechanical properties must be known in order to derive proper values [63].
- The contact models are summarized in Appendix A.

2.4 Adhesion

Adhesion can be regarded as a state of minimum energy that is attained when two solids are brought into intimate contact. This means that a certain force is required to separate the solids, as shown in Fig. 2.11. The interaction between two materials is often described in terms of the work of adhesion W_a related to the thermodynamic interfacial free energies:

$$W_a = \gamma_1 + \gamma_2 - \gamma_{12} \, . \tag{2.30}$$

This simple relation – known as the Dupré equation – combines the surface free energies of both materials γ_1, γ_2 with the interfacial free energy γ_{12}. Unfortunately, it is quite difficult to obtain experimental values of γ for solids [64].

Adhesion is always small, if interdiffusion, interdigitation and/or reorientation of molecules or molecular groups can be neglected. When a surface is covered by a liquid film then adhesion is also small, since the molecular rearrangement of the liquid is faster than the receding motion of the probe that contacts the surface [65,66]. Adhesion increases with decreasing roughness, showing that adhesion has a distinct range of action. In addition to low roughness, soft and flexible materials can also show strong adhesion, since these materials replicate the roughness profile of the counter surface, leading to intimate contact. This underlines that material deformation – increasing with increasing normal force [67] – contributes significantly and has to be considered by appropriate contact models as introduced in Sect. 2.3.4.

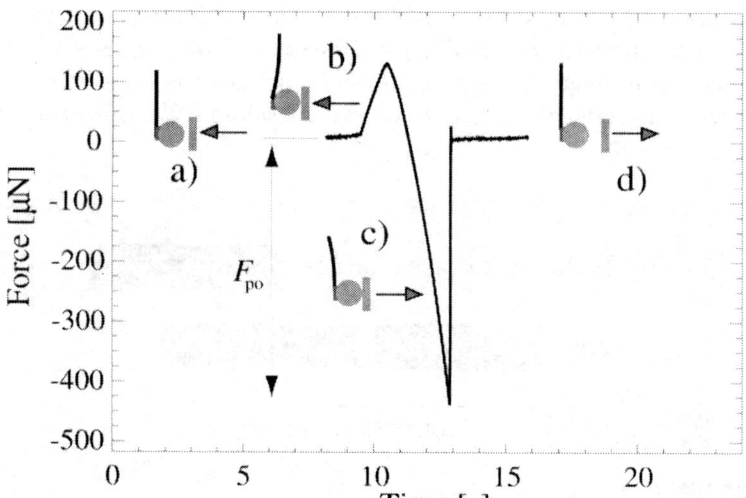

Fig. 2.11. Force–distance curve obtained from the SILICON model system. The ball is attached to the spring, while the flat sample can be moved. (**a**) Ball and flat are not in contact. (**b**) The flat sample contacts the ball and the spring becomes deflected. (**c**) After reversal of the motion the spring deflection decreases and becomes zero. Due to adhesion the two samples stick and the spring becomes deflected in the opposite direction. (**d**) When the back driving force of the spring equals the adhesion, the ball jumps back to its zero position.

Figure 2.12 shows the magnitude of different contributions to adhesion for the SILICON model system. Four main contributions can be identified. The weakest influence arises from molecular interaction followed by electrostatic attraction, capillarity and the action of excess charges. Whereas the action of capillarity and excess charges are continuum quantities, molecular attraction and electrostatic forces are caused by quantum mechanical interactions, e.g., van der Waals forces and/or interlocking forces of molecules.

While molecular forces are direct attraction forces, capillarity acts indirectly via a liquid. The interaction between liquid molecules (intermolecular forces), which usually leads to mutual attraction, is called cohesion. There are also interactions between a molecule in the liquid and all neighboring molecules. The resulting force is on average zero and the molecule is in equilibrium. If we consider a molecule at the liquid surface, then a part of the interaction with the other molecules is missing. An interaction with the molecules of the gas phase takes place. These forces are in general weaker. The molecule experiences a resulting force directed inwards, perpendicular to the surface. In order to bring a molecule from the inside to the surface, work must be carried out against all of the intermolecular forces, i.e., additional energy is required. The reverse process is connected with a gain in energy. This energy is called the surface energy.

In the macrorange the term adhesion is mainly connected to wear or thin film adhesion to a substrate [68]. Adhesive wear results from the rapid formation of strong chemical bonds during abrasive friction [69]. Both high temperature and increasingly intimate contact due to asperity breaking lead to a joining of the materials, which is also called cold welding. Continued shear

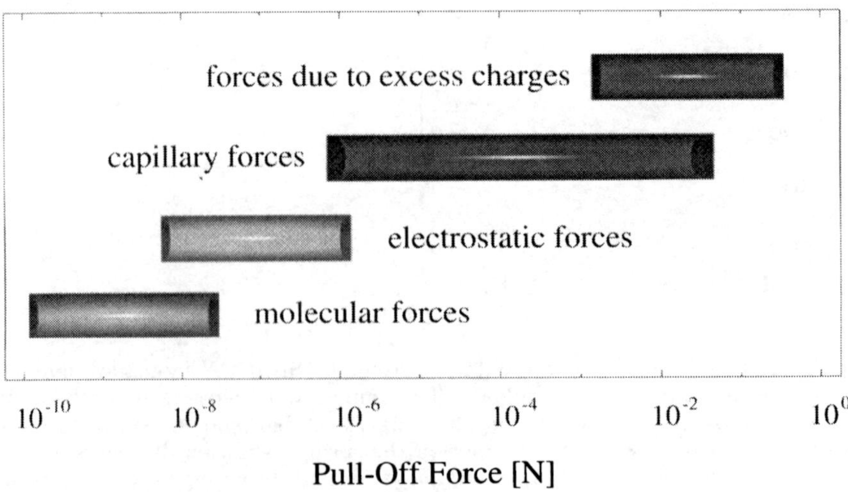

Fig. 2.12. Forces contributing to adhesion. The range for each contribution refers to the assumptions of the SILICON model system.

breaks particles out of the surface and degrades the material. The strength of adhesion depends on the number of formed microcontacts, the magnitude of the binding force acting between the partners, and the electron configuration of the involved bodies. When low forces in the microrange are accompanied by smooth surfaces, then wear is negligible and the interaction of both solids is confined to the mutual film of adsorbates. This film is also responsible for adhesion.

In biology, adhesion contributes to the coagulation of cells [70–73] and supports the formation of a bio-film [74]. This formation begins when a cell settles at a boundary surface and starts replicating itself. For adhesion of the cells to the boundary surface different mechanisms can be responsible. Of particular importance are van der Waals forces, electrostatic attraction as well as hydrogen bonds [75]. Each boundary surface offers adhesion potentials for micro-organisms. However, the attachment is favored if the boundary surface is already occupied by organic polymers (e.g., polysaccharides). Such polymers are usually of biological origin.

Probing the epidermis of the bacterium *Staphylococcus* it was found that donor-acceptor interactions play an important role in the adhesion process [76]. Besides molecular attraction roughness also influences adhesion. It was possible to correlate surface roughness and liquid contact-angle to the adherence and removal of bacteria from fluorinated polymers, stainless steel, glass and polypropylene [77,78].

Bowen et al. [79] studied cell surface adhesion in the presence of a liquid environment using single, living, immobilized cells. The probing cell was replaced by other cells allowing the fundamental study of cell adhesion, in particular the effect of the substrate (type, roughness, preparation, coatings), the strength of cell/surface interaction, effects of cell life cycle and growth conditions, the time of development of adhesive contacts, the influence of pH, or the effect of weakened adhesion.

Whereas these kinds of adhesion aim at permanent attachment, in the Case Studies of this book the authors are mainly concerned with repeatable attachment/detachment processes such as, for instance, the adhesion of the great green bush cricket to a smooth surface.

2.4.1 Molecular Forces

The shortest range of interaction is governed by molecular forces. To induce strong attraction the spacing between the solids must be reduced to a distance lower than about 10 nm. Mostly this is not the case, since natural and technical surfaces are not atomically smooth.

Attractive interactions between electrically neutral particles can occur in diverse ways, e.g., by the fact that the particles carry fixed and static electrical dipole moments, as for instance in water molecules. In addition, between rotation-symmetric particles, which are not polar, an attractive interaction caused by quantum fluctuations in the electronic structures of two closely neighboring particles leads to dipole moments, which attract mutually. This type of attraction is called the van der Waals interaction.

Van der Waals forces consist of a repulsive and an attractive contribution. The repulsive part is due to the effect that two molecules cannot penetrate each other, while the attractive part occur between neutral molecules, which do not carry a fixed dipole moment. Thus, quantum fluctuations induce transient dipole moments, which lead to mutual attraction. In quantum mechanics it has been shown by disturbance theory that the attractive part is proportional to $1/d^6$, where d in this case is the distance between the molecules. Both parts, repulsion and attraction, are often described by the Lennard–Jones model for the interaction potential $\Psi(d)$ between two molecules:

$$\Psi(d) = \Psi_0 \left[\left(\frac{d^*}{d} \right)^{12} - \left(\frac{d^*}{d} \right)^{6} \right] . \tag{2.31}$$

Here the repulsive part of the interaction is modelled by a potential ($\approx 1/d^{12}$). The parameter d^* describes the range of the interaction, the parameter Ψ_0 its strength. At $d = d^*$, $\Psi(d)$ becomes zero. For $d < d^*$, the potential increases very strongly. This part of the potential describes the impenetrability of molecules, which possess a finite radius d_0. By means of the Derjaguin approximation [48] a force law for the SILICON model system can be derived:

$$F(d) = \frac{AR}{3} \left(\frac{1}{d^2} - \frac{d_0^6}{d^8} \right) . \tag{2.32}$$

Here A is the Hamaker constant, R the radius of the ball and d is now the distance between ball and flat.

A necessary precondition for the generation of a dipole moment is that the molecule can be polarized. The polarizability is a measure of the ease with which the electron cloud of a molecule can be distorted. It remains the basis for the calculation of the non-specific attraction (London dispersion force), which occurs during the approximation of two molecules. Each molecule influences the electron cloud of the other molecule and, thereby, induces a dipole moment. The induced dipoles attract each other. Dispersion forces are weak and are most important in non-polar solvents, where no solvation forces act (see Sect. 2.5.1). They become stronger when the electron cloud is large.

Under ambient conditions most surfaces are covered with water. The interaction can be separated into an attractive van der Waals force and an electrical double-layer repulsion force. The total free energy of the interaction can be taken as the sum of these two contributions. The interaction of identically charged materials is described by the Derjaguin-Landau-Verwey-Overbeek (DLVO) theory. When dissimilar metal oxide surfaces interact across water the van der Waals contribution will generally be attractive. However, in the case of similar charges, the electrical double-layer contribution is repulsive or attractive for small separations. Moreover, the interaction can even change from one state to the other state through charge regulation with changes in separation.

Surface forces can be measured. In principle this is accomplished by a surface enlargement that requires energy. The surface energy is also called surface tension. Both terms are equal for liquids. For the macroscopic measurement a metal ring with diameter D is pulled slowly from the water and the necessary force F is measured. The surface energy is then calculated from:

$$\gamma = \frac{F}{2D} . \tag{2.33}$$

The literature gives the value for pure water as 72 mN/m [25]. The magnitude of the surface tension depends on the type of molecule and is, thus, different for each material.

2.4.2 Electrostatic Forces

Electrostatic forces can be divided into two different contributions. First of all bulk excess charges q present on the surface induce the classical Coulomb attraction (see Fig. 2.13). This force vanishes after proper grounding of the samples.

The second contribution arises from the electrostatic contact potential resulting in the electrical double-layer force that is the scope of further discussion.

Unlike Coulomb attraction, the double-layer force remains constant after grounding. The mechanical interactions in the microrange should always be treated as a combination of solid and liquid properties. With respect to sliding under low normal forces the interface layers have to be examined carefully. If solids are in contact with water – and this is always the case for experiments

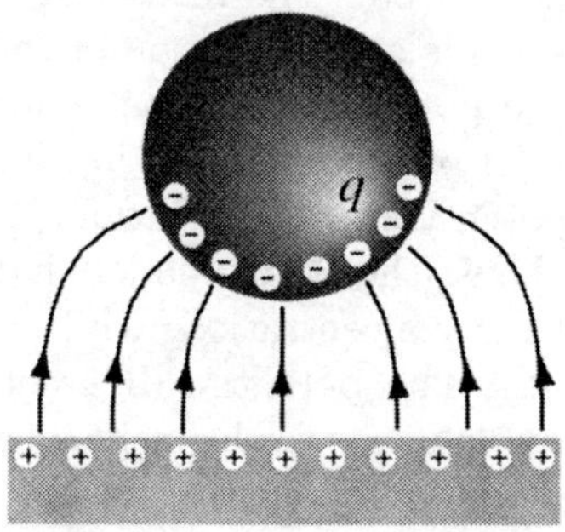

Fig. 2.13. Electrostatic force generated by surface charges. The accumulated charges of one solid are equilibrated by the opposite charges of the other solid. When both samples are electrically connected and grounded this force contribution vanishes. q is the sum of the charges.

in air – the interface has a different distribution of electrical charge carriers than the liquid bulk phase or the solid [80]. The enrichment of charge carriers at the interface is predominantly caused by adsorption of OH^- or H^+ of dissociated water. An additional source of charges are impurities (in the oxide for the case of the SILICON model system). Most biological macromolecules become charged when brought into an aqueous environment, since the molecules expose weak acidic and basic functional groups. These phenomena are properly reflected by the DLVO theory; see [81] for a review. The DLVO theory, developed in the years 1939 to 1945 by Derjaguin and Landau (Soviet Union) and Verwey and Overbeek (the Netherlands), supplies an explanation for the stability of dispersion colloids. On the surface of the particle, anions or cations are adsorbed forming a rigid film that shields the charges in the solid. Several layers of scattered ions follow, which form symmetrically around colloidal particles. With consideration of the electrical neutrality condition (in the time average) a symmetrical charge cloud develops. Excessive charges cause the repulsive characteristics of colloids, if these approach a defined distance in the dispersion. From this the stability of dispersion colloids results. The DLVO theory provides an approximate solution to the Poisson-Boltzmann equation. Very often the Poisson-Boltzmann equation is linearized leading to the Debye-Hückel theory and the following force law:

$$\Psi(d) = \Psi_0 \mathrm{e}^{-d/\lambda_\mathrm{D}} \ . \qquad (2.34)$$

Here d is the distance from the solid surface, λ_D is called the Debye length and Ψ_0 is the potential at $d = 0$ (see Fig. 2.14).

From here it is a small step from small colloidal particles to larger surfaces. The available surface charges at the solid surface attract oppositely charged ions of the liquid adjacent to the surface. A part of these solvated ions becomes so strongly attached to the surface that they can be regarded as rigidly bound to the solid described by the Stern model [82,83]. The remaining ions form a film of scattered ions. The model of Stern is a combination of the double-layer model by Helmholtz – in which the potential decreases linearly with distance – with the Gouy Chapman model of the scattered ion film, which applies an exponential decline of the potential to the solution. The Stern model contains both models as limits. Thus, the Stern model changes into the Helmholtz model for high counter ion concentration, while the Gouy Chapman model holds for high dilutions. The potential difference at the boundary between rigid and scattered double-layers to the inside of the liquid is called the Stern potential. Refined models differentiate within the rigid section of the double-layer between an inside and an outside Helmholtz layer. Here, the inner Helmholtz layer is formed by the centers of charge of the specifically adsorbed, i.e., partially desolvated, ions. The outer Helmholtz layer represents the centers of charge of specifically adsorbed solvated ions, which are predominantly held by electrostatic forces. In both layers partly aligned water molecules are available. If the particles in the liquid move, then

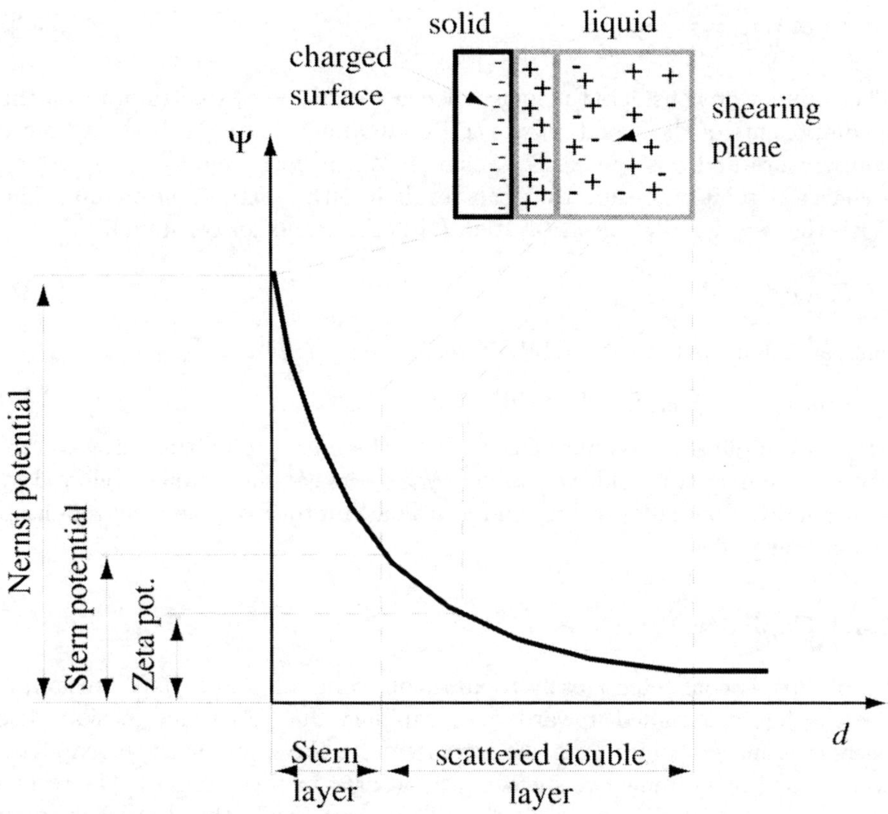

Fig. 2.14. Electronic potential in the vicinity of a charged solid surface. Adjacent to the surface an ordered layer of water molecules assembles to balance the charges in the solid. With increasing distance to the solid the state of order decreases until the water shows bulk properties.

a shearing plane can be assigned. The difference of potential of this shearing plane to the inside of the liquid is defined as the zeta potential. To a first approximation the shearing plane is located at the boundary between rigid and diffusive double-layer. Therefore, the zeta potential is often equated to the Stern potential. During the approach of two solids the double-layers begin to overlap. For low surface potentials the electronic double-layer force F_{edl} between two planar surfaces can be calculated by [25]:

$$F_{\mathrm{edl}}(d) = \frac{2\xi_1\xi_2}{\epsilon_{\mathrm{e}}\epsilon_0} A_{\mathrm{c}} \mathrm{e}^{\frac{-d}{\lambda_{\mathrm{D}}}} , \tag{2.35}$$

where ξ_1 and ξ_2 are the surface charge densities of the surfaces with the area A_{c}. ϵ_{e} is the dielectric constant of the liquid adjacent to the solid and ϵ_0 is the dielectric constant of vacuum. λ_{D} is the Debye length describing the thickness of the electrostatic double-layer as

$$\lambda_{\mathrm{D}} = \sqrt{\frac{\epsilon_{\mathrm{e}}\epsilon_0 k_{\mathrm{B}} T}{e^2 \sum_i^n c_{\mathrm{i}} \chi_{\mathrm{i}}^2}} \; . \tag{2.36}$$

The summation considers the ionic valence χ_{i} and the concentration c_{i} of the n components of the liquid. k_{B} is the Boltzmann constant, T is the absolute temperature and e is the electron charge. When both double-layers overlap a charge regulation occurs that can result in either attraction or repulsion. With the van der Waals contribution (A is the Hamaker constant),

$$F_{\mathrm{vdW}}(d) = A_{\mathrm{c}} \frac{-A}{6\pi z^3} \; , \tag{2.37}$$

one can calculate the total DLVO force:

$$F_{\mathrm{DLVO}}(d) = F_{\mathrm{edl}}(d) + F_{\mathrm{vdW}}(d) \; . \tag{2.38}$$

As an example of a complex DLVO force–distance profile the silica–water–vapor system shows repulsive van der Waals force at short-range, followed by an attractive double-layer force and a weak asymptotic double-layer repulsion at long-range [84].

2.4.3 Capillary Forces

Capillarity is connected closely to adhesion, cohesion and surface tension. A wetting liquid is pulled upwards in a capillary due to surface tension. The necessary energy comes from the interaction of the liquid with the capillary wall. In sliding systems two surfaces are brought in close contact. The resulting slits and pores act as capillaries. The water inside the slits leads to an increase of the normal force and to resistance against shear. This means that the friction force is directly proportional to the common area that is wetted. If the liquid does not wet both surfaces, a capillary bridge is not formed. The deformation or the breaking of the capillary bridges can be regarded as the dominant energy loss mechanism in the microrange [85]. Capillary action depends strongly on the interfacial properties of liquid and solid as well as liquid and vapor. Figure 2.15 shows the formation of a capillary bridge.

The force induced by a liquid bridge can be calculated by multiplication of the Laplace pressure Δp,

$$\Delta p = \gamma \left(\frac{1}{r_1} + \frac{1}{r_2} \right) \; , \tag{2.39}$$

with the area that is immersed in the liquid ($A_{\mathrm{l}} = \pi x^2$ for the SILICON model system) [86]. One obtains

$$F_{\mathrm{c}} = \Delta p A_{\mathrm{l}} = \gamma \left(\frac{1}{r_1} + \frac{1}{r_2} \right) \pi x^2 \; , \tag{2.40}$$

where γ is the interfacial energy (see Sect. 2.4.1). It is assumed that the curvature of the meniscus is spherical. The variable x is the radius of the wetted portion of the ball, and r_1 and r_2 are the radii of curvature. r_1 swings in plane and r_2 out of plane. One plane is positioned perpendicular to the other. The Laplace pressure equals the surface tension divided by the spherical meniscus curvature.

For the configuration of the SILICON model system (and only for this configuration) the capillary force can be derived from:

$$F_c = 2\pi\gamma R(\cos\Theta_1 + \cos\Theta_2) , \tag{2.41}$$

where R is the radius of the ball and Θ_1 and Θ_2 are the contact-angles as shown in Fig. 2.15. The Laplace pressure can be reduced to $\Delta p = \gamma/r_1$ when $r_2 \gg r_1$. For hydrophilic surfaces ($\Theta \approx 0$, i.e., $\cos\Theta \approx 1$) one obtains for identical material ($\Theta_1 = \Theta_2$):

$$F_c = 4\pi\gamma R . \tag{2.42}$$

This approximation holds for the majority of experiments shown in this book. The derivation of (2.42) is given in Appendix B. Note that F_c does not depend on the thickness of the adsorbed water film. The validity of (2.42) has been probed by theoretical investigations formulating a set of equations to calculate the meniscus profile with emphasis on surface interactions and liquid evaporation [86]. It was found that thicker liquid films cause larger attracting forces [87]. This observation was confirmed by our experiments probing the friction force as a function of atmosphere. Examples are given

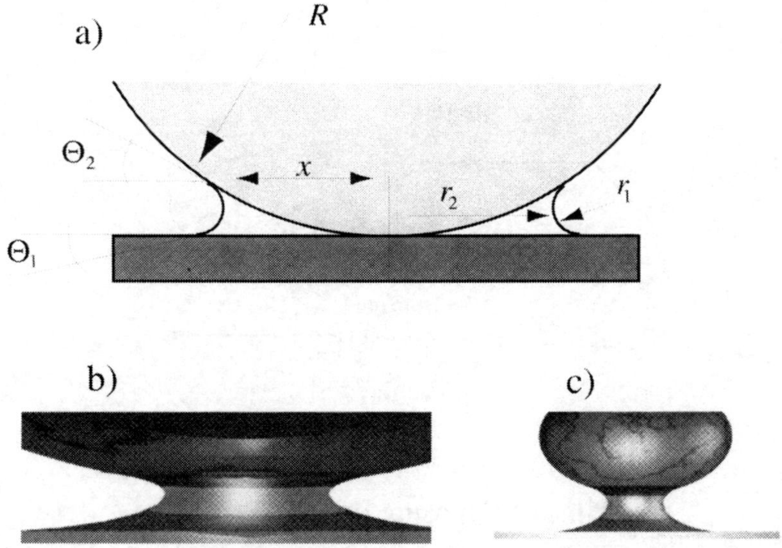

Fig. 2.15. Formation of a capillary bridge. (**a**) By means of two circles the capillary neck is approximated. This is a special case and requires identical contact angles between ball and liquid and surface and liquid. (**b, c**) show two more realistic cases.

for vacuum conditions in Sect. 2.5.5 and for different relative humidities in Sect. 2.6.6.

The reason for the increased capillary force is the increased area that is wetted. For this case, (2.42) does not hold anymore, since it contains no term to describe the influence of water film thickness. In addition, any change of the vapor phase leads to non-equilibrium conditions that could also be accompanied by different contact-angles or an even modified free energy of the liquid.

Furthermore, the strength of the capillary effect depends on capillary condensation in the wedge between ball and surface because of increased water vapor pressure in narrow slits [88]. The condensation is stronger for balls with larger diameter, since the wedge is longer (compare Fig. 2.15b, c). Besides the quantity of water, the capillary force also depends on the time spent in contact, since the capillary rise is not an instantaneous process [89].

In order to investigate the influence of a changing water coverage, tests with chemically modified Si surfaces were performed. A Si(100) sample was etched with hydrofluoric acid just before the first pull-off test to make the surface hydrophobic and then exposed to the laboratory environment (relative humidity $\approx 50\%$) for the following tests. The initial tests showed no pull-off force (see Fig 2.16). After about 24 hours the pull-off force increased significantly and saturated. In the range between zero coverage and beginning saturation (2.42) does not apply.

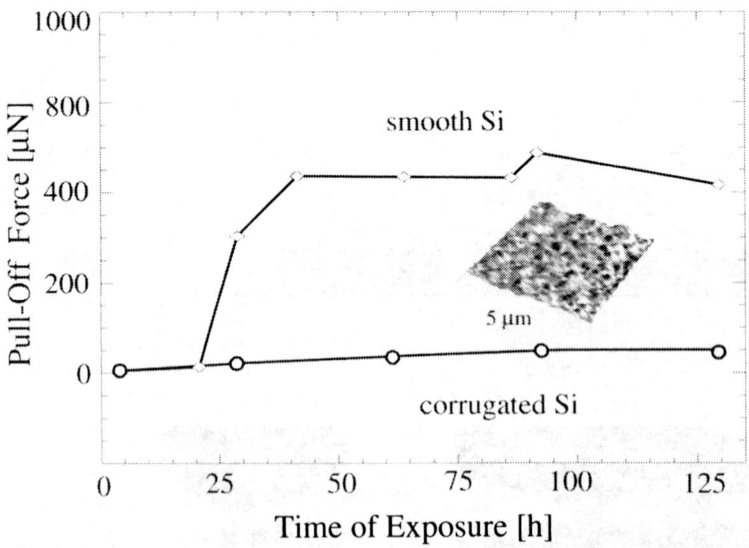

Fig. 2.16. Pull-off force of a hydrophobic polished and corrugated Si flat. Reoxidation of both samples increases the water coverage. Because of the reduced contact area between the ball and the corrugated Si flat, the friction force shows no dependence on the time of exposure to humid environment.

Due to exposure to air the surface eventually became hydrophilic caused by oxidation [90–92]. Adsorption of atomic oxygen occurs even at room temperature [93]. Initially oxygen attacks the backbonds of silicon [94], leaving the hydrogen termination unchanged [95]. The oxide starts to grow when the hydrogen termination has vanished [96]. The initial stages of oxidation indicate a uniform growth of SiO_2 [97]. The oxidation process is promoted by the presence of water vapor [98,99]. This effect changes the thickness of the water coverage. For all further tests a constant force was determined. The magnitude of the saturated pull-off force corresponds very well with the results obtained with (2.42) (F_c=445 μN).

To evaluate the influence of topography, the capillary force was measured for a corrugated silicon surface ($R_{a,corr}$=7 nm). The corrugated surface was prepared by sputter epitaxy of additional Si, generating a sinusoidally shaped topography (see inset of Fig. 2.16 [100]). Most obvious was that the pull-off force measured on the corrugated surface was hardly affected by the time of exposure. The explanation for this behavior is based on two competing sub-mechanisms. According to (2.39) the Laplace pressure increases with decreasing contact area, since radius r_2 decreases and cannot be neglected as in (2.42). This is illustrated in Fig. 2.17a. While the contact of the ball with a smooth surface results in a single capillary bridge [(2.42) applies], the contact with the rough surface leads to multiple sub-contacts accompanied by separated capillary bridges. For this situation both radii must be considered and F_c will increase.

On the other hand, the curvature of the corrugated surface leads to an inhomogeneous distribution of water due to minimization of surface tension, Fig. 2.17b. The drive to minimize surface tension leads to an increased water film thickness in the profile minima. Accordingly, the thickness on the tops

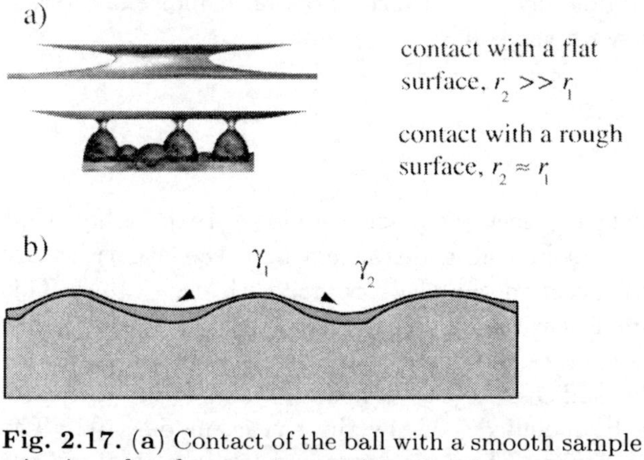

Fig. 2.17. (**a**) Contact of the ball with a smooth sample and a rough one. (**b**) Minimization of surface tension resulting in inhomogeneous thickness of water coverage. r_1 and r_2 are the radii of curvature according to Fig. 2.15.

is reduced, thus decreasing the wetted area and with this the capillary force. The magnitude of this decrease depends strongly on the size and lateral distribution of the asperities forming the corrugated surface. When the size of the asperities is small compared with the contact area – for the case of a ball with a large diameter – the resulting capillary force increases with increasing roughness, since many capillary bridges can form. In the opposite case when the size of the asperity is comparable to the size of the contact area (small ball) a single asperity forms a capillary bridge and F_c is small .

Summarizing this section one can state:

- Of all the attractive forces in the microrange, capillarity has the strongest impact provided that electrical excess charges can be neglected due to grounding.
- The action of capillarity can be properly described by simple approximations. However, the actual geometry (e.g., asperity size, sample corrugation) has to be identified carefully in order to minimize computational errors.
- The liquid monolayers that are in direct contact with the solid are subject to electrical double-layer enforced ordering.
- Ions in the liquid balance the charges present on the surface of the solid.
- Molecular forces (van der Waals contribution) act in the nanonewton range. They are responsible for the mutual attraction or repulsion of liquid molecules and their interaction with the confining solids.
- The measurement of the attracting forces reflects mainly capillary force hiding subtle molecular interactions.
- The magnitude of the measured forces depends, thereby, crucially on the shape and size of the probe.
- Blunt, ball-shaped probes detect capillary forces, while sharp, needle-like probes penetrate the double-layers and also experience molecular forces due to intimate contact with the solid.

2.5 Lubrication

Lubrication is one of the key aspects of macrotribology [101]. A lubricant is mainly used to keep two solids at a distance where the asperities are prevented from getting in direct mechanical contact with each other. This requires that the lubricant has to be sufficiently viscous in order for it not to be squeezed out of the contact. To describe lubrication effects, a Newtonian fluid model normally suffices. As the dimensions and forces decrease, non-linear effects have to be included, e.g., by the Eyring model [102]. The Eyring model considers the effects of temperature and strain rate of viscous flow, which develop from consideration of the activation energy required for a molecule to jump from one position to another. This model is very simple and does not apply to polymer non-linearities. Friction and adhesion forces with magnitudes lower than about 1 mN acting on contact areas in the micrometer range are strongly affected by the action of adsorbed liquids. The influence depends on the thickness of the liquid film and on the roughness of the contacting surfaces. As the thickness of the lubricant decreases below about 10 nm molecular influences become notable. For polymers sliding performance depends strongly on the molecular weight [103]. Lubrication effects were observed for films with a thickness of just two layers [104] or even one single layer when dealing with Langmuir–Blodget films. Ultra-thin films of fluorocarbon polymers are applied to lubricate magnetic storage disks. Significantly altered physical properties (e.g., disjoining pressure) were found in the range of a few monolayers. Whereas in the microrange capillarity has the strongest impact, confinement processes determine the behavior in the nanometer range. The main effect in thin film lubrication is solidification.

To analyze extremely thin films of lubricants in terms of coverage or ordering, special equipment like electrical conductivity measurement of point contacts [105] and simulation combining thin liquid films and surface roughness [106] were introduced in recent years. In this book the authors focus on the analysis of confined simple liquids and water investigated with the Surface Forces Apparatus (SFA) and an especially designed microtribometer to analyze the effect of water.

2.5.1 Simple Liquids – Experiments

The study of lubricant confinement was initiated by experiments using simple globular liquids such as, for example, octamethyltetracyclosiloxane (OMCTS). The experiments were carried out with the SFA [107]. For the purpose of understanding, this section begins with a general description of the apparatus. Further instrumental details are explained in Sect. 5.2.5.

The apparatus can be thought as two glass cylinders arranged as a cross located in a chamber that contains the liquid under investigation. On top of the cylinders thin sheets of mica are glued to serve as the actual samples, contacting each other in a single point with an area of a few square micrometers. Mica was chosen, since the sheets are atomically smooth over a range that is many times larger than the probed area. Since mica is a transparent material, interferometry can be used to determine the spacing between the cylinders and, thus, the thickness of the confined liquid film. By means of an oscillatory drive and tangential force detectors friction can also be measured. In experiments where both cylinders were approached without oscillation it became evident that a distinct force is necessary to squeeze out the liquid. The force (solvation force, for the case of water hydration force or DLVO contributions due to dissolved electrolyte ions) increased with decreasing liquid film thickness. From a certain distance the force started to oscillate between attraction and repulsion. For the case of water no oscillation between attraction and repulsion was found. Instead, only the repulsion force was modulated. The

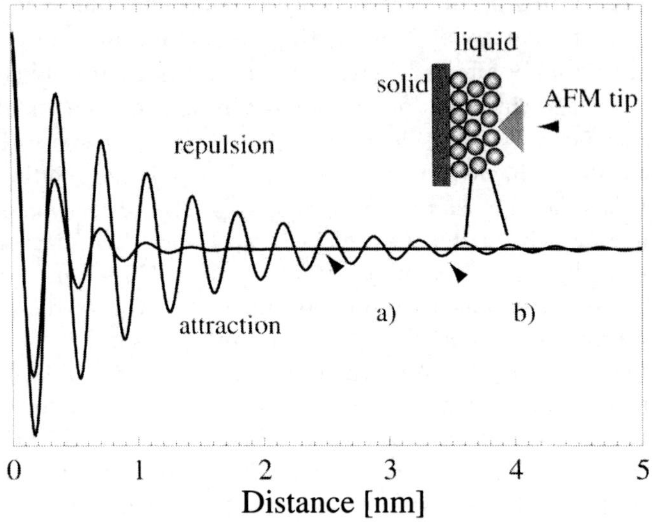

Fig. 2.18. Schematic solvation force curves of a molecular stack with short-range order (**a**) and of a molecular stack with long-range order (**b**). The curves reflect the molecular resistance experienced by an approaching probe and can be measured by SFA and AFM (see inset).

period length was found to be equal to the molecular diameter as shown in Fig. 2.18. These findings – obtained experimentally [108] and by computer simulation [109–113] – suggest that the molecules of the film are ordered in a layered structure. The magnitude of the in-plane order is described by the Debye–Waller factor, ranging between 0 and 1. Zero corresponds to a liquid decoupled from the wall, i.e., without any order, and 1 corresponds to the crystalline state at $T = 0$ K [114]. The squeezing process thus removes the liquid layer by layer. For highly viscous liquids the solvation force oscillates, while for low-viscous liquids a purely exponential decay was found by simulation. The structure of the solvation force depends on the physical nature of the surface, i.e., its roughness and the termination of chemical bonds [115].

Another phenomenon arising from confinement is the reduction of mechanical degrees of freedom of the molecules and the appearance of fluctuations during sliding. Israelachvili [116] found that films with a thickness lower than 10 molecular diameters become more viscous and that films of four molecular diameters undergo a phase transition into a liquid-crystalline phase. Friction experiments showed periodic stick/slips. However, friction experiments cannot prove crystallization of the liquid, since the change from stick to slip also occurs in less-ordered systems due to viscosity increase, as shown later. During the stick regime both samples adhere to each other. When the back driving force of the spring is large enough, spontaneous sliding releases the accumulated energy.

In contrast to a thick lubricant film, very thin films are able to sustain a certain amount of shear stress. Granick et al. [117] showed that complete crystallization of the liquid film is not a necessary precondition in order to observe a critical shear stress. Depending on the molecular structure various liquid films behave differently. While globular molecules tend to acquire higher ordering, chain or branched molecules show a lesser degree of order. Films of chain molecules and hydrogen-bonded liquids undergo a glass transition instead of crystallization [114, 118, 119]. However, a critical shear stress was also found for glass-forming liquids, far above their glass-forming temperature. Heuberger [120] performed SFA experiments and introduced vibrations perpendicular to the direction of sliding. The vibrations were tuned to the range from Hz to kHz and to distances comparable to the molecular diameters. It was shown that vibrations prevent the molecules from ordering, although the oscillation frequency was far away from molecular relaxation (10^9 to 10^{12} Hz). As a result, ultra-low friction was achieved.

The main results of the experiments with confined liquids can be summarized as follows [117]:

- Systems showing a disordered arrangement of the liquid molecules between two solids exhibit no critical stress when subjected to shear.
- The apparent viscosity increases with increasing normal force [121].
- The critical shear stress increases gradually as a result of time-dependent ordering mechanisms.
- The solid-like properties are caused by confinement and not by the material properties of the lubricant.
- The smoother a surface, the more drastic is the influence of normal force on the critical shear strength due to enhanced ordering of the film.

2.5.2 Simple Liquids – Simulation

Basically, the structural and dynamical characterization of interfaces belonging to a confined structure is difficult. There is always the question of whether the order is real or induced by the mechanical contact. Hence, analytical *ab initio* predictions of the surface structure with a given type of molecule are not possible [122]. It follows that confinement properties cannot be fully explained by macroscopic terms; they demand molecular analysis. Furthermore, continuum mechanics loses its applicability at very small separations of the bodies [123].

To describe the phenomena theoretically it was found that the continuum van der Waals theory fails to predict the interaction of molecules under confinement [25, 115, 124]. Therefore, sophisticated simulation techniques were applied. To interpret the friction experiments, the ordering of the molecules was intensively analyzed by either dynamic Molecular Dynamics (MD) or by quasistatic Monte Carlo (MC) simulation. MD simulations try to model dynamic properties on the molecular scale by computing the trajectories of thousands of atoms in the phase space by solving their equation of motion [125–130]. For a comprehensive treatment of the basic physical properties of the sliding friction of thin liquid films the reader is referred to [131]. One of the main problems of the dynamic approach is that MD simulations are based on the equation of motion. To solve those equations, the mass of the confining wall must be inserted. However, there is no reasonable criterion to specify the mass and, therefore, the relevance of MD simulations seems to be questionable according to [132]. Atomic molecular dynamics simulation has also inherent limitations because of the short time and length scale that are not sufficient for the consideration of slow relaxation processes [133,134]. It has to be mentioned that some authors [135] state that MD simulation should not be used to study sliding dynamics because of the short times. This reduces the range of application to quasi-static processes.

MD calculations of confined hexamers [135] revealed that the molecules order themselves under shear. The molecules adopt conformation with the top of the confining walls. The atomic corrugation leads to a pinned solid adsorbate structure. As a result, the shear rate increases as a function of temperature, pressure or wall symmetry. The film can solidify and form domains with crystalline order. Shear may affect the order by melting the microcrystals [136,137]. In non-equilibrium MD simulations of confined oligomer melts, it was found that the velocity profile normal to the confining walls is in strong correlation with the density profile in the slit [138]. The velocity profile is constant for the middle section. However, much higher shear rates have to be applied to induce the flow. A slip occurs between the first layer and the rest of the system [139]. For solids showing low surface energy it is likely that the slip occurs at the solid wall and not by solid–liquid transition [140].

MC simulations employ a quasi-static approach to calculate the equilibrium properties of the confined film during a stepwise change of its thermodynamic states. The stick/slips are conceived as continuous phase transitions. Schoen [132] found that mere confinement forces the molecules to arrange themselves in layers parallel to the walls. In the simulation the walls were treated as Einstein solids where the single atom moves in an effective mean field along with the other atoms in the wall. The atoms of the lubricant were treated as harmonic oscillators described by Lennard–Jones interatomic potentials. The indentation of atoms in the wall is prevented by a repulsive background potential. The impact of thermally induced corrugations on the film's structure is small.

Fuchs et al. [141] performed quasi-static MC simulations of the static friction behavior of ultra-thin liquid films. Solidification of the film was attributed to the template effect imposed by the atomic structure of the confining walls. The solvation force curve for globular as well as for elongated molecules was in very good agreement with experiments. The stick/slips are interpreted as thermodynamical phase transformation. It was deduced that a yield point separates the stick from the slip regime. The very simple Lennard–Jones model gives a good description of confined thin films. In addition, the experimentally observed yield stress does not depend on the formation of a completely solid-like film. This statement is very important, since no solid-like film is formed when the walls are incommensurate with the prospective film, which is the case for most of the technical systems. Moreover, it was found that for a liquid-like confined film the shear stress also follows Hooke's law initially. For larger stresses a yield point exists that is a factor of three lower than for a completely solid-like film. It was concluded that any sufficiently thin confined film is probably capable of sustaining shear strain [142].

The following conclusions can be drawn from simulations:

- Solidification appears to be a universal property of confined liquids [143].
- It is not necessary for the confined phase to have a perfect crystalline structure to sustain shear.
- The shear stress for confined liquid-like films is about one order of magnitude lower than for solid-like films because of the lower degree of order.
- To determine whether a confined phase is solid-like or liquid-like, the microstructure must be investigated thoroughly.
- The physical state of a confined phase cannot be determined solely from its thermodynamic properties.

2.5.3 Water – Bulk Properties

It was shown in the previous section how simple liquids in thin film configurations influence the tribological behavior. This section extends the study to water, since all surfaces are covered by it, unless they are hydrophobic or dried. Water is of great importance for insects, since it influences the constitution of biological structures as the common solvent for all biological activities [144]. Biological processes are strongly influenced by the structure and intermolecular hydrogen bonding of water, shown by vibrational sum spectroscopy to analyze the OH stretching mode that is sensitive to structure and bonding information [145]. The impact of the water film thickness on friction can be directly seen in the sliding behavior. Figure 2.19 shows three friction force curves: curve (a) was measured with a system that is covered with more than 10 nm of water.

The overall sliding behavior is liquid-like, characterized by a smooth transition from static to dynamic friction. The friction force curve shows no rapid fluctuations. In contrast, curve (b) shows the sliding behavior of a film covered with just a few monolayers. Due to reduced capillary action the friction force is lower. The periodic fluctuations are caused by the increased viscosity of the confined water film. When the water film is reduced to one last monolayer (due to vacuum desorption or annealing) the periodic stick/slips vanish (curve (c)). Friction is low until this monolayer becomes penetrated. Then friction increases again due to cohesion.

Table 2.2. Atomic radii of hydrogen and oxygen

Radius	Hydrogen	Oxygen	Ref.
atomic radius (empirical) [pm]	25	60	[146]
atomic radius (calculated) [pm]	53	48	[147]
van der Waals radius [pm]	120	152	[148]

Throughout the centuries water has been the subject of intensive research. A comprehensive collection of data on bulk water can be found in [149]. Although water is a small molecule with low-viscosity [116] (see Table 2.2), its main difference from the simple liquids discussed before is that water is a dipole molecule that is able to form hydrogen bonds to neighboring water molecules or to solid surfaces [150] (see Fig. 2.20). Protons are the dominant charge carriers in hydrogen-bonded systems [151]. Water shows even in the bulk state many special properties and anomalies [152,153]. A

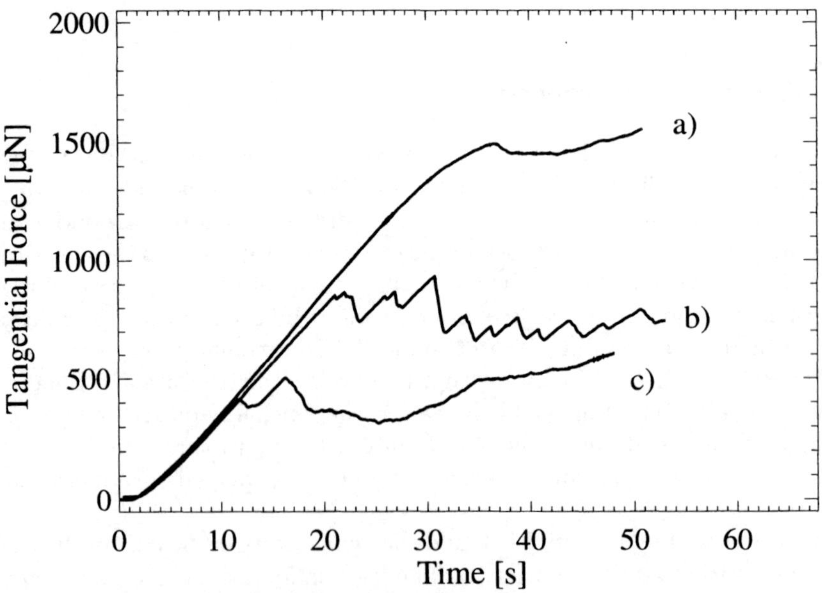

Fig. 2.19. Tangential force curves for different water coverages. (**a**) Thick water films ($\gg$ 10 nm) induce liquid-like sliding and high friction force due to capillarity. (**b**) For a coverage of a few monolayers the sliding behavior changes from liquid-like to solid-like. (**c**) When the water film is reduced to one monolayer the friction is low and almost fluctuation-free.

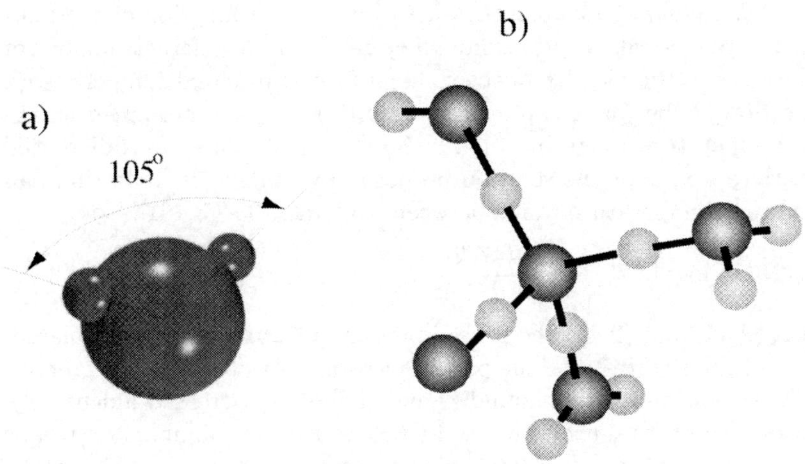

Fig. 2.20. (**a**) Water molecule. (**b**) Tetrahedral structure formed by water molecules.

certain degree of order can be found in water resting in an equilibrium state, where hexagonal structures – similar to ice [110, 154–157] – were observed. These structures are not stable and vanish as the temperature increases. Bulk water can also be arranged in a tetrahedral structure with a high degree of short-range order. The connectivity of water molecules caused by hydrogen bonds determines physical properties such as freezing point, viscosity η and dipole relaxation frequency f_D [158] as described by the Eyring–Debye model:

$$\frac{1}{\eta} \sim f_D \sim e^{\frac{-\Delta\Gamma}{k_B T}} , \tag{2.43}$$

where $\Delta\Gamma$ is Gibbs potential, T is the temperature and k_B is the Boltzmann constant. The increase in dielectric relaxation is a property that corresponds to the viscosity of water [159]. Viscosity increases due to a higher degree of connectivity. With decreasing amount of water, especially when water forms adsorbate films, these properties change significantly.

2.5.4 Water – Molecular Film Properties

Leaving the area of bulk water it is interesting to investigate the interface between water and solid [160] and how water is adsorbed on a surface. Water vapor is always present in air. The content of water is described by the partial pressure of water, or expressed with the more common term of relative humidity (r.h.) as the ratio of the actual partial pressure of water p to the saturation pressure of water $p_s(T)$ at the prevailing temperature:

$$\text{r.h.} = \frac{p}{p_s(T)} . \tag{2.44}$$

For a preset humidity the partial pressure of water is a function of the temperature. In other words, a preset humidity results in a different number of water molecules in the vicinity of the solid surface at different temperatures. The saturation of the water vapor pressure $p_\mathrm{s}(T)$ can be determined within a certain temperature range by solving the Clausius–Clapeyron differential equation. Here we adopt the solution proposed by Magnus in 1844 that can be applied to a temperature range between 0°C and 100°C [161]:

$$\log[p_\mathrm{s}(T)] = \log(4.525) + \frac{7.4475T}{T + 234.69} . \tag{2.45}$$

Combining (2.44) and (2.45) the partial pressure of water can be calculated. Figure 2.21 demonstrates that the partial pressure of water depends on temperature as well as on relative humidity and cannot be increased indefinitely.

To adsorb water, the surface has to be able to keep the liquid. This means that the surface has to have hydrophilic surface properties in order to be wetted. Wetting is an interfacial process dominated by long-range interactions [162]. Surfaces become wetted after exceeding a critical temperature. In ultrahigh vacuum – where clean surfaces can be prepared – the initial adsorption takes place at step edges, which are energetically preferred locations [163–165]. Detailed studies exist for the initial processes of water adsorption on semiconductor surfaces [166], e.g., the adsorption and reaction of hydrogen and water on Si(100) [167–174]. Köhler et al. [175] investigated the water adsorption on Si(100) by STM and found that water chemisorbs via molecular precursor preferentially at terrace dimers. It was shown that the following

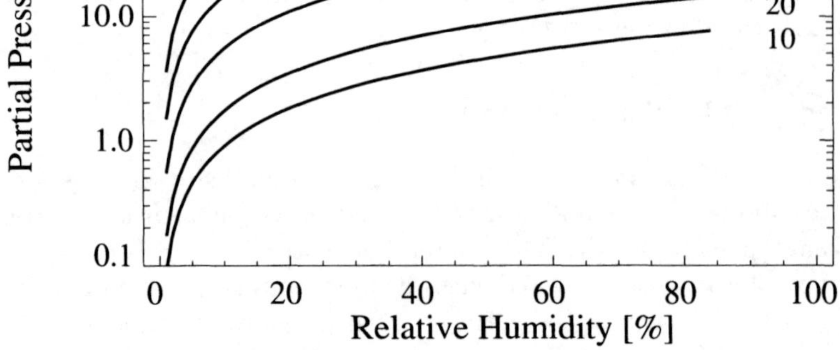

Fig. 2.21. Partial pressure of water versus relative humidity for different temperatures calculated using the Magnus equation.

adsorption of water leads to the dissociation of the water molecule to OH and H, both passivating the dangling bonds of the Si(100) surface.

In air after forming the first monolayer, surface diffusion is obviously suppressed by ordering processes [176]. This order is also manifested in a polygonal shape of the water island with edges making angles of 60° and 120°. In addition, water islands on mica showed an alignment with its principal crystallographic directions [177]. The inclination of a surface to adsorb water depends on its polarity. A hydrophilic surface is a polar surface while a hydrophobic surface is non-polar. Hirata et al. [178] studied the structure of water by the reference interaction site method for the solute–solvent system. The solute was assumed to be a periodic two-dimensional array. For a negatively charged surface the hydrogen atoms of the water molecule point towards the surface. The angle between the surface normal and the vector of the dipole moment changes from the first to the second layer. The second layer has a hydrogen atom pointing towards the vapor phase; thus more water can be adsorbed. In our SILICON model system silicon represents a negatively charged surface that induces ordering of the water molecules with the hydrogen atoms contacting the surface (see inset of Fig. 2.22). The orientation of the water molecules depends strongly on the surface atomic structure.

Continued adsorption leads to a closed monolayer of water. It was shown by AFM experiments on mica that the water film thickness varies from one to several molecular monolayers (0.15 nm to 0.4 nm for 45% and 70%, respectively [179]). The film is formed in equilibrium with the vapor and shows uniform structure [180]. Upon mechanical contact with the AFM tip, two-dimensional islands form. After formation of a uniform layer the increase in humidity leads to the growth of two-dimensional domains with a clear orientation. An ordered structure is assumed [89,181,182]. A full ice-like ordered monolayer is formed above 90% relative humidity. MD simulations confirmed two-dimensional ice-like water bilayers [183]. According to [184] ice-like water layers form with a saturation at 90% relative humidity at 296 K. The layer does not contain any free OH groups, indicating the well-ordered structure.

On SiO_2 a water film thickness of 0.3 nm to 0.6 nm was measured [185]. Other experiments (by ellipsometry [186]) showed that the water film on SiO_2 ranged between 2 nm and 8 nm [187]. In addition, measurements with AFM have shown that the water film thickness is about 10 nm for a relative humidity of 92% [188,189]. However, when the surface is cleaned with extreme care, water films with a thickness of about 150 nm were measured [190].

The structure of adsorbed water can be imaged at low temperatures. As shown in [191] the interaction of water with single-crystal MgO(100) leads to an ordered (3×2) commensurate solid monolayer at temperatures higher than 185 K. The measurements were performed by LEED (see Sect. 5.3.4) and helium atom scattering. Freund et al. [192] investigated the formation of water films on gold with a specially designed STM for relative humidities between 10% and 90%. Water droplets with a diameter of 30 nm to 50 nm and

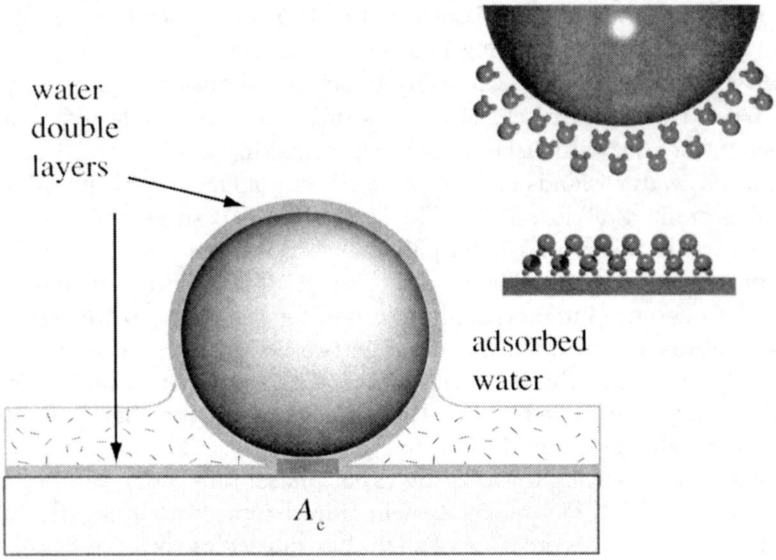

Fig. 2.22. According to the electrical double-layer model both ball and flat sample are covered with two layers of water. When the solid surfaces carry negative charges, the oxygen atoms of the water molecule point away from the surface. Additional water leads to the formation of a capillary bridge. A_c denotes the actual contact area. Only water under this confinement is subject to viscous interaction.

a height of about 10 nm are formed on a particularly flat surface with small irregularities at humidities above 20%. Increasing humidity (above 55%) leads to a continuous water film with a thickness of about 10 nm. For titanium the water film thickness increases rapidly with humidity. A thickness of 50 nm was measured at r.h.=20%. Even in quite dry conditions a closed layer of water or droplets was found, even in dry nitrogen atmosphere.

A further significant change in physical properties appears when the water is subjected to geometric confinement enhancing the order [193]. For instance, the viscosity increases exponentially toward the surface when two parallel surfaces confining water are approached [194]. However, the change in viscosity is rather continuous [195] and the structure and polarization of the interlamellar water are reinforced.

By means of Newton's law the viscosity can be estimated for a Couette flow geometry, which is characterized by two parallel walls separated by a liquid film of molecular dimension. This idealization applies well to the SILICON model system (see Fig. 2.22). The area where the friction force is dominated by viscosity effects (F_{fv}) is confined to the real contact area as determined by the Hertz model:

$$F_{\mathrm{fv}} = A_{\mathrm{c}} \eta \frac{\mathrm{d}v}{\mathrm{d}(d_{\mathrm{H_2O}})} , \tag{2.46}$$

where d_{H_2O} is the thickness of the water film and v is the sliding velocity. For low coverage the derivative can be replaced by v/d_{H_2O} and after supplying reasonable experimental values for $F_{fv} = F_f$=100 μN, A_c=90 μm^2 (see Sect. 2.3.4) and v=40 μm/s the viscosity can be calculated for a water film thickness of, for example, 4 nm:

$$\eta = \frac{F_f d_{H_2O}}{A_c v} = 111\,\text{Pa} \cdot \text{s}\,. \tag{2.47}$$

In comparison with the calculated value, the viscosity of water in the bulk state is 10^{-3} Pa·s. This underlines that at a certain thickness the water looses its bulk properties.

While in wider slits water still exhibits its tetrahedral structure, clusters and networks occur in slit-like pores until linear chains in quasi two-dimensional slits are formed [196].

Whereas non-polar surfaces do not wet, polar surfaces tend to build a water double-layer as shown in Fig. 2.22 with distinctively different properties than those of bulk water [197, 198]. The water dipole layers have a thickness of about 0.3 nm. In addition to the already ordered double-layer the geometric confinement further reduces the degrees of freedom (translation and rotation) of the water dipoles. As a result, the molecules become arranged in an ordered solid-like structure. Thus, when two polar solids are brought in contact, a stable arrangement of four water layers may form. In bulk water Brownian motion and rotation of water molecules disturb the ordering process, whereas in the compact part of the double-layer the water dipoles are stabilized and remain at their positions.

The ordering effect is also reflected by the hydration force curve (see Fig. 2.18). Hydration force curves are caused by the tendency of water to form discrete ordered layers [199]. The observed curves show an exponentially declining oscillation as a function of the distance to the solid surface. The period length corresponds to the diameter of the water molecule. The hydration force measured between a conical AFM tip and a flat surface in water induces a pressure of 10 MPa at a spacing of 2 nm [200]. The hydration force can be caused by the ordering of water due to the presence of two confining walls or due to molecular undulation forces [201].

An example of water confinement in biological systems is the membrane. A biological membrane is a matrix formed by a lipid bilayer. Water is necessary to build the bilayer [158]. The hydrophobic or hydrophilic interaction of biomolecules with water is caused by their functional groups. Hydrophilic head groups compensate for deficient hydrogen-bonds in the water film due to the possession of hydrogen bond donors. The inter-water hydrogen bonds are optimized [201]. Surfactants affect the ordering of water molecules at the air/water interface, with cationic surfactants present. Water molecules in double-layers will align with the oxygen atom pointing away from the surface [202].

Summarizing the statements it can be concluded that:

- Water shows already in the bulk state a certain degree of order.
- The order increases in the vicinity of a solid that allows itself to be wetted.
- Polar solids form a water double-layer.
- The ordering is increased tremendously by confinement.

2.5.5 Lubrication and Water Film Thickness

With respect to the findings in the previous section the following experimental results will be discussed. The experiments investigate the interactions of the hydrophilic SILICON model system to determine the dependence of the friction force on the water film thickness in a vacuum chamber [17]. In these experiments, the authors used the fact that water desorbs as the pressure drops, finally forming a single monolayer or less. The friction force was measured initially in air and subsequently as a function of the total pressure of the vacuum system.

The ball and the flat sample were cleaned by rinsing in isopropanol and methanol (5 minutes) followed by a two-minute dip in distilled water. Due to the native oxide (1.2 nm) this surface is terminated by OH groups, which attract water molecules [203,204]. The surface is charged by adsorbed ions and by charges moving into the oxide film [205]. It is a known fact that SiO_2 becomes negatively charged (charge density of 6.7×10^{18} cm^{-3}) when exposed to water vapor [206]. Parallel to the friction measurement the thickness of the water film was measured by STM, as explained in Sect. 6.1.3. With decreasing pressure the water film decreases in thickness. Figure 2.23 shows two distinct ranges. The range of high water coverage – caused by the rinse in water immediately before the test – is followed by the range of low water coverage. Both ranges are separated by a step at 0.1 mbar due to different pumping speeds. In the range of low coverage the pumping speed is constant. The measured water film thickness is the sum of the water film on the Si sample and on the tungsten STM tip, which is also oxide-covered. At a pressure below 10^{-7} mbar the last monolayer of water desorbed. This was confirmed by photoelectron spectroscopy, which indicated a binding energy of oxygen caused by the bond with silicon. In the case of water coverage the oxygen peak would have been shifted to the value caused by the bond with hydrogen.

Because of low normal forces (F_n=500 µN) and low surface roughness (R_a < 1 nm) mechanical interlock phenomena as a cause of friction were neglected. The results suggest that the sliding behavior is strongly influenced by the thickness of the water film.

As the pressure dropped, the equilibrium point between liquid and vapor constantly changed. Lower water film thickness leads to lower capillary force and, thus, to lower friction (see Fig. 2.24). To learn more about the dynamics of the process the authors also took the friction force curves as a function of time (period of 3 minutes) into consideration (see Fig. 2.25). When the water

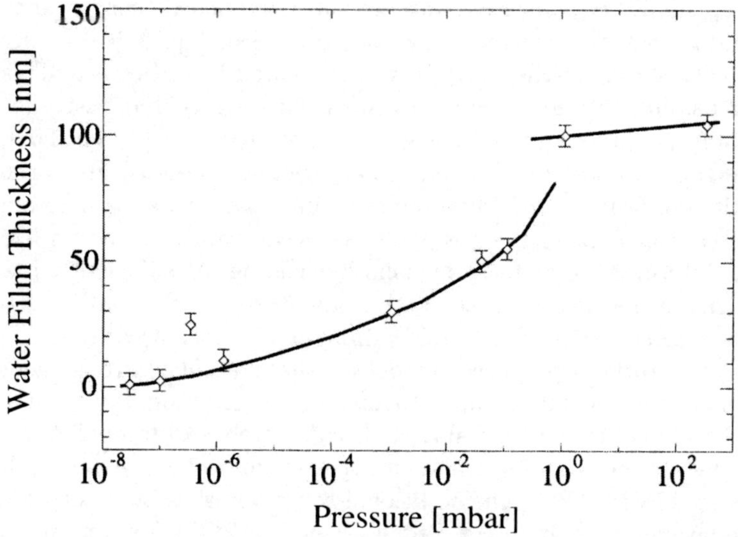

Fig. 2.23. Water film thickness measured by STM in the spectroscopy mode as a function of the pressure in the vacuum chamber. The approaching STM tip experiences an ionization current upon contact with the water film. The distance between the onset of the current and the exponential increase due to the contact with the solid is interpreted as water film thickness.

film is thick the frictional behavior appears to be liquid-like. The magnitude of the friction force is solely dominated by capillarity. As the pressure contin-

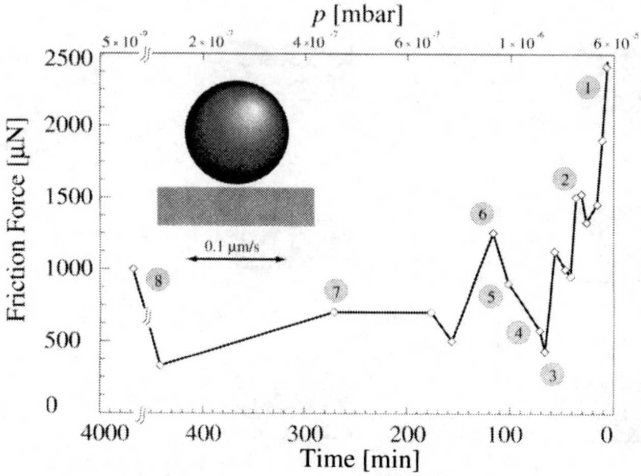

Fig. 2.24. Friction was measured as a function of time while evacuating the vacuum chamber. The exposure of the water-covered silicon surfaces to vacuum leads to desorption and, therefore, to continuously decreasing water film thickness.

ues to decrease, capillarity vanishes and more subtle influences, for instance increased viscosity can be detected. After some time of liquid-like sliding behavior, solid-like sliding suddenly appeared, indicated by stick/slips. This occurred after passing the first minimum. Stick/slips are characteristic for either dry sliding conditions or, as shown for carbohydrates [25,117,207], for ordered quasi-crystalline liquid layers. Then, friction increased up to the first local maximum. From this point two more minima and maxima followed where the decrease in friction force was always accompanied by liquid-like sliding and the intermediate increase by solid-like sliding. At sufficiently low pressure, the friction became constant with a low degree of fluctuation. Finally, the friction force curve left its main minimum and increased.

In Fig. 2.25 the authors propose a model for the molecular arrangement between the ball and the flat sample to explain the friction force curve of Fig. 2.24. At point (1) ball and flat stick completely. Additional water molecules are present surrounding the double-layers (indicated by the capillary bridge shown in Fig. 2.22). The additional water vanishes by desorption but the double-layers remain in direct contact. At point (2) the first minimum is already passed. The sliding is solid-like, showing a distinct stress value of static friction. Points (3) to (6) demonstrate the transitions from liquid-like

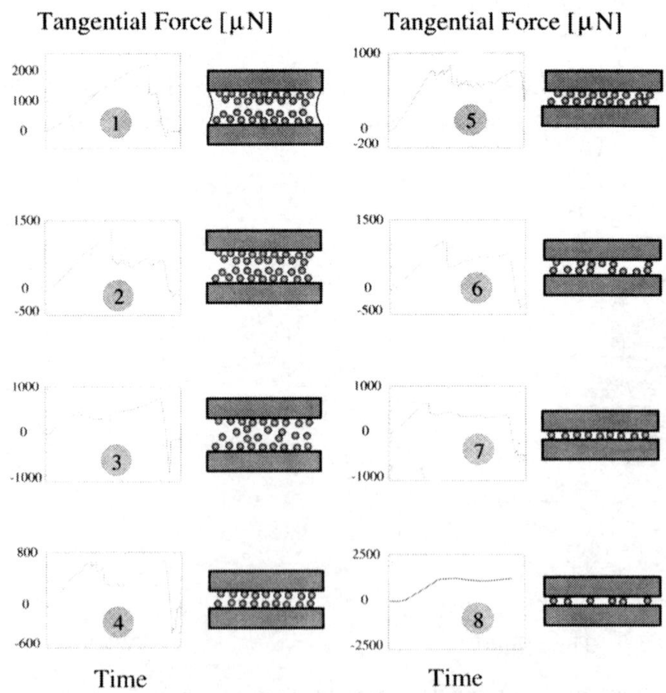

Fig. 2.25. Friction force curves over a period of 3 minutes and proposed molecular arrangement of the water molecules in the slit.

sliding in one of the minima to solid-like sliding in the increasing part of the curve. The static friction and the stick/slip amplitude increase. At point (7) liquid-like sliding is dominant. Point (8) depicts the sliding behavior of the ball when a single monolayer of water is left.

It can be concluded that the first transition – from liquid-like to solid-like sliding – appears when the water film has reached a stable arrangement of four monolayers. In this state the interdiffusion of double-layer water and bulk water ceases. Viscosity-induced friction is now the dominating factor. According to (2.46) the friction force is proportional to the water film thickness. This explains the further decrease of the friction force curve.

Continued desorption seems to destroy the stable four-layer arrangement and the sliding takes on a more liquid-like property. In the transition period between this stable arrangement and the next stable state – probably the three-monolayer arrangement – the friction decreases, since the viscosity is lowered due to the disordering of the water molecules. When the stable three-layer arrangement is achieved, the stick/slips appear again, indicating that solid-like sliding was re-established. Viscosity is again high, though on a decreased level. Finally, the same scenario takes place for the transition from three to two layers.

The final increase of the friction force curve can be explained by the formation of a sub-monolayer water film. The last monolayer is eventually penetrated, giving rise to true electronic adhesion dominated by charge transfer and covalent bond formation [203, 208]. Thus, in this range water acts as a lubricant [209], since it shields the van der Waals forces of the silicon atoms.

When the same experiment is started from vacuum conditions (Fig. 2.26), this means with two dry surfaces in contact, the friction started from large values. By means of an electronically controlled valve water vapor is allowed to enter the chamber. The chamber rests at this pressure for half an hour before the friction is measured. When water is present, the friction decreases at first. With increasing pressure the friction shows similar maxima and minima as in the experiment shown above. This experiment supports the assumption that the changes in viscosity are caused by water in monolayer arrangements. The formation of a slit between the ball and flat sample obviously supports the adsorption of water, leading to somewhat thicker films than on the bare Si surfaces.

Additional support for our results comes from studies performed on a pairing of planar SiO_2 samples as a function of the residual pressure [210]. A similar vacuum system was used to decrease the water film thickness by desorption. A silicon sample (18 mm × 15 mm × 0.4 mm) was placed on a second silicon sample (76 mm × 32 mm × 0.4 mm). AFM roughness analysis showed that the real contact area is reduced to a few contacting spots. By tilting the lower sample the upper sample was allowed to start sliding at a certain tilt angle α. By means of the relation $\mu = \tan\alpha$ the coefficient of friction was determined. The experiment clearly showed decreasing friction

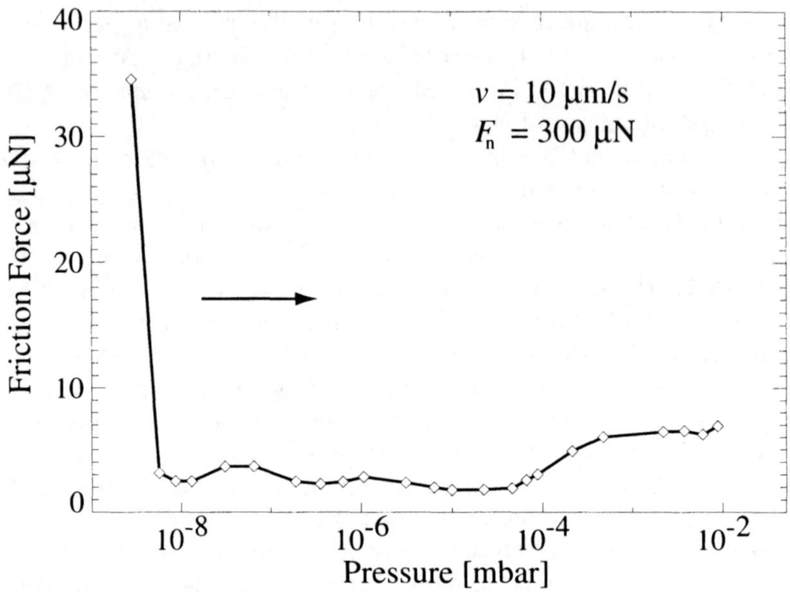

Fig. 2.26. Friction force measurement in a vacuum chamber starting from low pressure. Stepwise the content of water is increased, leading to the formation of water layers between the ball and flat sample.

as the result of decreasing capillarity followed by an increase due to increased viscosity. The final increase due to cohesive action was reproduced also.

Similar experimental results were obtained for a single crystalline Ni(100) surface where friction was measured for a decreasing thickness of an ethanol film. Smooth sliding at a coverage of about 300 monolayers changed into the stick/slip regime around the monolayer thickness, followed by strong sticking in the sub-monolayer range [211].

Our results disagree with the findings obtained with the SFA. It was shown in [65] that confined water shows low friction and no fluctuation in comparison with hydro-carbons. However, since the fluctuations in our tests appear in a force range that is three orders of magnitude lower, the authors assume that the SFA analysis is not sensitive enough to detect those small scale stick/slips.

The results can be summarized as follows:

- Friction decreases with decreasing water film thickness.
- When capillary action ceases to act, subtle influences can be measured.
- Ordering effects in the water film lead to increased viscosity.
- When the water film is removed completely, cohesive interaction (van der Waals forces, chemical bonds) leads to the increase of friction.

2.5.6 Solid Lubrication

Among others (hard coatings [212]) solid inorganic materials such as graphite or molybdenum disulfide (MoS_2) are the most widely used materials for solid lubrication [213–221]. These materials have a layered crystalline structure and show strong anisotropy in their response to shear leading to the sliding of individual layers. In addition, this physical effect is also accompanied by a chemical change of the material (e.g., oxidation of the matrix material [222]).

Another way to obtain solid-like lubricants involves the careful selection of molecular properties leading to a robust lubrication film [223].

2.6 Friction

Friction is an everyday experience. Friction is necessary, e.g., for a clutch to operate properly or for an insect to initiate motion. In this case friction is a desired property. On the other hand, friction means loss of energy, and when friction is accompanied by wear it also means severe damage or destruction. Friction acts in an external way but also internally. External friction is the friction between solid bodies. It defines the resistance affecting a solid body, which retards or even obstructs the movement of the body. Internal friction describes the loss of energy inside of a body, for instance in a liquid [224] or inside a crystal when electrons driven by the electric field interact with the metal lattice (electromigration) [225,226].

Friction works against the relative motion of contacting bodies and expresses itself as friction force, which opposes the acting force, or as friction energy. Macroscale friction is mainly caused by mechanical interlock caused by the roughness of the contacting surfaces. In order to maintain the motion of a body against the friction force, it is necessary to perform work. This process represents, in energy terms, a transformation of mechanical energy into heat energy. The heat that is released during friction is called frictional heat. Its occurrence is one of the reasons why the mechanical energy delivered by a machine is always smaller than the supplied energy. But not only a moving body experiences a friction force. Force is also necessary to overcome inertia and static friction. One has, therefore, to differentiate between static, sliding or rolling friction, according to whether a body adheres, slides or rolls, respectively. Rolling friction appears when wear is involved and debris form a third body.

In approximately 2500 B.C. the ancient Egyptians discovered the fact that their carriages slid better on damp sand and therefore poured water on their pathway. In about 1000 B.C. the Greek philosophers Aristotle and Plato thought about friction and found that there are materials, that "love themselves" and such, which "hate themselves", i.e., materials that attract or repel each other. In the 16th century Leonardo da Vinci dealt with friction and found that the form of a body has obviously no influence on the friction force, but probably does its weight. Two hundred years later Amonton cast da Vinci's approach in the following form, which for a long time remained universally valid:

$$F_{\mathrm{f}} = \mu F_{\mathrm{n}} , \tag{2.48}$$

where F_{f}, F_{n} and μ denote friction force, normal force and coefficient of friction, respectively. Today one knows that the coefficient of friction is not a material constant but that it very strongly depends on the boundary conditions such as temperature and humidity. Additionally, in practice adsorption and/or oxidation change the properties of the pure materials, which has a dramatic influence on the frictional behavior. Charles-Augustin de Coulomb justified the simple laws of sliding friction and examined in 1801 the internal friction of liquids by means of torsional motion.

The history of microfriction analysis is much shorter. Although it was known for a long time that the macroscopic behavior of friction is determined by microscopic origins, the equipment to probe the materials in this particular length and force range was not available until the mid-1980s. Table 2.3 summarizes the main differences between the macro- and microfriction phenomena. The normal force ranges are clearly different. However, since the

Table 2.3. Differences between the macro- and microfriction

Parameter	Macrorange	Microrange
normal force range	> 1 N	< 1 N
ratio of real to geometric contact area	$A_{\mathrm{c}} \ll A_{\mathrm{g}}$	$A_{\mathrm{c}} \leq A_{\mathrm{g}}$
impact of wear	large	small, often negligible
heat generation	large	negligible
impact of surface properties [a]	negligible	dominant

[a] with respect to the upper atomic layers

real contact area in macrosystems is only a fraction of the geometric contact area, the microtribological pairings can be regarded as model systems for the macrocontact. All tribo-systems, no matter on what scale, experience identical physical influences. However, the impact of these influences on the overall behavior of the system strongly depends on the investigated length and force scale. While wear is a common process in a macrosystem, microsystems show negligible wear when the solids are separated by adsorbed liquid films. Therefore, the impact of physical and chemical influences depends on the level of activation, as shown for silicon [227]. Since heat generation in microsystems is low, chemical changes of the adsorbates can be neglected. As shown in Sect. 2.2, the contact area of our SILICON model system of 90 μm^2 and a normal force of 500 μN result in a contact pressure of about

5 MPa. This value is too low to initiate severe change in the material or in the chemistry of the adsorbate coverage. The only change that is conceivable is a rearrangement of the molecular structure [228].

One question that is still under intensive discussion is the actual atomic origin of friction [229–232]. Friction means that in order to maintain the sliding of a body a permanent force has to act on this body. Without this force the motion would stop. Along a distance x a certain amount of energy has to be spent:

$$W = \int_{x=0}^{x=x_1} F_t \mathrm{d}x \ . \tag{2.49}$$

Because of energy conservation this energy input is transformed into energy of motion and an energy loss

$$W = \frac{m}{2}v^2 + W_{\mathrm{ph}} + W_{\mathrm{el}} \ . \tag{2.50}$$

The energy loss is caused by atomic lattice vibration (phonons) W_{ph} [233, 234] due to shear-induced displacement of atoms with respect to their equilibrium position. When an unstable configuration is reached the excited atom jumps back to its initial position and thus energy is dissipated in the form of vibration transferred either continuously by acoustic waves or quantized by phonons, finally becoming transformed into heat [235]. The other contribution to friction is of electronic origin W_{el} [236] caused by electron scattering, the generation of electron/hole pairs or the formation and breaking of adhesive bonds [237,238]. In general, the electronic contribution is considered weak. However, if there is a significant electronic coupling between the molecules and the substrate, electronic friction can dominate phononic friction [239]. All contributions act on a timescale between pico- and nanoseconds [240].

The described mechanisms apply very well to unlubricated systems. However, real-life microprocesses show that the friction between two interacting bodies is always accompanied by a third body, the adsorbate film between both surfaces. Simulations demonstrated that this third body is responsible for an energy minimum without deforming the surface atoms, leading to non-zero static friction [241]. The only case for zero static friction is when the surfaces have similar lattice constants (registry of surfaces=commensurability) [242]. Thus, the number of atoms that lead to energy loss is equal to the number of atoms that gain energy from the sliding process.

Friction measured on different length and force scales very often shows instabilities expressed in periodic stick/slip cycles. Moreover, it seems that stick/slips occur on any length scale [243]. On the largest scale earthquakes are accompanied by stick/slips [244–246]. Machine chatter, squeaking doors or violin playing are examples of stick/slips on the macroscale [247]. In the microworld stick/slips appear not only in micromechanical systems [248,249] but also in biological systems. Finally, at the atomic level stick/slips were revealed with the AFM [250,251]. Whereas the measured curves look similar, the origins are very often different [252], as shown in the following sections.

2.6.1 Macroscale Stick/Slips

In macrosystems roughness is the main controlling factor of friction [253, 254]. Due to mechanical interlock followed by an abrupt release of the force stick/slips appear. Here, frequency and amplitude of the stick/slips are a function of the mechanical parameters as for instance mass and stiffness of the system [255]. If no wear is involved, then the asperities of one solid have to slide over the asperities of the other solid. This effect can hardly be measured, since the macroscopic surface consists of many asperities, averaging this effect. However, the simulation of a one-asperity contact sheds some light on these hidden processes. To model such a contact the interaction of a diamond pyramid – attached to a spring – and a corrugated surface was investigated, (Fig. 2.27a, b). The use of a diamond pyramid reduces the capillary action due to minimization of the contact area, thus confining the interaction to topography aspects only. It can be seen that each time when the spring deflects

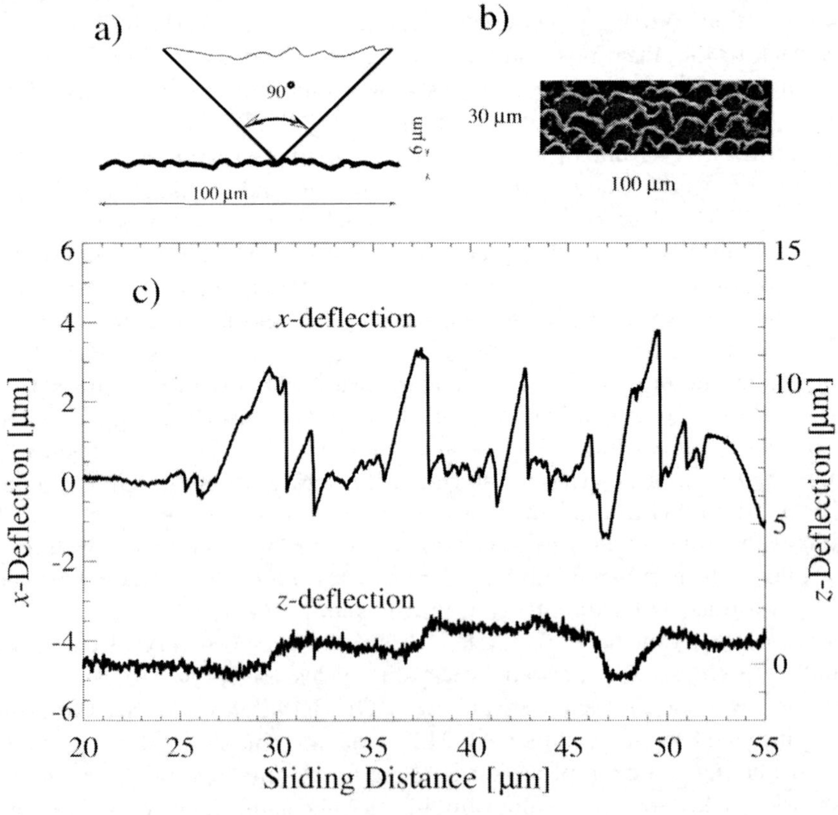

Fig. 2.27. (**a**) Test setup comprising a diamond pyramid scanning a corrugated SiC sample. (**b**) SEM image of a 30 μm × 100 μm sample. (**c**) Deflection of the spring in the sliding direction (friction) and perpendicular to the direction of sliding.

in the x-direction, a z-deflection also occurs. This means that the diamond has to climb up the asperity. Once the summit is crossed the slip releases energy that is stored in both the lateral as well as in the vertical directions. Figure 2.27c shows that the train of stick/slips is as random as the shape of the profile [256]. This effect averages for systems with many asperities and the train of stick/slips resulting from the overall mechanical properties of the system can be periodic.

When wear is involved, stick/slips tend to vanish. This is caused by the onset of rolling friction as part of a three-body effect. Due to the breaking of asperities debris is formed – for the case of hard and brittle surfaces. Softer surfaces show the effect of ploughing [257]. Here one of the sliding partners digs into the surface of its counterpart. This effect can also be accompanied by stick/slips.

2.6.2 Microscale Stick/Slips

In microsystems the occurrence of stick/slips depends both on interlock as well as on viscosity phenomena. The occurrence of friction instabilities is explained with the help of the SILICON model system. As the flat sample wafer silicon with different roughness profiles was used. In the first experiment the ball contacts a polished silicon flat (set I) while in the second experiment a lapped Si flat was used (set II). With this combination a significant difference in roughness was achieved, as shown in Table 2.4. Before the experiments the ball and flat sample were etched with hydrofluoric acid (HF) for 2 minutes to remove the oxide film followed by a rinse in distilled water. The HF dip leads to hydrogen termination and thus to a hydrophobic surface (see Sect. 5.3).

Table 2.4. Sample sets I and II

Set	Description	$R_{\mathrm{a,ball}}$ [nm]	$R_{\mathrm{a,flat}}$ [nm]
I	ball on polished Si wafer	0.2	0.17
II	ball on lapped Si wafer	0.2	1200

Figure 2.28 shows the tangential force curves for both sets. The left-hand column shows the friction force curves obtained from set I, while the right-hand column depicts set II. The upper two graphs, (a) and (b), were taken immediately after the chemical treatment. Graphs (c) and (d) were taken after one week and graphs (e) and (f) after two weeks. Between the measurements the samples were stored in humid air (r.h.=80%). All curves show similar values with regard to static and dynamic friction and in all curves stick/slips are visible.

A significant change in the sliding properties of set I was observed as a function of the storage time in the humid environment. Starting with random stick/slips at the beginning, the fluctuations became periodic after one week. After two weeks the stick/slips almost vanished. Set II showed no dependence on the storage time. The stick/slips are randomly distributed with respect to time and amplitude.

To explain the stick/slips the roughness argument does not hold for set I. For this case interlock phenomena can be neglected, as AFM roughness analysis proved. Due to the initial HF etching the Si surfaces were made hydrophobic, as shown by the large contact-angle of $\Theta > 90°$. Thus, both surfaces get in intimate contact and either van der Waals forces or chemical bonds pull both solids together. The stick/slips occur, since the shearing force breaks the bonds and frees the ball for a short time until a new attraction is generated when the ball (and the spring) returns to its initial position.

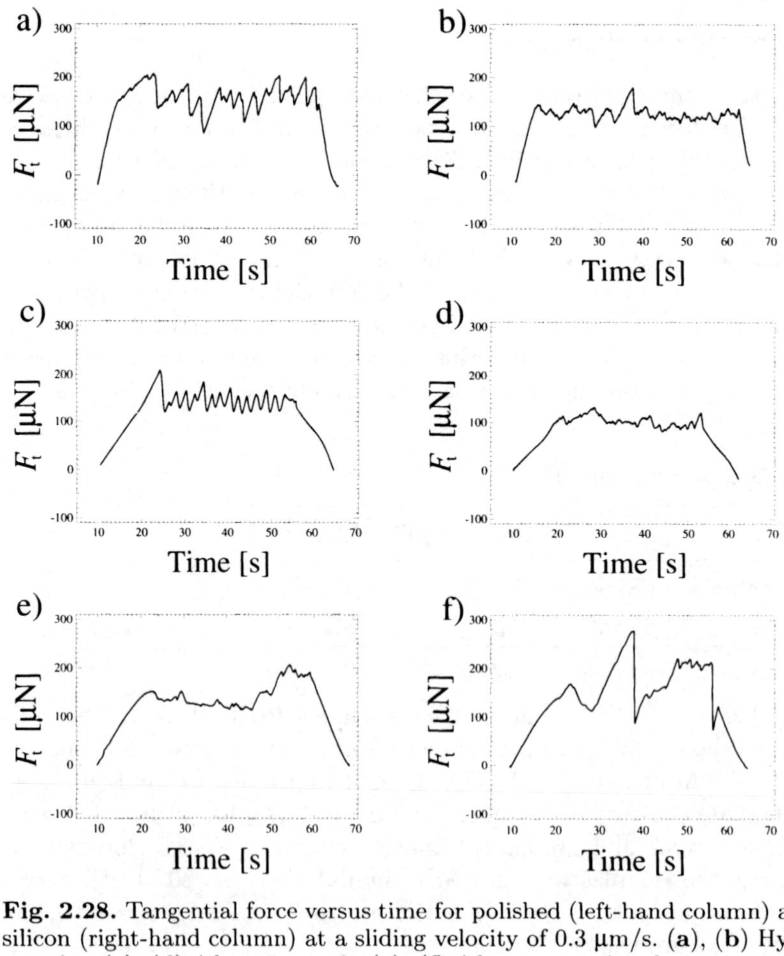

Fig. 2.28. Tangential force versus time for polished (left-hand column) and lapped silicon (right-hand column) at a sliding velocity of 0.3 μm/s. (**a**), (**b**) Hydrophobic samples. (**c**), (**d**) After one week. (**e**), (**f**) After two weeks of storage in humid air.

After one week the force curve (c) showed that the ball periodically sticks and slides. The contact-angle measurements indicated that the surface is not completely hydrophobic anymore, but is rather in an undefined state between hydrophobic and hydrophilic. Thus, the surface was wetted by a certain amount of water. It was shown in Sect. 2.5 that the viscosity of a thin liquid film increases dramatically when the film has a thickness of just a few monolayers. Furthermore, when this liquid is confined between the sliding partners, this effect is amplified. As a consequence, the sticking mechanism is caused by the increased viscosity of the liquid.

After two weeks of storage in a humid environment the water coverage of the ball and flat sample increased further. As a result, the friction force curve of the polished sample showed no drastic stick/slips anymore. The sliding changed from solid-like to liquid-like behavior. This is also visible from the smooth transition from static to dynamic friction.

In contrast to the phenomena described above the stick/slip behavior of set II can be explained mainly by roughness effects. In micro- and macroscale experiments mechanical interlock is a function of roughness in combination with normal force. In Fig. 2.29 both the ball and flat sample are displayed in scale. It is evident that the ball can be pressed into the grooves of the lapped Si surface. The higher the normal force, the more tangential force must be exerted to maintain the sliding. The stick/slips are, therefore, as

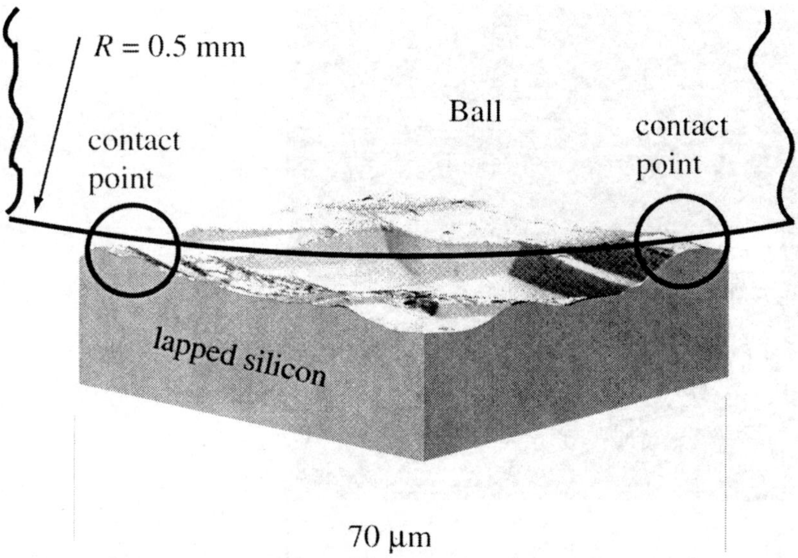

Fig. 2.29. AFM image of the topography of the lapped Si surface in contact with the ball. The actual contact region is reduced to two singular spots.

random as the profile of the lapped sample. With increasing water coverage the stick/slips are stronger due to additional capillary action.

2.6.3 Nanoscale Stick/Slips

Nanoscale interaction can be investigated with the AFM. Here, the contact is reduced to a few atoms, e.g., the top atom of a cantilever tip and the surface atoms of the sample [258]. Stick/slips appear, since the AFM tip has to climb up the potential hill of surface atoms. The increase of the force is followed by the sudden drop caused by the down motion at the other side of the potential hill. According to the atomic arrangement, the strength and frequency of the stick/slips depend on the direction of sliding [259–261]. To measure the stick/slips on this scale an ultra high vacuum AFM was used. Figure 2.30 shows a 5 nm × 5 nm image of freshly cleaved highly oriented pyrolytic graphite (HOPG). Instead of the expected hexagonal structure a trigonal symmetry of the potential maxima was found. According to [262–264] this behavior is caused by the interaction of the tip and surface potential. The tip tries to keep a path of lowest energy. This leads to a zig-zag course from one potential minimum to the next, resulting in stick/slips. As a consequence, only a fraction of the surface atoms takes part in the scanning process. This also means that the tip rests most of the time in the potential minima. For increased sliding velocity the stick/slips decrease in amplitude,

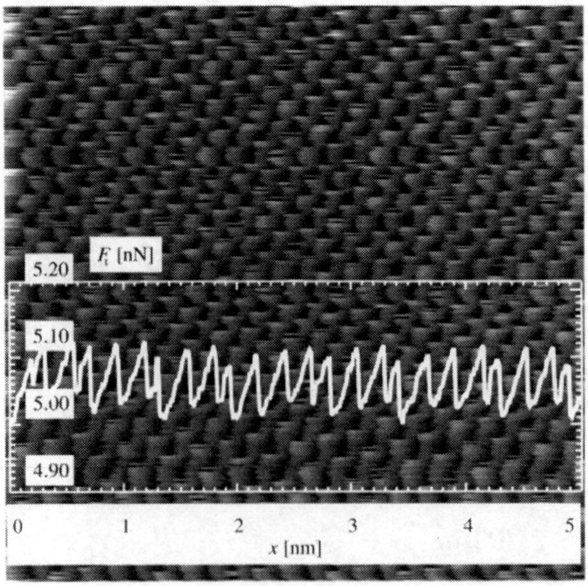

Fig. 2.30. Topography and stick/slips measured on graphite by AFM in vacuum. The lateral-force signal shows stick/slips that can be related to the location of the potential maxima and minima.

since the relaxation time, i.e., the time to bring the tip back to its initial position, decreases.

It has to be pointed out that – as in microsystems – the HOPG surface has to be dry. Adsorption of water plays a far more important role than on the microscale. Due to capillarity the stick/slips are damped, leading to a blurred image.

2.6.4 Stick/Slips and Sliding Velocity

In this section friction instabilities will be investigated as a function of the sliding velocity. While the stick/slips occurring in model sample set II (Fig. 2.28b) were more or less erratic, a characteristic frequency can be assigned to the stick/slips of set I (Fig. 2.28c). Therefore, the stick/slip range was Fourier transformed to identify the characteristic frequency f_{ss} (for details see Sect. 5.1.7). The obtained frequency is characterized by the mechanical parameters of our SILICON model system. For sliding velocities greater than 5 µm/s a characteristic frequency could hardly be obtained, since the stick/slips were erratic and the spectra became increasingly noisy. With increasing sliding velocity the frequency increased and the stick/slip amplitude (ΔF_{f}) became smaller.

Based on the transition model, describing liquid-like to solid-like sliding transitions of systems confining thin liquid films [265], a critical velocity was determined by

$$v_{\mathrm{crit}} \approx \frac{\Delta F_{\mathrm{f}}}{5 k_x \Delta t} , \tag{2.51}$$

where Δt is the slip time and $k_x = 50$ N/m is the spring constant in tangential direction. Using Fig. 2.28c the following values were inserted: ΔF_{f}=70 µN, Δt=700 ms. This resulted in a critical velocity of about 0.4 µm/s, meaning that a system sliding with $v > v_{\mathrm{crit}}$ experiences no stick/slips. This is in good agreement with our SILICON model system experiments.

It turns out that in addition to the static and dynamic friction forces, the curves can be characterized by the stick/slip amplitude and the stick/slip frequency, (see Figs. 2.31 and 2.32). Static and dynamic friction are plain numerical values. ΔF_{f} and f_{ss}, however, reflect the type of sliding. Solid-like sliding is characterized by stick/slips. If the sliding is liquid-like, no stick/slips can be detected and an initial stress overshoot $\hat{\sigma}$ (see inset of Fig. 2.31) is followed by a smooth and almost fluctuation-free friction curve. The velocity-induced change from sliding with stick/slips to sliding without is assumed to be dependent on the decrease in viscosity of the thin confined water film (in transition between hydrophobic and hydrophilic). With increasing sliding velocity the ordering of the water film is disturbed, leading to lowered viscosity. The experiments showed a critical velocity below which the sliding shows stick/slips. Above v_{crit} the stick/slips disappear. For higher velocities it is assumed that the ordering of the water layers is destroyed. As a result,

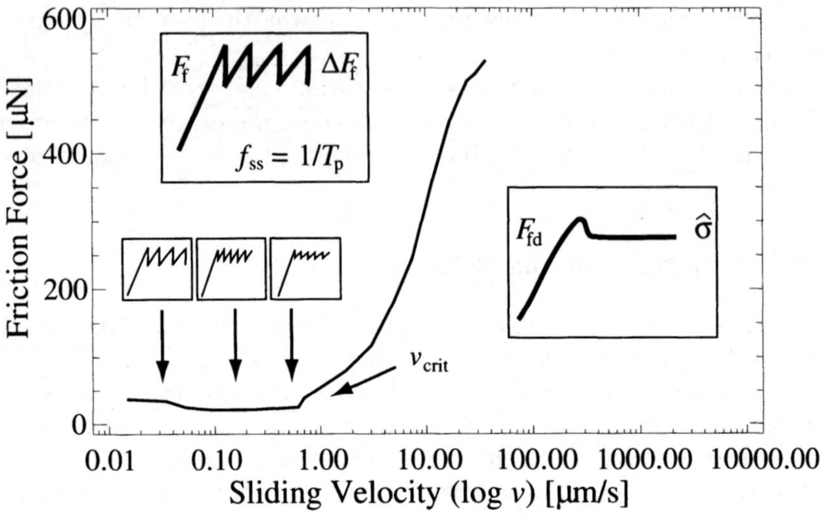

Fig. 2.31. Friction force as a function of sliding velocity for the SILICON model system. The insets depict the sliding behavior for velocities lower and higher than v_{crit}. For increasing sliding velocity below v_{crit} the amplitude of the stick/slips decreases while the frequency increases.

the water film is not able to separate the ball and flat Si sample. Both solids contact, leading to higher friction as a function of the sliding velocity, as explained for dry surfaces.

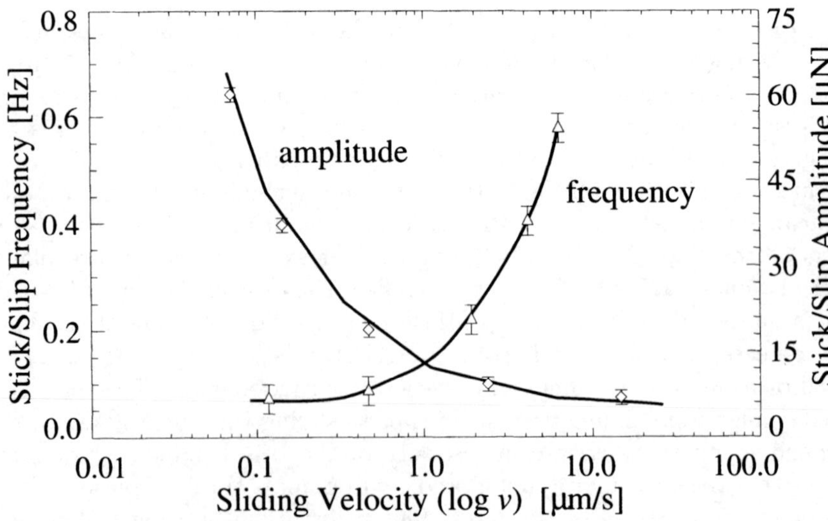

Fig. 2.32. Frequency f_{ss} and amplitude ΔF_f of stick/slip events as a function of the sliding velocity. For the definition of ΔF_f see Fig. 2.31.

In summary:

- For dry sliding conditions stick/slips are mainly caused by roughness effects; this means by mechanical interlock.
- Ploughing the material as well as the periodic breaking of adhesive contacts can cause stick/slips too.
- In macrosystems stick/slip sliding is isotropic, since both the large normal force and the large contact area average the friction force.
- With shrinking dimensions, the variety of influences increases. Besides interlock also capillary forces, wettability, surface termination and liquid layer confinement are crucial factors for both the magnitude of the friction force and the instabilities.
- Stick/slips can be caused by increased viscosity due to confinement of liquids in the contact area.
- Stick/slips appear below a critical sliding velocity v_{crit}.
- Nanoscale stick/slips are caused by atomic roughness (potentials).
- The strength and the frequency of nano-stick/slips depend on the direction of sliding. This behavior is caused by the atomic surface configuration.

2.6.5 Friction and Sliding Velocity

In this section we consider the sliding behavior of dry- and water-lubricated surfaces. The differences at the macroscopic and microscopic levels are discussed. Water as the lubricant was selected, so that conclusions could be drawn for systems with an identical intermediate liquid.

Macro- and microtribological systems show a dependence of the friction force on the sliding velocity (the so-called Stribeck curves) for dry as well as for lubricated systems (see Fig. 2.33). There are velocity ranges where the sliding is fluctuation-free and others with strong fluctuations, stick/slips or other instabilities. Generally, a decrease of the friction force down to a minimum followed by an increase with increasing sliding velocity is measured. Depending on the sliding velocity, mechanisms of dissipation are activated or deactivated. The stronger the loss mechanism works, the more energy must be delivered subsequently, in order to keep the sliding process upright. That means that with constant normal force a larger tangential force must be exerted. As mentioned before, the results obtained from macro- and microscale experiments are very similar although the force range clearly differs [266].

While macrosystems can be lubricated actively with water [267], generally a liquid film (mainly water) exists due to adsorption on the surfaces of microsystems, provided that those surfaces can be wetted. In macrosystems the sliding velocity dependence can be described very well phenomenologically by differential equations, since the sliding partners are coupled over oscillating systems to the environment [255]. In the equations of motion viscous damping is commonly used as the loss term. Computer simulations and measurements have shown that fundamentally four ranges can be differentiated in the friction force/sliding velocity curve (see Fig. 2.33a). Within the range of small velocities one finds approximately constant friction (range I, boundary lubrication). The friction falls with rising sliding velocity (range II, mixed lubrication) to a clear minimum (range III, elastohydrodynamic lubrication). Then the friction again rises (range IV, hydrodynamic lubrication). One assumes within the range I and II that the friction partners squeeze the lubricant partly or completely out of the contact and get into direct mechanical contact with their asperities. The mechanical interlocks are then loosened periodically as a function of the mechanical constants of the sliding system, and one registers stick/slips. The inclination to stick/slips vanishes up to the minimum. Due to the water in the contact, the mechanisms of dissipation, e.g., the generation of acoustic waves in the solid, are blocked, and the friction is small (range III). In this range the system is unstable and can fall into self oscillations. The increase of friction within the range IV is caused by the velocity dependence of viscous friction. This property is valid even for liquid thicknesses of just a few nanometers [268]. With higher sliding velocity a closed liquid film between the friction partners forms with increasing

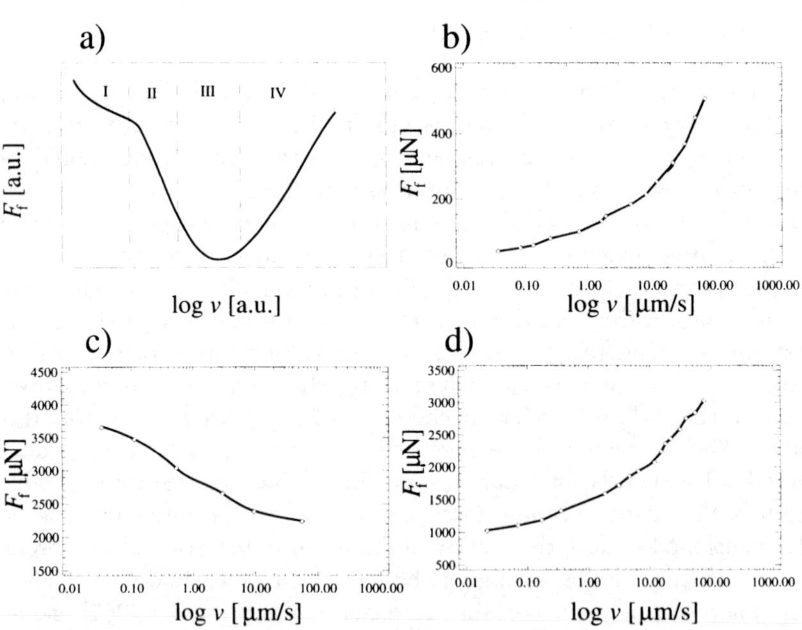

Fig. 2.33. Different types of Stribeck curves. (**a**) Schematic for a lubricated macrosystem. Four areas of lubrication can be identified. (**I**) boundary lubrication, (**II**) mixed lubrication, (**III**) elastohydrodynamic lubrication and (**IV**) hydrodynamic lubrication. (**b**) Dry microsystem. (**c**) Water-lubricated microsystem. (**d**) Microsystem in vacuum.

mechanical resistance. With water this increase is very small [269], since the viscosity relative to oil is small.

Dry Systems

The fundamental dissipation mechanism in dry systems is the generation of phonons induced by mechanical interlock of asperities with sizes of micrometric order [270]. This is suggested in the Frenkel–Kontorova–Tomlinson model [271–273], which is very similar to models describing earthquake mechanisms [244,245]. Nanoscale vibrations are excited by the contact of the surface atoms, which relax over phonons very effectively. If the path of dissipation is blocked, e.g., by phonon resonance [274], the loss of energy and therefore also friction is small. In addition to phonons friction on metal surfaces also induces low-energy electron–hole pairs in the substrate. When the metal surface is covered by a single adsorbed layer of Xe, as shown in numerous experiments by Krim et al. [233,240,242,275,276], friction can also be caused by the generation of phonons within the adsorbate layer [277].

In macrosystems generally the friction force increases with decreasing sliding velocity. This is justified by an enlargement of the contact area with increasing retention time in the contact at small sliding velocities. This effect has negligible influence on microsystems, especially on our SILICON model system, since the contact pressure is too low to enlarge the contact area. In addition, ploughing through asperities as a source of dissipation [278] hardly applies to the microrange because the forces are too small to induce plastic deformations.

Dry systems rarely occur in reality, since adsorption usually forms a liquid film on the surface of the solid. However, there are conditions, in which friction pairings can be understood as dry. In the macro-area one speaks of quasi-dry systems, if the normal force is so large that the liquid (this means the applied lubricant) is pressed out of the contact. In this case the surfaces come directly into contact. In macro- as well as in microsystems the dry state can be generated artificially in a vacuum chamber (Fig. 2.33d). Alternatively, hydrophobization of the ball and flat sample leads also to dry conditions. Figure 2.33b (hydrophobic surfaces) and Fig. 2.33d (vacuum) show that the friction force increases with increasing sliding velocity. Although in both cases the normal force was equal, the friction force showed higher values in the vacuum experiment. This means that besides inertia (the ball and spring try to remain in the deflected position during sliding) adhesive components are involved as loss mechanisms. It is known that adhesive junctions can form by adhesion avalanches on a time-scale that is orders of magnitude lower than the time-scale of sliding [69,208], meaning that adhesion is always present as a source of increased normal force. Whereas in the hydrophobic state the Si surface is water-free but terminated by hydrogen, in the vacuum experiment the bare Si surfaces contact, leading to higher adhesion due to van der Waals interaction or the direct formation of Si bonds. Therefore, the friction force in vacuum is higher than for the H-terminated Si surface.

Nanoscopically, at low sliding velocities there is always the chance for elastically deformed asperities to return to the undeformed state. Stick/slips can lead to a momentary loss of contact, and thus to relaxation [279]. Higher sliding velocities reduce the time for the asperity to return to the undeflected (i.e., zero friction) state, leading to permanent a deflection. This is even more pronounced at higher sliding velocities. A similar behavior was also found in a single-asperity AFM experiment [280], showing that the cantilever remains in a deflected state (torsion) when the sliding velocity is high enough.

Confined Systems

When the thickness of the water film is in the particular range where confinement-induced effects of molecular ordering occur, then the friction force – velocity curve has the form shown in Fig. 2.31. As in dry systems, the friction also increases. However, the sliding behavior is different, showing a transition from solid-like sliding to liquid-like sliding with increasing sliding velocity. As long as the sliding velocity is small, the shearing force acts on the ordered molecules and experiences the increased viscosity. Above v_{crit} the water film remains disordered. This may lead to direct contact of the solids, resulting in a behavior like dry systems, i.e., increasing friction with increasing sliding velocity.

Lubricated Systems

Unlike lubricated systems on the macroscale, microsystems show only decreasing friction without any minimum as a function of the sliding velocity (Fig. 2.33c). The main dissipational mechanism is generated by the capillary effect. Increasing sliding velocity reduces the time to re-establish a capillary bridge; thus the wetted area becomes smaller and according to (2.40) the friction decreases. This effect is more pronounced when in addition to sliding velocity also the relative humidity is changed. The hydrophilic surfaces of the ball and flat sample have a common water film with a thickness as discussed in Sect. 2.5.4. Higher relative humidities lead to increased friction and to a decrease of the slope of the curves (see Fig. 2.34). Due to shear the capillary bridge is deformed. The faster the ball slides, the less time is available to restore the capillary bridge. This effect is more pronounced for low relative humidities and the curve falls strongly with increasing sliding velocity.

Although there are some monolayers of water between the ball and the flat sample, the increase of friction due to viscosity effects is masked by capillary action. The water that is subject to viscosity influences is located only in the contact region of the double-layers (see Fig. 2.22), whereas the area that is involved in the capillary effect is much larger and therefore stronger, covering the effect of viscosity.

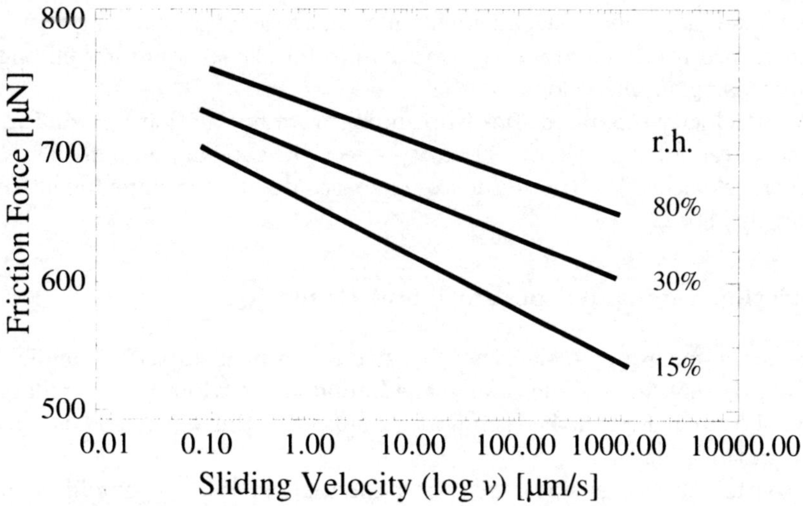

Fig. 2.34. Linearly fitted Stribeck curves of the hydrophilic SILICON model system. The parameter is the relative humidity.

The results can be summarized as follows (see also Fig. 2.35):

- The amount of water available on the surface as well as the state of order determine the velocity dependence of friction.
- Completely dry surfaces show an increase of friction due to solid–solid interaction and reduced mechanical relaxation time (i.e., time for the spring to return to its undeflected state).

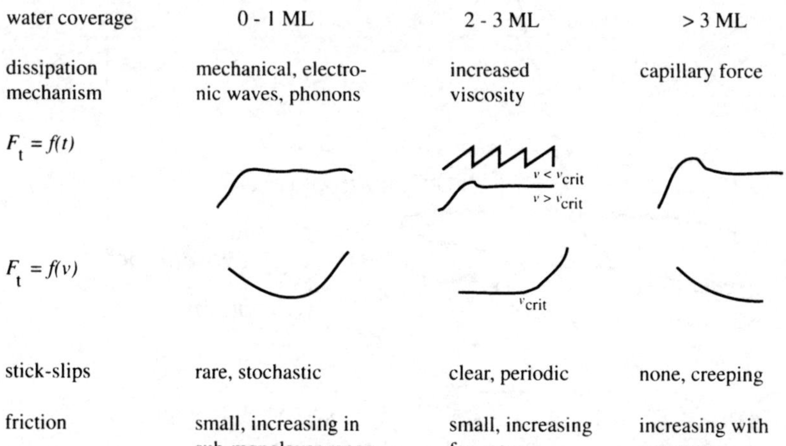

Fig. 2.35. Dissipation mechanism, sliding behavior, velocity dependence and occurrence of stick/slips as a function of the water coverage for the SILICON model system. ML = monolayer.

- Water films in a confined arrangement show increased viscosity and, therefore, increased friction and a transition from solid-like to liquid-like sliding with increasing sliding velocity .
- A critical velocity separates stick/slip sliding from fluctuation-free sliding.
- Thicker water films mask the viscosity effect by capillary action. Thus, at higher velocities, the friction force decreases due to the impediment of capillary bridging.

2.6.6 Friction versus Temperature and Humidity

Whereas in the macrorange the humidity strongly influences tribo–chemical reactions [281–285], in the microrange the humidity contributes to capillary action as shown in Fig. 2.34. Temperature supports this effect and is also discussed in this section.

The friction force was measured as a function of relative humidity at different temperatures for the hydrophilic and hydrophobic SILICON model system. The temperature was taken as the variable parameter. The tester was placed in an environmental chamber. Before each experiment the humidity in the chamber was set to 85% and then decreased sequentially to 20% by purging the chamber with dry nitrogen. Figure 2.36 shows the results for low and high temperatures. For each cycle the normal force was kept constant at F_n=2000 μN.

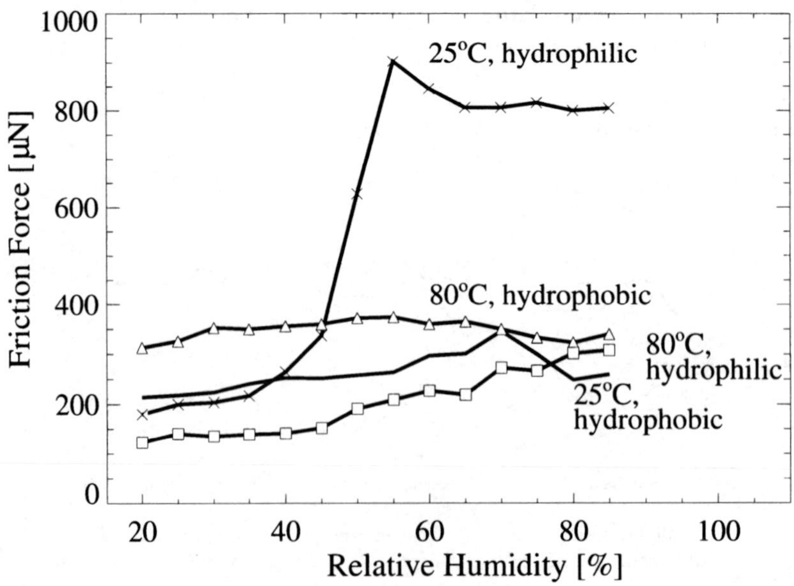

Fig. 2.36. Friction force versus relative humidity for the hydrophilic and hydrophobic SILICON model systems. The parameter is the sample temperature, showing higher adsorption for lower values.

A hydrophilic sample surface can retain a large amount of water. Thus, the more humid the environment is, the more water can be adsorbed, and the higher is the friction force. When the temperature is raised, desorption becomes stronger than adsorption and the friction decreases. At each temperature the surface finds its equilibrium point in a constant ratio of adsorption and desorption. A change in sample temperature shifts this equilibrium point. The higher the temperature of the sample is, the more energetic are the water molecules, and thus the more readily they can leave and return to the surface. This explains the weak dependence of the friction force on humidity changes at high temperatures. At low temperatures it is obvious that the sample surface can adsorb more water. At a certain point the thickness of the water film stays constant and the friction force saturates.

In contrast, hydrophobic Si shows a weak dependence on humidity. When the temperature is raised the friction force increases slightly. It is possible that at higher temperatures the hydrophobic termination by hydrogen is not stable, leading to a partly H-terminated surface. The desorption process may also be enhanced by friction. As a result, the solids come into more intimate contact, allowing short-range van der Waals forces or chemical bonds to act [286].

2.6.7 Friction and Contact Geometry

It was shown in Sect. 2.4.3 that capillarity strongly contributes to adhesion and also to friction. Of great importance is the magnitude of the real area of mutual contact and the separation of the surface atoms not in direct contact with each other [88,287–289]. Besides roughness these values also depend on asperity deformation [67,290] and the elasticity of the bulk material.

In order to achieve a gradual modification of the surface profile, the polished silicon sample was replaced by corrugated material. A profile analysis performed with a tactile profilometer is shown in Fig. 2.37. The lateral resolution of the profilometer was 32 nm for a scanning range of 1 mm. As probe, a cone-shaped diamond was used with a radius of 1.5 μm. Images obtained by a scanning electron microscope showed that the specification of the profilometer is adequate for the surface to be investigated. This means that the size of the diamond does not introduce artifacts (see Sect. 5.2.4). The surface is characterized by a sequence of bumps. From sample S1 to S7 the bump diameter and height increase; however, the height distribution is not uniform.

The friction tests with ball against corrugated sample were performed in air as well as in high vacuum. The normal force was set to 800 μN. The sliding distance was 35 μm at a sliding velocity of 4 μm/s. Figure 2.38 shows friction force and roughness values.

It can be shown that the friction force decreased with increasing roughness. This behavior is caused by the decreased effect of capillary bridging. The rougher the surface, the smaller is the contact area and thus the area

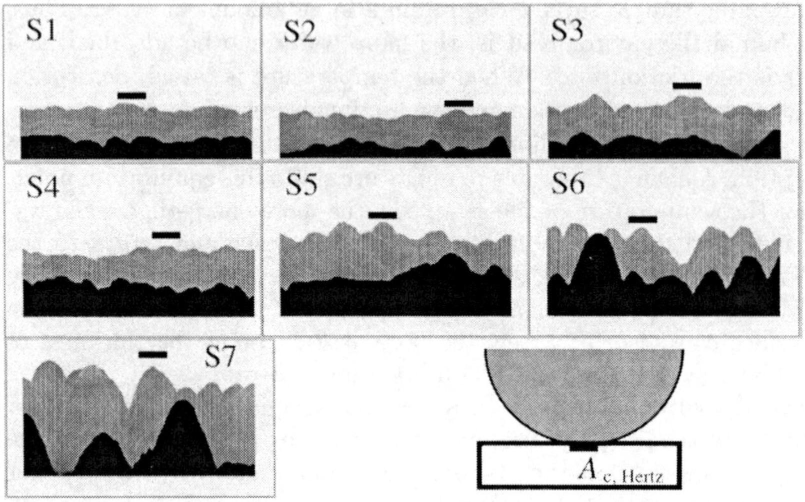

Fig. 2.37. Profiles of corrugated Si samples measured with a tactile profilometer. The scan range is 100 μm. The horizontal bar indicates the Hertzian contact area for the SILICON model system.

that can be wetted. The lower right-hand corner of Fig. 2.37 depicts the con-

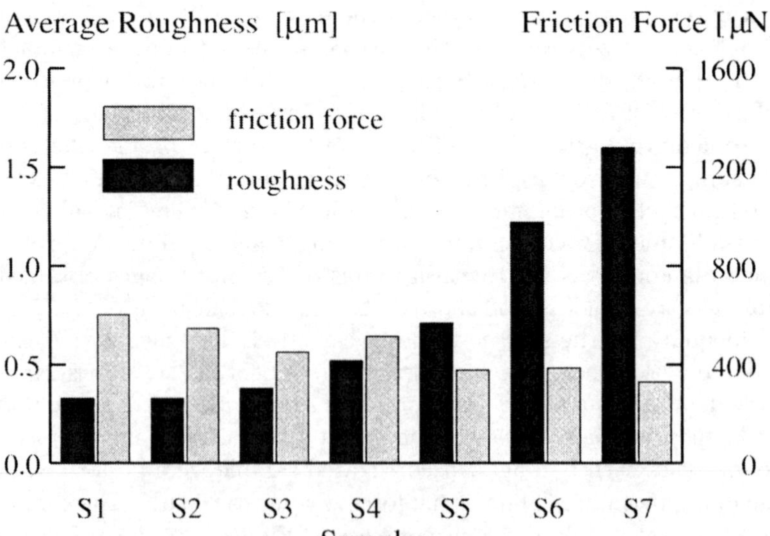

Fig. 2.38. Roughness measured by tactile profilometry of the corrugated samples and friction measured in air. Increased roughness leads to reduced capillarity and, thus, to lower friction.

tact area that would be formed between the ball and a flat sample according to the Hertz model. It is evident how the corrugation affects the common area of real contact.

In addition, the effect of decreased water film thickness as a result of minimized surface tension (see Sect. 2.4.3) leads also to reduced capillary force. Water bridges can only be formed at locations where large asperities come in contact with the ball. As documented in the profiles the difference in height between the largest and smallest asperities is sufficiently high to prevent any further capillary bridging. Accordingly, the friction force is low.

To underline these findings experiments with different humidities were performed (see Fig. 2.39). The tester was placed in the environmental chamber where the relative humidity was varied between 20% and 85% (for details see Sect. 2.6.6). Whereas the rough sample S7 is nearly unaffected by humidity changes, sample S1 shows a strong increase. To exclude the influence of water completely, the experiments were then carried out in the vacuum chamber at a pressure of 2×10^{-6} mbar. By desorption most of the water was removed. As a result, the friction force increased with increasing roughness due to mechanical interlock phenomena. This behavior is very similar to that in macroscale experiments, where capillarity plays a minor role.

A second effect that influences friction is the curvature of the asperities. With increasing roughness a larger number of individual asperities are visible.

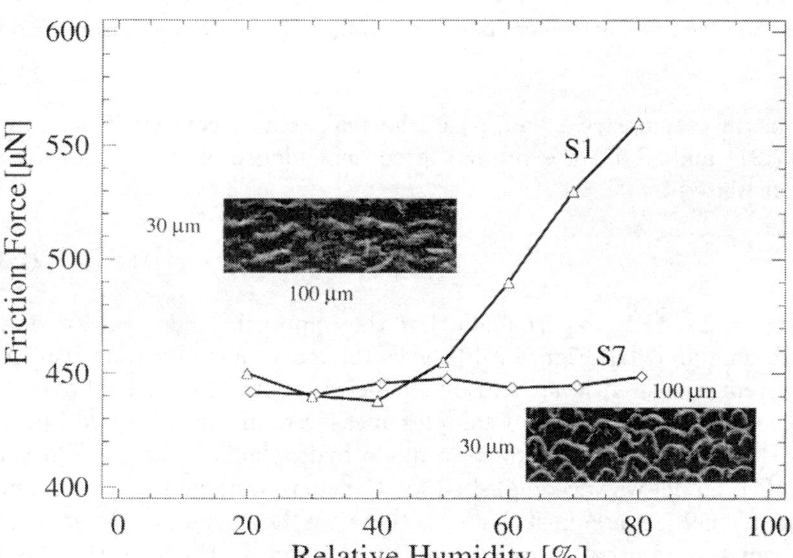

Fig. 2.39. Friction force versus relative humidity. The two insets show the surface of a smooth (**S1**) and a rough (**S7**) surface. The rougher the surface, the smaller is the impact of increased humidity, since the contact of the ball and rough sample occurs only at the tops of the asperities.

These asperities have an increased curvature. This is shown in Fig. 2.37. While sample S1 makes a large contact with the ball sample S7 has only punctual contacts. Since capillary force – increasing normal and friction force – is proportional to the radius of the asperity (see Sect. 2.4.3) a higher curvature leads to decreased friction. This was also shown for nanometer-sized contacts between an AFM tip and a silicon surface with regular asperities generated with a focused ion beam. Both pull-off force and friction decreased with increasing asperity curvature [291].

The findings can be summarized as follows:

- Decreasing contact area decreases friction due to lower probability of liquid bridging.
- Increasing roughness is connected with increasing asperity curvature. The capillary force decreases.
- If the water is removed from the contact, the friction force behaves as in a macrosystem and shows higher values with increasing roughness.

2.6.8 Normal Force Dependence

The normal force dependence of our SILICON model system can be described by a combination of the Bowden Tabor (BT) model (2.52) and Hertzian contact mechanics (2.20) for low water coverage. When the normal force increases, the friction force increases accordingly, but with decreasing slope:

$$F_{\mathrm{f}} = \tau A_{\mathrm{c}} \,, \qquad (2.52)$$

where τ is the shear stress and A_{c} is the real area of contact. Combining (2.20), (2.21) and (2.52) the normal force dependence of the friction force can be calculated by

$$F_{\mathrm{f}} = \tau\pi \left(\frac{RF_{\mathrm{n}}}{K}\right)^{\frac{2}{3}} \,. \qquad (2.53)$$

It was shown by AFM experiments that the approach holds also for small contact diameters [292]. Figure 2.40 shows the load curves for the SILICON model system measured in air and under high vacuum (10^{-7} mbar) and the curve according to (2.53). The samples measured in air were hydrophilic whereas the samples in vacuum were made hydrophobic. The curve fitting with the BT model was accomplished for the curve obtained in the vacuum experiment, since it is assumed that only there are the silicon samples in direct contact, not the adsorbate films or the oxide layers. In this case the elastic constants (Young's modulus and Poisson ratio) for Si as given in Table 2.4 were used. The best fit was obtained for $\tau = 50$ MPa. This value is similar to data obtained with the SFA [65].

The curve measured in air shows a similar load behavior, with the exception that at zero normal force there is still friction. The offset is caused by

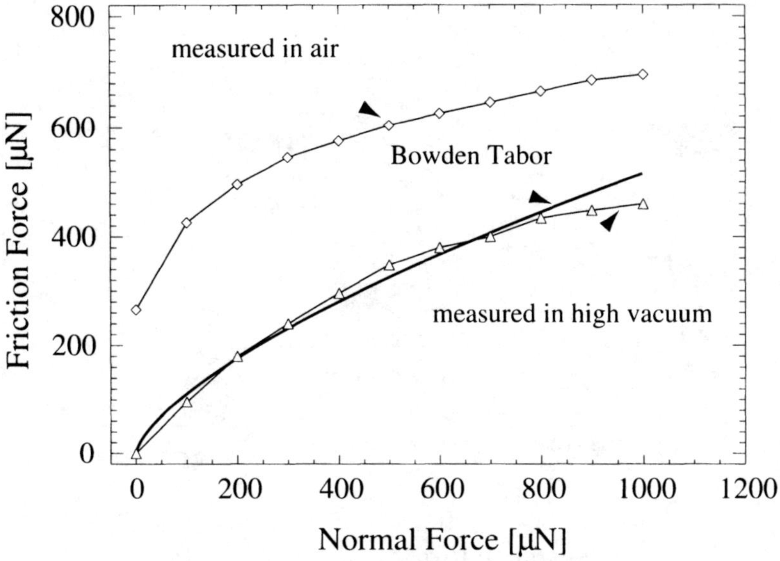

Fig. 2.40. Friction force as a function of normal force for measurements in air and in high vacuum. The thick line indicates the load behavior according to the Bowden Tabor model.

the capillary action. It was already mentioned in Sect. 2.4.3 that the capillary force acts in the same direction as the normal force, creating a resulting normal force $F_n' = F_n + F_c$. When no external normal force F_n is applied, then the resulting normal force is equal to the capillary force. By taking $F_c = 4\pi\gamma R$, the capillary force has a value of about 445 μN. With the values of τ=50 MPa, K=108 GPa and $F_n' = F_c$ the friction force according to (2.53) was calculated and a value of 254 μN was obtained. This is the offset shown in Fig. 2.40.

To illustrate the action of water Fig. 2.41 shows the friction force measured as the ball approaches the flat sample. When both liquid films make contact, a capillary bridge is formed and the friction force shows a jump. If the distance is now decreased further, the two solids approach until both water double-layers make contact. The friction force increases further but saturates, Hertzian behavior is observed as discussed above. When the vertical motion of the flat sample is reversed (from approach to retraction), the friction force decreases due to decreasing normal force. In the range where the capillary bridge is stretched the friction force is nearly constant. In this range the ball follows the flat with a small time lag. Finally, the capillary bridge breaks and the force curve returns to its zero position.

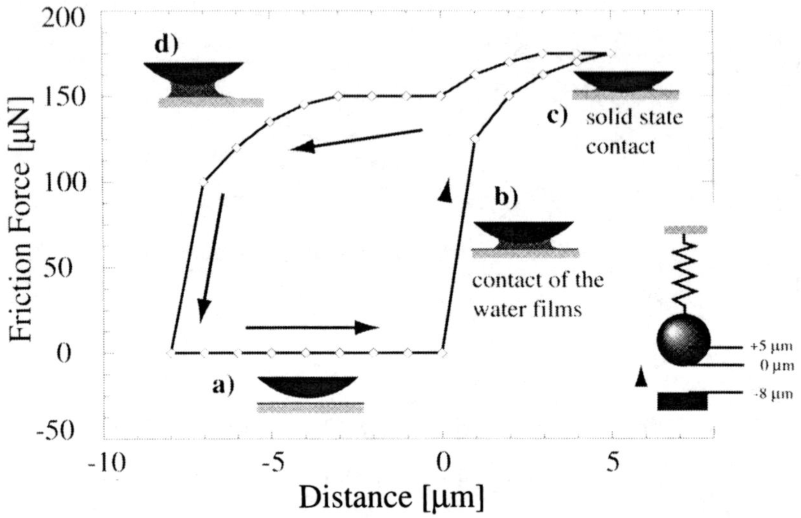

Fig. 2.41. Friction force versus distance between ball and flat sample. In (**a**) the spacing is large; thus the ball and flat sample have no contact. The friction force is zero. At (**b**) the liquid films make contact and form a capillary bridge. At (**c**) the distance is further reduced and the normal force increases. When the ball is withdrawn the capillary bridge is stretched and then finally breaks, leading to zero force, (**d**).

2.6.9 Vibrations

Mechanical vibrations can be self-excited [293] or can be introduced deliberately to reduce friction and adhesion. The degree of friction decrease depends basically on the frequency and amplitude of the vibration. Since mechanical systems have multiple degrees of freedom, the systems show a complex response, which is sometimes chaotic [294]. Bhushan and co-workers [295] have investigated friction and vibration using an AFM. They found that the friction is zero for ultrasonic frequencies above 500 kHz. Amplitudes in the nanometer range were sufficient to evoke a super-lubricating effect. This effect is caused by a periodic loss of contact between tip and sample and by squeezing the film leading to elastohydrodynamic lubrication. Similar results were obtained with the SFA [120, 296], showing that vibrational amplitudes of 0.1 nm lead to ultra-low friction (see Sect. 2.5.1). It was also found that the friction reduction is connected with a coupling of the mechanical resonance of their system and the molecular-scale relaxation. This involves solid-like and liquid-like transitions. Hess and Soom [297] investigated friction for unsteady sliding conditions by looking at individual asperities. They found that a loss in contact is necessary to reduce friction. Vibrations were also used for microassembly applications [298]. Böhringer et al. [299] used ultrasonic

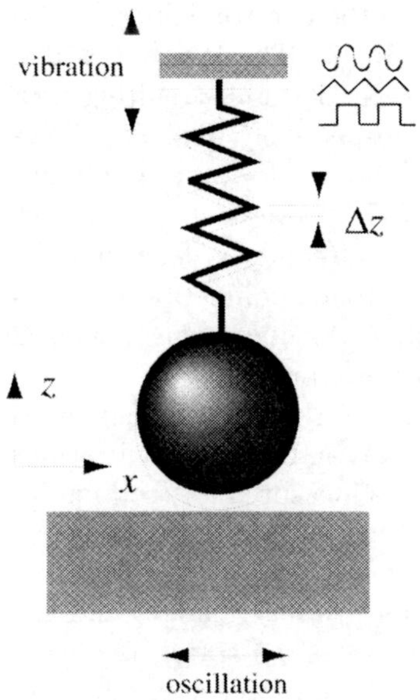

Fig. 2.42. Schematic of the SILICON model system. The flat sample oscillates at constant frequency, while the vibration Δz of the ball is increased from zero to 400 Hz. In addition, different waveforms – sine, sawtooth and rectangle – were applied.

vibrations to eliminate friction and adhesion in micropart handling. It was demonstrated that ultrasonic vibrations (20 kHz range) can be used to overcome adhesive forces, leading to low coefficients of friction. So far, mainly ultrasonic vibrations were used to reduce friction and adhesion. Here the authors extend the investigation to the effect of low frequencies combined with amplitudes in the nanometer range on static and dynamic friction.

Our simple SILICON model system was used to measure the friction force while vibrating the ball. For this experiment both the tangential and the vertical motion were generated by piezo-transducers. The control unit of the transducer was connected to a waveform generator. In this manner, sine, sawtooth and rectangular vibrations could be excited with an amplitude, Δz. Figure 2.42 shows the setup. The flat sample was set to constant sliding with a velocity of 3 μm/s by applying a sawtooth signal. The applied normal force F_n was set to either 800 μN or 1600 μN. The tester was placed in the environmental chamber, where both humidity and sample temperature were controlled.

In the experiments the vibration amplitude, waveform, temperature, humidity, mass of ball and normal force were used as parameters while driving the ball with a frequency from 0 Hz to 400 Hz. Figure 2.43 shows a representative set of friction force curves as a function of the vibration frequency with the amplitude of the vibration as the variable parameter. By means of a rectangular voltage signal an offset was added to the control voltage of the piezo-transducer. The normal force varied accordingly. The lowest possible change in amplitude was 6 nm and the highest was 300 nm, resulting in a change of normal force ΔF_n of 0.6 µN and 30 µN, respectively. The plot shows that the smallest amplitude had no influence on friction. However, any larger value reduced the friction force significantly. Higher vibration amplitudes shifted the friction curves to lower values and the frequency f_0, when the friction force becomes zero, to lower values. The starting points of the curves have higher values for larger vibration amplitudes Δz due to an increased normal force resulting from the offset caused by the vibration.

The friction force is influenced by the waveform of the vibration signal as well. Almost no influence was measured for sine waves. Sawtooth vibrations showed only a weak influence. The largest effect was measured for rectangular signals with an instantaneous switch between the high and low state, as shown in Fig. 2.43. When the ball and flat sample come into contact the capillary bridge is formed immediately. This means that for gradual changes and low frequencies – as caused by sine or sawtooth signals – the spring is always able to follow the vibration and no decrease in friction force was detected. However, if the change is sudden, then obviously the capillary bridge breaks, leading to a momentary loss of contact and thus to a low friction force. It follows that the inertia of the ball/spring assembly is responsible for the reduction of the friction force. This means that for a higher mass of the ball the sticking time is shorter. Larger balls with higher mass cannot follow the induced up and down motion, i.e., they are most of the time in a floating state.

Hydrophobic silicon showed lower friction and f_0 is lower than for hydrophilic silicon. A similar behavior was obtained for tests in a dry atmosphere and by increasing the sample temperature. Low relative humidities shifted f_0 and the set of curves to lower values. By increasing the sample temperature the set of curves was shifted to lower friction values as well. Higher temperatures support the desorption of water, reducing the size of the capillary bridge. Water is a coupling medium. If the capillary bridge is small, then the stiction is small too, leading to a reduced transmission of the tangential motion from the flat sample to the ball. The transmission of the tangential motion is also influenced by the magnitude of the normal force. Higher normal forces shift f_0 and the whole set of curves to higher values.

It is interesting to note that experiments in ultra-high vacuum showed a similar behavior. Although water was removed by desorption, thus excluding capillary force, the friction decreased in the same manner as in the experiments in air. The attraction of the ball and flat samples was caused by van der Waals interaction and chemical bonds. In addition, since the experiments were performed in vacuum, air as a damping medium was excluded. It was, therefore, easier for the spring to follow the forced up and down motion.

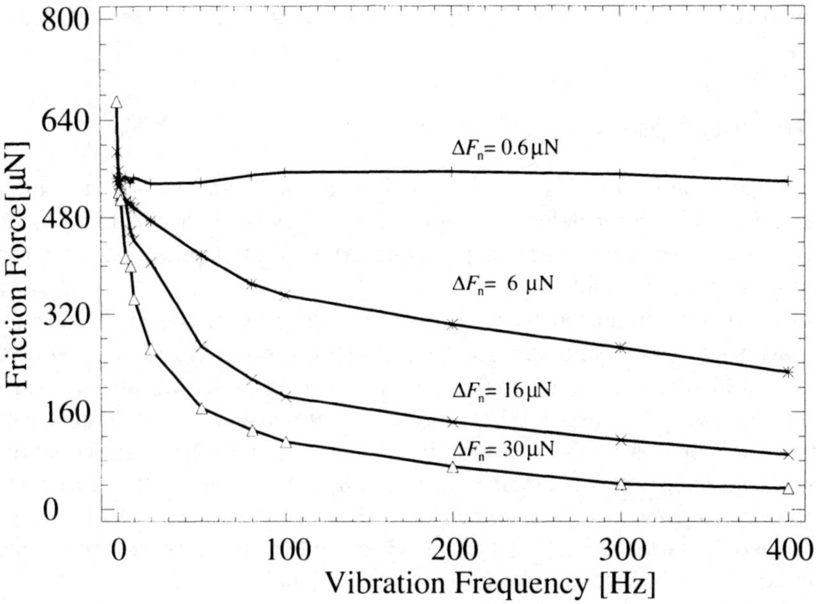

Fig. 2.43. Friction force versus frequency of rectangular vibration for different vibration amplitudes ΔF_n. The lowest change in normal force (0.6 μN) corresponds to a change in vertical direction of 6 nm.

2.7 Wear

Like adhesion and friction, wear can also be divided into macro-, micro- and nano-events. On the macroscale repeated plastic deformation and the action of heat during friction lead to degradation of the material that is called wear. If the impact is strong enough, grains can be removed from polycrystalline materials. However, wear is not only influenced by mechanical attack. It was shown in [300] that combined chemical and mechanical stressing leads to an exponential decrease of the shear stress from 15 MPa for 3% relative humidity to 3 MPa for 80% relative humidity. Thus, water vapor accelerates the debonding of particles. Corrosive wear of single-crystal brushite ($CaHPO_4 \cdot 2H_2O$) was measured in aqueous solution [301]. Wear was a strong function of the normal force and crystallographic orientation. Depending on the bond strength the material showed a distinct anisotropy of wear. As a result of fracture and fatigue, debris form that can remain in the interface or can be transferred from one solid to the other [302].

2.7.1 Microscale Wear

Microscale wear analysis was performed on hip-replacement materials. In this study a third-body wear mode was identified as the cause for higher roughness. Wear debris form as a result of fatigue of the metal as well as the polymer surface. The diameter of rounded debris was about 0.4 μm to 0.8 μm [303].

2.7.2 Nanoscale Wear

Wear on the atomic scale is accompanied with the formation of crystallographic defects like point defects or kinks. Starting with the removal of single ions the wear process proceeds to the generation of etch pits [304] and the flattening of asperities [305].

AFMs can be configured to measure wear. Very often the Si cantilevers are replaced by stiffer cantilevers with a sharp diamond tip [306]. In order to obtain sufficiently high normal forces the spring constant in the vertical direction can be adjusted by changing the dimensions of the cantilever. One drawback of this method is that increasing the normal spring constant decreases the sensitivity in the lateral direction, i.e., for the friction measurement. A way to prevent this is shown in [307] where two double-leaf springs were applied in series. The first spring serves to set the normal force while the second spring is used to detect the friction signal.

2.7.3 Wear of Organic Monolayers

Whereas the pressure in microtribological studies of silicon is very often not high enough to induce wear, especially when the solids are separated by a water film, wear is observed for organic monolayers.

Self-assembled monolayers (SAMs) form spontaneously by immersing a substrate into a solution of an active surfactant in an organic solvent. They are stable due to the fact that the monolayers are held onto the substrate surface by a chemical bond.

The SILICON model system was modified to measure octadecyltrichlorosilane (OTS, $C_{18}H_{37}SiCl_3$) self assembled monolayer films. Therefore, the silicon flat was first cleaned with piraña solution (mixture of 3:1 v/v of H_2SO_4 and 30% H_2O_2 at ≈ 80°C) and then thoroughly rinsed with distilled water.

The first step of applying the SAMs consists of creating a hydroxylated Si(100) surface. There are two ways of achieving this. In the first case, the sample is immersed in freshly prepared piraña solution for 1 h, is then thoroughly rinsed using distilled water and finally blown dry using nitrogen. This procedure removes any organic contaminants present on the silicon surface but retains the native oxide that is already present on the Si(100) crystal.

The second method consists of first removing the native oxide layer and then reoxidizing the sample. The method for this is detailed in [7]. It consists of first etching the sample in a 40% v/v solution of HF for 10 minutes to remove the native oxide, followed by immersion in distilled water for 10 minutes and immersion in H_2O_2 for 15 minutes and a final water rinse for 15 minutes. This procedure results in a 0.65 nm thick oxide coating [7]. The authors

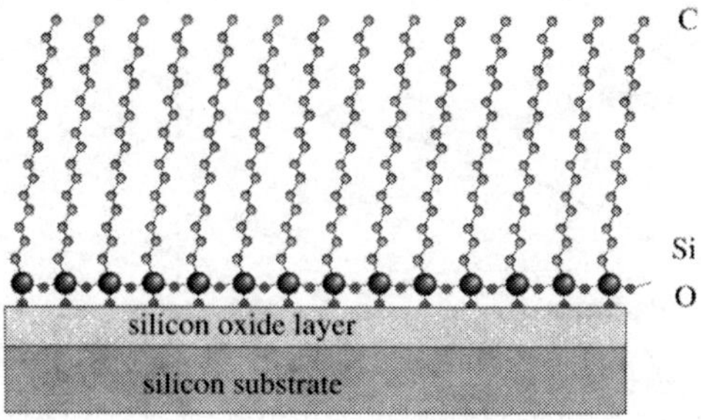

Fig. 2.44. Structure of OTS self-assembled monolayer film.

used both methods to prepare SAMs and did not observe any significant differences between these two methods using our SILICON model system and AFM measurements. Film preparation was accomplished on the oxide-coated Si(100) crystals by immersing them in the solution containing the OTS. The samples were immersed in 50 ml of a 1 mM solution of OTS in bicyclohexyl contained in sealed brown glass bottles for 24 h at 25°C. Afterwards, the samples were removed and rinsed with distilled water, immersed in chloroform and then blown dry with nitrogen. The structure of the film is shown in Fig. 2.44.

By means of Auger Electron Spectroscopy (AES) the coverage was probed; this is shown in Fig. 2.45. The spectrum of OTS showed a strongly decreased

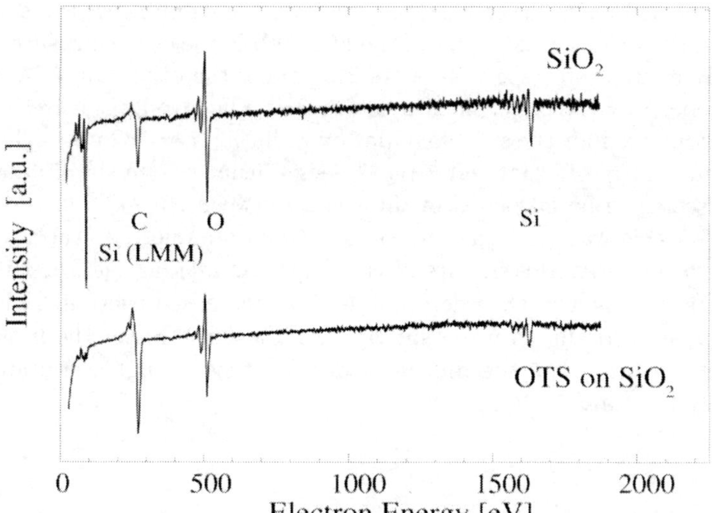

Fig. 2.45. AES spectrum of an OTS self assembled monolayer film.

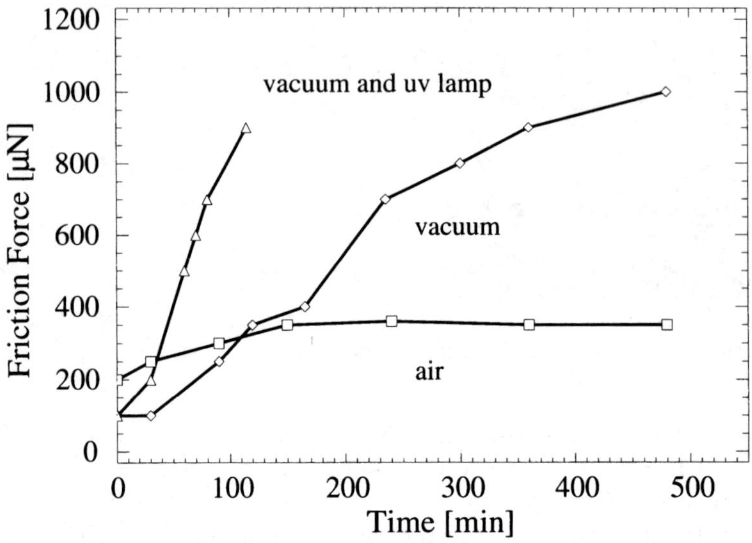

Fig. 2.46. Wear study of OTS stressed in air, high vacuum and high vacuum combined with uv radiation.

contribution of silicon compared with the spectrum of oxidized silicon. This means that the OTS film homogeneously covers the oxide.

Wear was detected during sliding. Tests were carried out in air, high vacuum and in high vacuum in combination with ultraviolet (uv) light stressing. The results are shown in Fig. 2.46. The curves taken in vacuum ($p = 2.5 \times 10^{-7}$ mbar) show lower friction force at the beginning of the experiment, caused by the desorbed water of the ball. While the curve measured in air shows only a small increase with increasing time, the other curves are drastically influenced by the experimental conditions. The strongest increase of friction was monitored for the sample under uv radiation (λ=182 nm). The influence of vacuum suggests that water plays a significant role in the sliding process. Water seems to be adsorbed in the space between the OTS chains. Due to desorption this water is removed. The authors assume that the water contributes to the ordered structure of the SAM. As long as the order is maintained the friction is low. Disordered films show increased friction. This effect becomes magnified when during sliding uv radiation stresses the film. Here the authors assume that the film degrades, leading to higher friction and consequently to wear.

Part II

Biological Friction Systems

3. Biological Frictional and Adhesive Systems

The biological world is part of the physical world and, therefore, the rules of mechanics also apply to living systems. The mechanics of living motion systems is the subject of an integrative discipline called biomechanics [39]. Living creatures move on land, in the air, or in water. There are complex motions inside their bodies to provide fluid circulation or to generate forces for locomotion. The resistance against motion mediated by surrounding media and by the mechanical contact with various substrates was an evolutionary factor, which contributed to the appearance of many devices adapted to reduce such resistance. On the other hand, one always needs friction to generate the force to move on a substrate or to overcome the drag caused by friction elsewhere. A living motion system becomes optimized when it is capable of minimizing friction at one end of the system while maximizing it at the other end [308]. In other words, a living device needs a combination of maximum friction required for acceleration, deceleration and maneuvering, combined with minimum friction in joints for economic energy expenditure. Adhesion phenomena can also contribute to the functionality of such a system.

3.1 Biological Attachment

Living nature contains numerous examples of firm or releasable attachment systems whose design is based on the principles of adhesion. Sticky devices are used in locomotion and to hold firmly onto a variety of substrates. An additional type of adhesion-related system are those minimizing adhesion due to particular chemical composition on their surfaces and/or surface sculpturing. Those systems are responsible for keeping the surfaces clean.

Since many construction principles are simultaneously embodied in living nature due to particular developmental principles, many simple technical solutions will never become a reality in biological systems. On the other hand, many principles of natural design are very difficult to mimic in technological processes.

The biological diversity is immense, and it was impossible to put all interesting examples into one chapter. Attachment structures are found in great variety among animals and plants and include various sorts of barbs, hooks, suction cups, and sticky pads [309]. They operate by the use of various glues, surface tension, molecular adhesion, or mechanical grasping. However, existing assumptions about mechanisms should be treated with caution. Even selected examples show that the present state of knowledge is very poor. Numerous mechanisms have only been described and studied morphologically. Very few systems have been tested experimentally.

In conclusion:

- The biological world is a part of the physical world and, therefore, the rules of mechanics can be applied also to living systems.
- In the evolutionary process, living motion systems become optimized to minimize friction at one end of the system while maximizing it at the other end.
- The biological diversity is immense: attachment structures are found in a great variety among animals and plants and include various sorts of barbs, hooks, suction cups, and sticky pads. They operate by the use of various glues, surface tension, molecular adhesion, or mechanical grasping.

3.2 Systems with Reduced Friction

3.2.1 Fluids and Friction

There are two main types of active swimming systems: those generating propulsive force by rowing movements, using their extremities (water bugs, water beetles, Ciliata); and those using the entire body and/or specialized beating organs (fish, whales, dolphins, penguins, protozoan Flagellata) to undulate in the medium. Biological swimming systems have a variety of convergently appearing adaptations for decreasing their body density to prevent sinking and, by this means, saving energy for propulsive movements. Evolutionary processes have adapted nektonic animals to interact efficiently with the surrounding water. Reduction of drag due to friction in the boundary-layer close to the body surface is one of these adaptations.

Forces from the streaming medium act on the immersed body in two different ways. On one side, the fluid generates the pressure directed normally to the surface of the body and is responsible for inertial forces. On the other side, friction or shearing force acts tangentially to the surface of the body [310]. The resulting force depends on the geometry of the body, its relative dimensions to the flow and on such characteristics of the fluid as its density ρ, viscosity η and speed v. Analysis of the resulting force gives the Reynolds number *Re* showing the relationship between inertia and viscosity

$$Re = \frac{vl\rho}{\eta}\,, \tag{3.1}$$

where l is the body length. If two bodies of similar shape have the same Reynolds number, the flows around them will be similar. Whales swim at very high *Re* values ($Re \approx 10^8$). In this case, the viscosity influence on the body will be minimal. However, there is a so-called boundary-layer next to the solid wall of the body where the medium is viscous (see also Sect. 2.5.4). Outside the layer, the medium may be regarded as non-viscous. The unicellular organisms (protozoans) and plankton swim at a very low *Re*, where the viscosity of medium most influences motion. The boundary-layer mentioned above plays an important role in the frictional resistance of the medium against the moving body. Close to the solid wall, the velocity of the fluid or gas is close to zero. In the direction perpendicular to the body surface, the velocity gradually increases. At some distance from the surface it is similar to the velocity of the free flow. It is generally accepted that free flow is free of friction. The area that is most friction-active is the boundary-layer with a thickness lower than 1 mm. With increasing Reynolds number, the thickness of the boundary-layer increases.

The boundary-layer can flow in a laminar or turbulent way. In laminar flow, particles of the medium move almost parallel to the surface, whereas in turbulent flow, they rotate in one preferred direction. At the critical Reynolds number ($Re > 3.2 \times 10^5$), the laminar boundary-layer becomes turbulent [311].

Adhesive secretion (mucus), compliant skins, scales, riblets and roughness may influence the flow velocity gradient, the type of flow and the thickness of the boundary-layer around animals, and may seriously affect their drag in a positive or negative way. For example, transition of the boundary-layer from laminar to turbulent flow results in oscillations in the layer. If the system is able to suppress these oscillations before the critical Reynolds number is reached, the transition will not take place. In this case, it will occur at much larger Reynolds numbers or at much larger velocities of the flow. In other words, the critical Reynolds number will be shifted to significantly higher values. Such a boundary-layer damping is the result of elastic structures in the skin of some animals. Dolphin skin has a very special design. It is very smooth and relatively soft. Outer regions of the skin material consist of a covering layer, inner layer (corium papillae), corium grooves and underlying corium layer (Fig. 3.1, redrawn from [311]). The region of corium papillae is filled with a damping fluid.

Experiments with a dolphin model revealed a decrease of the frictional resistance (to 60%) due to a shift of the critical Reynolds number to higher values. A possible explanation of such a result is that viscoelastic properties of the outer part of the dolphin skin can damp short-wave turbulence appearing due to occasional surface irregularities. Under the pressure of microturbulence, the rubber-like outer layer deforms and changes the local shape of the skin surface due to a shift of the damping fluid located under the rubber-like layer. Such a material design is able to damp turbulence, which is 10–40 times larger than the thickness of the boundary-layer.

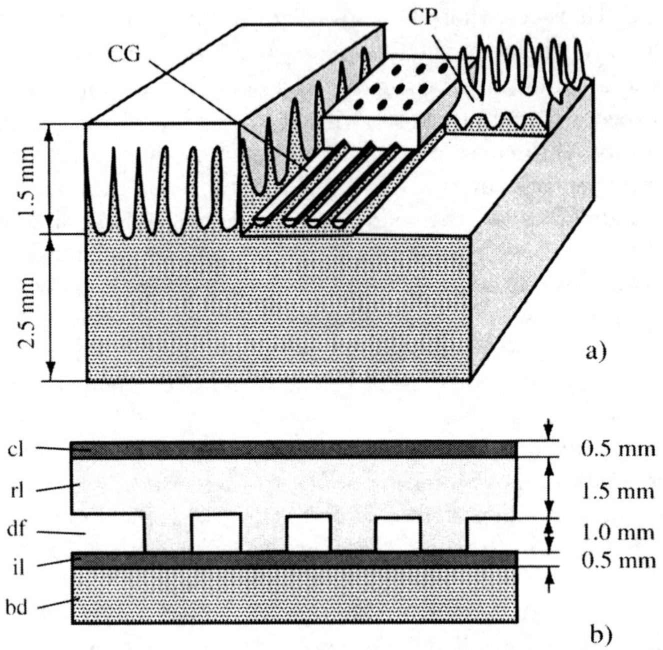

Fig. 3.1. Structures decreasing friction in fluids at high Reynolds numbers. (**a**) Schematic of dolphin skin. CG, corium grooves; CP, corium papillae; (**b**) Artificial coverage designed analogous to the dolphin skin. cl, covering layer; rl, soft rubber layer; df, damping fluid; il, inner layer; bd, hard underlying layer of the body.

Many fish have developed another mechanism to reduce friction in the boundary-layer, up to 60% in some species. They produce skin secretions, which are usually slightly soluble in water. In areas where microturbulence is strong, these substances can be locally dissolved. This results in the damping of microturbulence [311]. Due to skin secretions, fish can reach an extremely high speed in a short time. An alternative explanation of the mucus mechanism is that the long chain polymers found in mucus decrease the pressure gradient and considerably reduce friction. The effect is probably due to channeling of the flow particles in the direction of the main flow, resulting in a reduction of turbulence. Compliant surfaces could probably reduce drag by equalizing and distributing pressure pulses. However, the existing evidence that drag reduction actually occurs is not convincing. There is no indication that instantaneous heating, reducing the viscosity in the boundary-layer, is used by animals as a drag-reducing technique [312].

Fahrenbach and Knutson [313] have studied cell layers underneath the skin (epidermal surfaces) specialized for lowered or increased friction in representative animals having a backbone (vertebrates) by the use of scanning electron microscopy. They have reported that patterns with microridges and micro-outgrowths (microvillous) predominate in aquatic vertebrates. Friction in the boundary-layer of the body moving in the medium at high Reynolds numbers may be decreased due to such a sculpturing of the surface. The grooved scales of the shark skin is an example of such a system (Fig. 3.2, from [314]). Their size ranges from 200 μm to 500 μm. The surface of each scale contains parallel grooves between so-called riblets directed almost parallel to the longitudinal body axis. Interestingly, grooves and riblets of neighboring scales correspond exactly to each other so that the shark surface looks like a pattern of parallel stripes [315]. Experiments on flow resistance have been carried out with smooth-bodied models and with those covered with grooves of dimensions similar to original shark skin; see Fig. 3.3 (redrawn from [314]). The flow resistance measured in the grooved model was about 5%–10% lower than the resistance in the smooth model at a Reynolds number of 1.5×10^6. The geometry of the grooves and riblets can also influence results. The best results have been obtained with the surface profile shown in Fig. 3.3d.

An explanation of the shark skin effect has been proposed recently [314]. Presumably, in the boundary-layer, not only longitudinal microturbulence, but also cross-directed microturbulence occurs. The longitudinal grooves of the shark scales prevent the appearance of the cross-directed microturbulence. However, such an effect is only possible when the distance between the tops

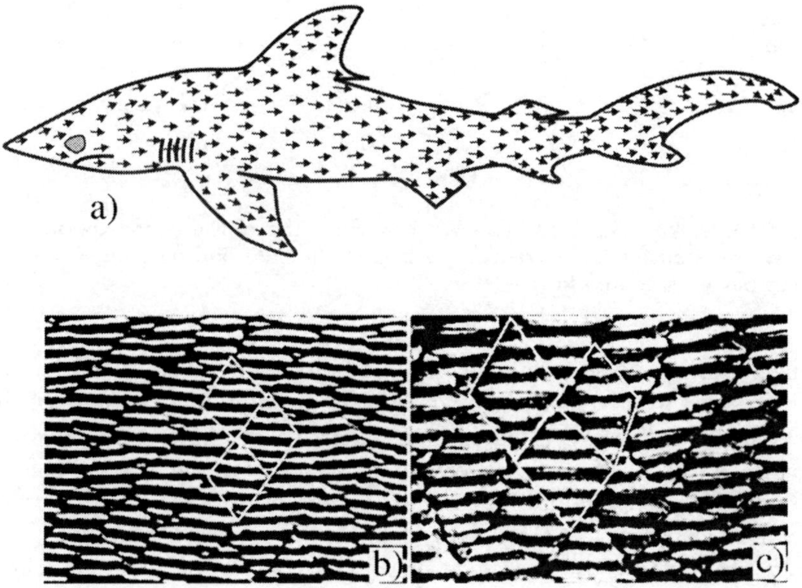

Fig. 3.2. Shark skin surface. (**a**) Directionality of the riblet groove pattern in the shark *Isurus oxyrhynchus*. (**b**) Scale surface of the shark *Sphyrna lewini*. (**c**) Scale surface of the shark *Etmopterus spinax*.

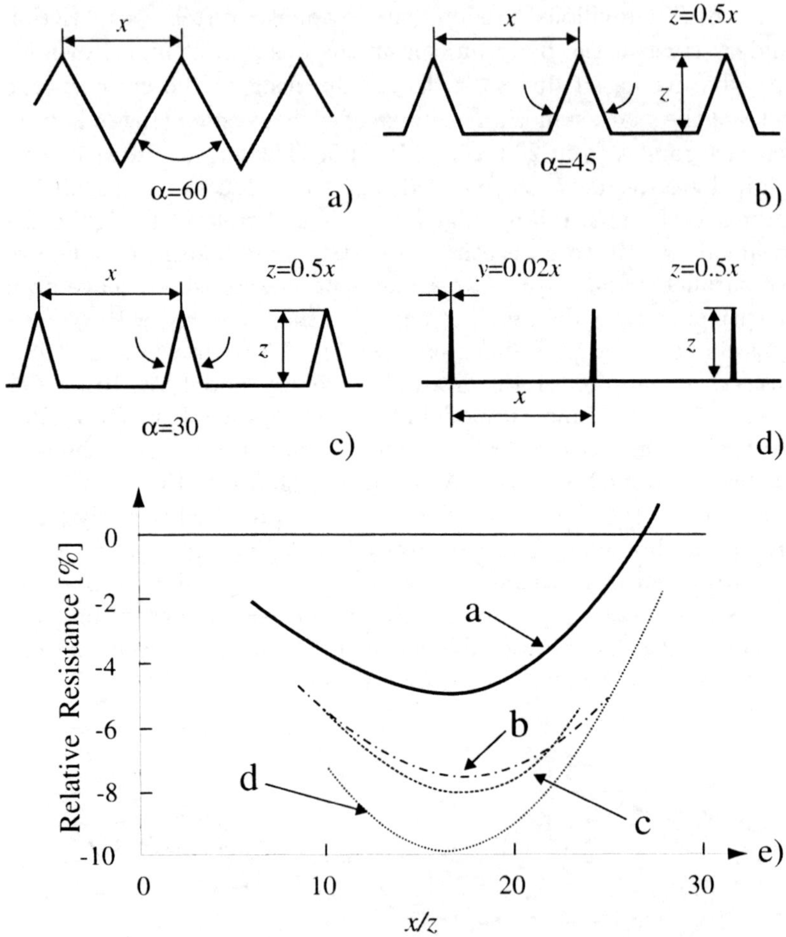

Fig. 3.3. "Shark skin effect". (**a**)–(**d**) Artificial riblet patterns used in experiments. (**e**) Relative resistance (shear stress) measured in the fluid flow on models covered with riblet patterns figured in (**a**)–(**d**).

of two neighboring riblets is smaller than half of the lateral wavelength. The riblet edges must also be sharp enough to prevent cross-directed flows of the boundary-layer. If flows directed at some angle to the riblet directions act on the surface, the riblets can redirect boundary flows in the direction of the riblet (or body) axis. The flow resistance measured in the direction parallel to the riblet axis was lower than the resistance in the cross-direction.

Effects similar to those described in this section can occur in air. Birds bear feather surfaces of microscale roughness, which can also contribute to a decrease of frictional resistance in the air. Fast flying insects are usually covered with hairs and skin outgrowths (acanthae) ranging from 10 μm to 100 μm, which have a similar function as previously suggested [316]. Industrial applications based on the design of the shark skin profile have been developed to decrease drag resistance in aircraft devices.

3.2.2 Joints and Articular Cartilage

Vertebrate bones, that are joined with each other, are covered by cartilage, which is the gliding surface of the joint. The coefficient of friction is very low (0.0026) [39].

Cartilage consists of two tissue elements: cells, and extracellular matrix, containing collagen fibers; see Fig. 3.4 (from [317]). In fact it is a fibrous composite material of collagen fibers embedded in a highly hydrated proteoglycan gel [318,319]. There are three main types of cartilage: hyaline cartilage, white collagen-rich fibro-cartilage, and yellow elastic fibro-cartilage. White fibro-cartilage is the most interesting one from a tribological point of view, because it is found mainly in invertebral and joint discs. This cartilage is

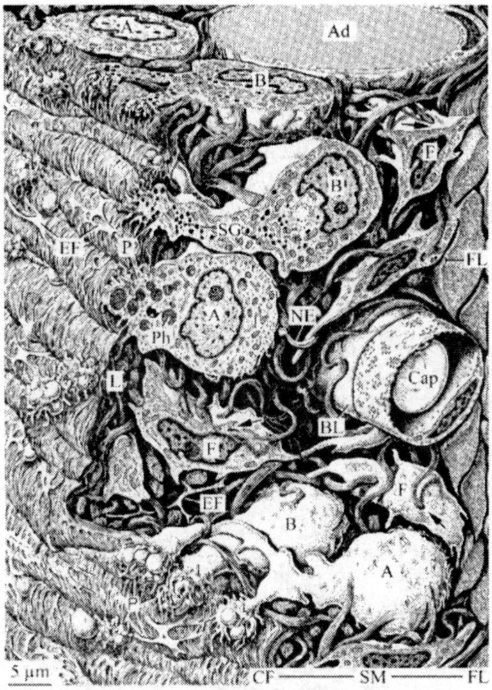

Fig. 3.4. Schematic of the synovial membrane (SM) of human cartilage composed of collagen fibers, elastic fibers, and synovial cells. A, macrophage-like A cells; Ad, white adipose cells; B, B cells responsible for secretion of proteoglycans and hyaluronic acid; BL, basal lamina; Cap, capillaries; CF, collagen fibers; EF, elastic fibers; F, fibroblasts; FL, fibrous layer; NE, nerve endings; P, processes of synovial cells; Ph, phagolysosomes; SG, secretory granules.

responsible for joint mobility. It provides lubrication of contacting surfaces [320] and serves as a kind of damper under dynamic loads.

The surface of cartilage is not ideally smooth. Its roughness can influence resulting friction. A roughness of 1 μm is typical of foetal cartilage, and 2.7 μm in healthy adults. The asperities have a slope of about 2° and a distance between them of 0.2 μm–0.9 μm. Their shape can change under load. In TEM studies, it has been revealed that surfaces of loaded joints were smooth at magnifications from ×35 to ×7,500, whereas the tibial surfaces of unloaded joints were irregular [321]. In osteoarthrosis, the roughness increases up to 5.3 μm [322]. This can decrease the lubrication quality of joint cartilage down to an average coefficient of friction ranging from 0.01 to 0.09. Cartilage is porous; its material is filled with synovial fluid [323,324]. The material of the white cartilage demonstrates viscoelastic properties under stress. However, under stress, fluid moves in and out of the tissue, and the mechanical properties of the cartilage change depending on the fluid content. Joint cartilage demonstrates unique lubricating properties, which surpass those of the best artificial anti-friction materials. Two different approaches to the tribology of human joints have been reported [39] (Fig. 3.5, redrawn from [39]). The first approach experimentally simulated intact synovial joints *in vitro*, whereas the second one tended to test the frictional behavior of cartilage on a glass surface. The first approach is a more natural one, but the geometric and kinematic complexity of such tests makes the results difficult to analyze. It is also almost impossible to compare such tribological data for various geometries. The second approach provided data comparable with results of standard tribological procedures, but has other limitations. In these experi-

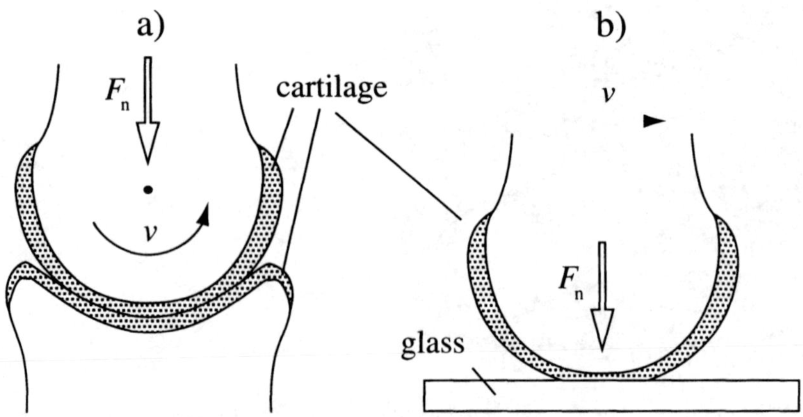

Fig. 3.5. Experimental design to study friction in vertebrate joints. (**a**) Concept for joint friction measurements *in vitro* on an isolated joint. (**b**) Concept for joint friction measurements between cartilage and glass.

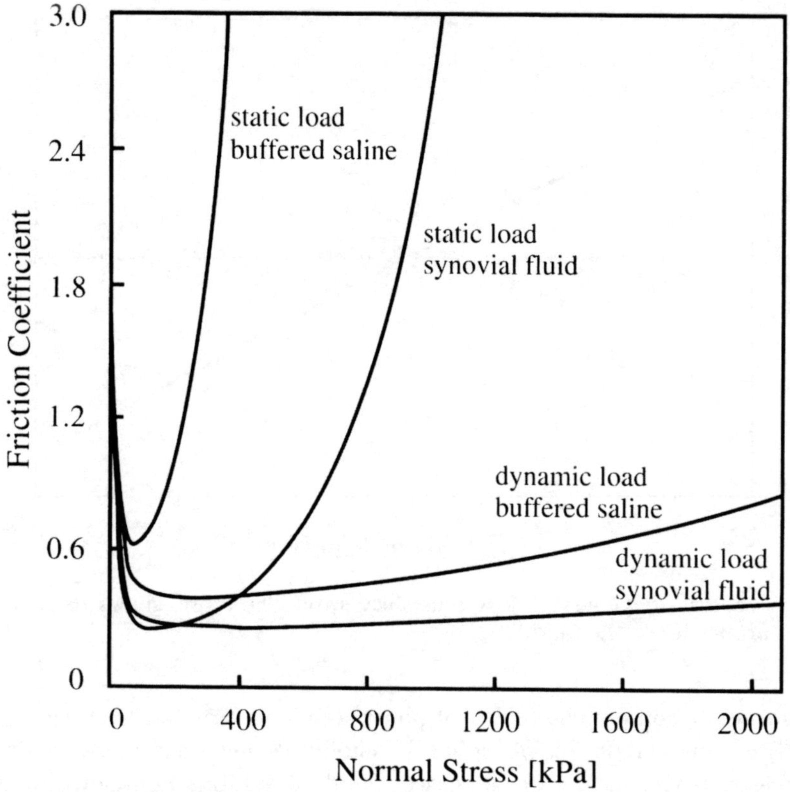

Fig. 3.6. Coefficient of friction between two cartilage surfaces as a function of the normal force between the surfaces. The coefficient of friction depends on experimental conditions (static or dynamic) and on the lubricating fluid (buffered saline or synovial fluid).

ments, an artificial material was used instead of cartilage as a corresponding material. The data are difficult to compare with natural conditions.

When cartilage was tested at different sliding velocities, it was found that the coefficient of friction at low sliding velocities (lower than 1 mm/s) is high compared to higher sliding velocities (1–400 mm/s). At static conditions, a considerably higher coefficient of friction was measured (Fig. 3.6, redrawn from [39]). The most efficient lubrication was obtained for a combination of a cyclic dynamic loading in a native synovial fluid. The advancing contact-angle (see Sect. 5.2.3) between saline drops and cartilage was found to be 100° ± 5°, indicating a highly hydrophobic surface. The friction of cartilage against hydrophobic surfaces is lower than the friction of cartilage against hydrophilic surfaces [325]. It seems that the secret of high lubrication abilities in joints is related to the synovial fluid composition.

Synovial fluid is a dialysate of blood plasma containing long chain protein molecules with many negative charges, hyaluronic acid, and phospho-

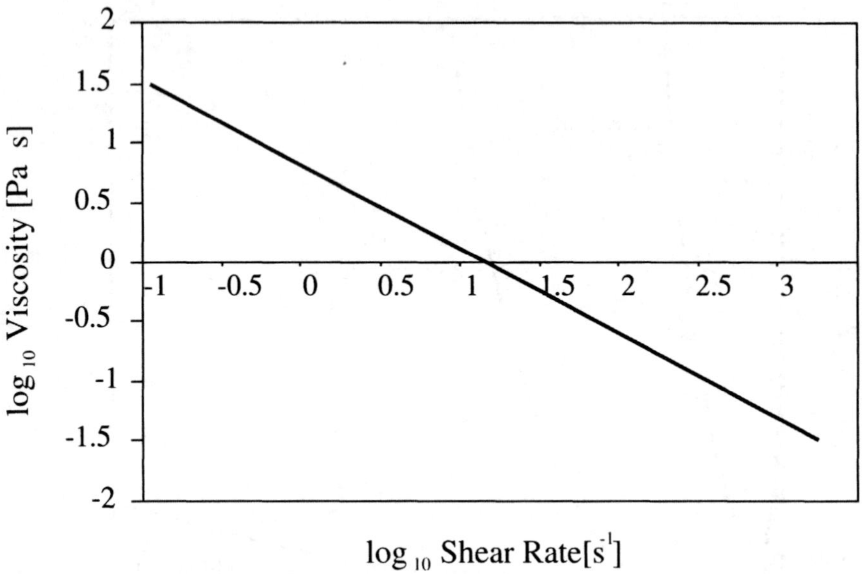

Fig. 3.7. Viscosity of synovial fluid versus shear rate. The graph shows the non-Newtonian properties of the fluid.

lipids. The major component (45%) of phospholipids extracted from canine (dogs, wolves) synovial fluid is phosphatidyl choline. The viscosity of synovial fluid decreases as the shear rate increases; the fluid exhibits non-Newtonian viscosity [322] (Fig. 3.7, redrawn from [322]). Synovial fluid glycoproteins, such as lubricin [326], lubricating glycoprotein (PSLF) [327], and lubricating glycoprotein–I (LGP–I) [328], have also been reported as highly effective lubricating substances.

The phospholipids extracted from synovial fluid have been shown to be surface-active in reducing the surface tension of water and to be readily adsorbed onto hydrophilic solids, whose surfaces then become hydrophobic. These adsorbed monolayers of synovial surfactant were then found to be excellent boundary lubricants *in vitro*, reducing the coefficient of kinetic friction in the dry state and under physiological loading by up to 97% for extracts and 99% for phosphatidyl choline alone, reaching μ=0.01. Surface-active phospholipid is put forward as the possible active ingredient in joint lubrication [329]. Phospholipids also show phenomenal anti-wear properties [330].

Interestingly, destroying hyaluronic acid, both friction and wear tests displayed no significant change in accordance with most previous studies of synovial fluid alone [331]. Hyaluronic acid could have carrier functions for the highly insoluble surface-active phospholipid, since hyaluronic acid has good wetting properties needed to promote hydrodynamic lubrication.

Synovial fluid as a whole can dramatically reduce frictional forces due to elastohydrodynamic lubrication, which usually occurs during rapid movements of joints. Hydrodynamic lubrication occurs when the fluid thickness between contacting surfaces is thicker than 1 μm. Since the cartilage material design also contributes to such a dynamic lubrication, the lubrication in vertebrate joints is usually called elastohydrodynamic. However, a TEM study has revealed that the fluid film in a loaded joint is much thinner than generally accepted. At magnifications of ×40,000 the joint surfaces are separated by a uniform 100 nm space containing fluid [321]. Four theories explaining cartilage lubrication mechanism have been reviewed by Fung [39]. These theories are contradictory, but some points are similar.

Fluid Transport Theory

The viscoelastic properties of the composite material of cartilage support loads mainly by the viscous response of the fluid rather than by the elastic response of the solid matrix material. Therefore, the friction remains low when the cartilage material is filled with a certain amount of synovial fluid. When the amount of fluid decreases, the friction tends to increase. This can happen, for example, when the fluid is squeezed out of the material. However, the time required to squeeze fluid out of the material is long enough to keep the surface lubricated for a short-time dynamic load. After the load is removed, the compressed cartilage sucks the fluid back into the pores [332–335]. The porosity of the cartilage depletes the lubricant film thickness, rather than increasing it, particularly when the lubricant film thickness becomes small [336].

Lubrication Layer Theory

At high pressures, the viscous fluid can presumably work as a layer, which supports loads [337]. Long-chained polymer molecules also aid in the support of loads [338]. Changes in the velocity profile of the lubricating layer can result in a lubrication effect. A high concentration of hyaluronic acid in the synovial fluid may contribute to this mechanism by increasing its viscosity, when the shear rate falls to a minimum [39].

Roller-Bearing Theory

The protein complex of the synovial fluid, but not hyaluronic acid itself, has a lubricating ability [339]. Electron microscopy observation revealed that hyaluronic acid was a 200 nm–500 nm formless mass and the protein complex was in the form of 20 nm–30 nm spherical particles. It was presumed that hyaluronic acid encloses the spherical protein particles in its network like the cage of a roller-bearing and that the particles can rotate freely. This mechanism can exist in parallel with elastohydrodynamic lubrication in animal joints.

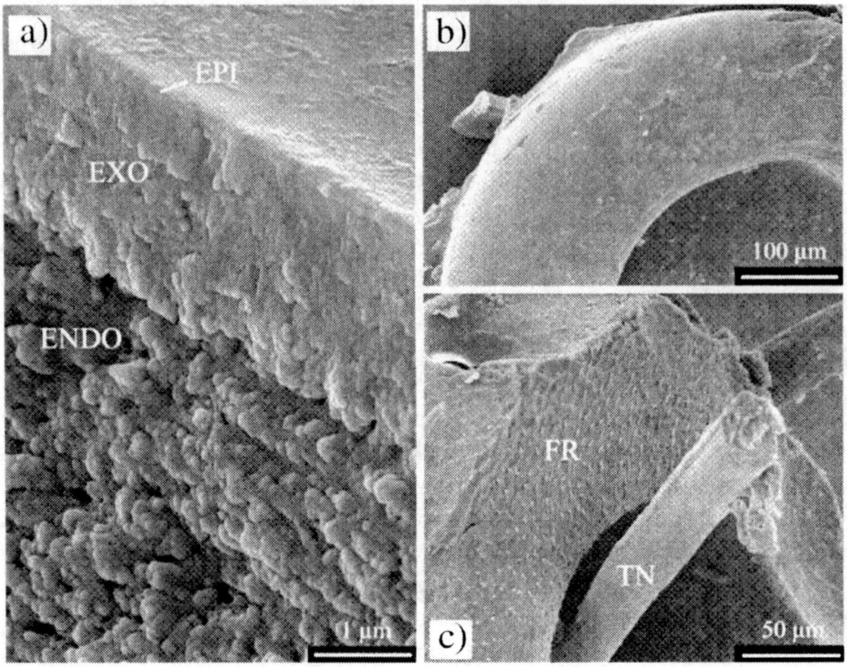

Fig. 3.8. Details of leg joints of the beetle, *Melolontha melolontha*. (**a**) Fractured femoral part of the femoro-tibial joint. (**b**) Surface of the femoral part of the femoro-tibial joint. (**c**) Frictional surface at the distal end of the second tarsomere. ENDO, endocuticle; EPI, epicuticle; EXO, exocuticle; FR, frictional surface; TN, tendon.

In insect joints, which are working under lower loading forces, but much higher frequencies than vertebrate joints [340–342], the joint surfaces are extremely smooth (Fig. 3.8a, b). Underlying tissues are penetrated with pore canals, which are presumably responsible for delivering lubricants in the contact area. Additionally, there are rough surfaces within joints which function in preventing movements in one particular direction (Fig. 3.8c). However, experimental studies testing friction in insect joints are completely missing in the literature.

Boosted Lubrication Theory

This theory [343,344] presumes that contact between two parts of cartilage in the joint will result in water absorption into the cartilage. After such a filtration procedure, proteins and other lubricating molecules remain in the contacting surface and provide lubrication.

Cartilage Material Theory

The viscoelastic mechanical properties of the cartilage result in local deformation under load [345,346]. These deformations can redistribute the load according to the contact stresses and local fluid pressures. An additional effect of the cartilage material may be caused by the chemical composition of the cartilage surface. Presumably, molecules of the synovial fluid interact with particular molecular species of the cartilage surface, and such an interaction acts as a kind of boundary lubricant coating surfaces [347]. It is suggested that at least part of the lubricating ability of LGP–I glycoprotein is dependent upon its ability to bind to cartilage [328].

The surface of the cartilage is covered by multi-layered discontinuous three laminar sheets, each 12 nm–14 nm thick. Each sheet is always separated from the cartilage itself and can be easily displaced. It consists of phospholipids, glycosaminoglycans and proteins. This membrane-like structure might have a protecting role in preventing direct contacts between the cartilage and toxic agents present in the synovial fluid and/or exerts a lubricating effect within the joint [348]. AFM studies confirmed the existence of an amorphous, non-fibrous surface, which may be vital for its normal lubrication and wear properties *in vivo* [349].

3.2.3 Muscle Connective Tissues

It is well known that skeletal muscles are made of muscle fibers held together in various configurations by an extracellular connective tissue framework [350]. The structure of the intramuscular connective tissue is functionally divided into three hierarchical clusters: the endomysium covering individual skeletal muscle fibers, the perimysium surrounding a group of skeletal muscle fibers called fascicles, and the epimysium enclosing the entire skeletal muscle unit (Fig. 3.9). The connective tissues not only provide stability to the muscle as a whole but also participate in the force transmission within the muscle. The contractile force is transmitted between neighboring muscle fibers by translaminar shear through the endomysium [351]. Shock-freezing experiments showed that reorientation of the collagen fibril network occurs with changing muscle length (Fig. 3.10, redrawn from [351]). Composite micromechanical models of connective tissues usually assume that direct interactions between fibers are negligible [352]. However, the mobility of collagen fibers in endomysium during reorientation may be influenced by friction in contacts between single collagen fibers and their bundles. In the case of the whole muscle, where active force and stiffness reflect the number of actin–myosin interactions and the shortening velocity reflects their turnover rates, molecular mechanisms that govern friction were only indirectly characterized by the rate of mechanical energy dissipation and the rate of actomyosin ATP (adenosine triphosphate) utilization. The amplitude of the frictional stress is proportional to the tissue stiffness [353].

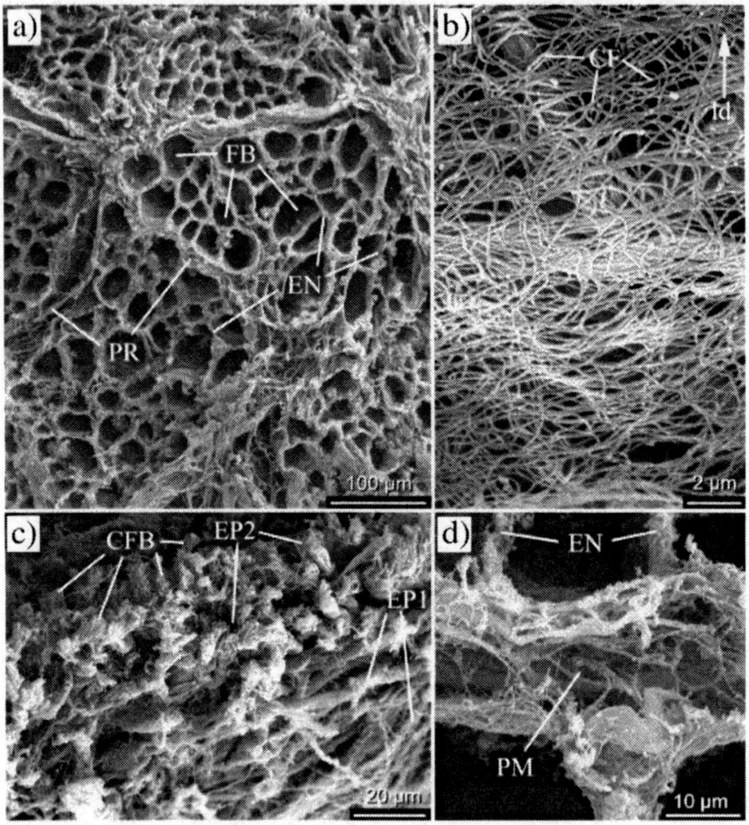

Fig. 3.9. Macerated triceps brachii muscle of the cui, *Galea musteloides.* (**a**) Cross-section showing endomysium and perimysium sheets. (**b**) Endomysium sheet, lateral aspect. (**c**) Cross-section through epimysium. (**d**) Cross-section through perimysium and endomysium. All muscle cell structures are removed to leave just the fibrous collagen network. EN, endomysium; EP, epimysium; PM or PR, perimysium. CF, collagen fibers. CFB, bundles of collagen fibers. EP1 and EP2, two layers of epimysium. FB, macerated muscle fibers. ld, longitudinal axis of the muscle.

It has been previously suggested that both endomysium and perimysium provide the necessary lubricated surfaces between muscle fibers and muscle fascicles, which enable the muscle to change its shape. This hypothesis, however, has not been proven experimentally. Muscle lymph could provide boundary lubrication for the sliding of connective tissue and for any relative movement of motor units in muscle contraction. Lymph was found to be surface-active at liquid and at solid surfaces. Extracted phospholipid was found to be an effective releasing agent as a monolayer, reducing the force of adhesion of 5% protein solutions by 76%, according to a standard test for tacky glues. The same monolayers were effective lubricants, reducing friction by 96% [354].

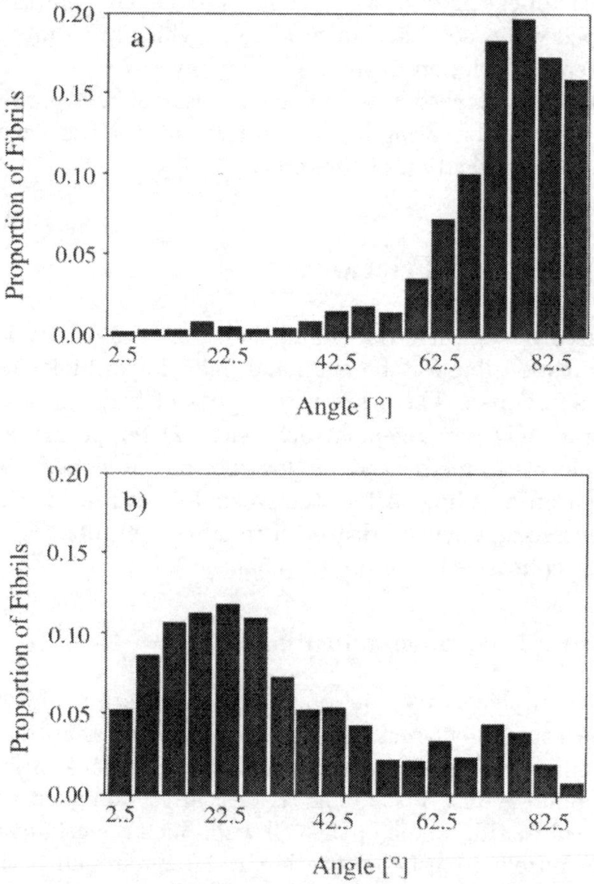

Fig. 3.10. The distribution of collagen fibril orientation in endomysium from highly shortened (**a**) and highly extended (**b**) muscles.

In summary:

- The resistance to motion from media and various substrates was an evolutionary factor, which contributed to the appearance of many devices adapted to decrease such a resistance.
- Evolutionary processes have adapted nektonic animals to interact efficiently with the water that surrounds them. Reduction of drag due to friction in the boundary-layer close to the body surface is one of these adaptations.
- Friction in the boundary-layer of the body moving in the medium at high Reynolds numbers may be decreased by sculpturing the surface.

- Viscoelastic properties of the outer part of the dolphin skin can damp short-waved turbulences appearing due to occasional surface irregularities.
- Many fish have developed water-soluble skin secretions, which are a mechanism causing the decrease of friction in the boundary-layer.
- The anti-friction mechanism of vertebrate joints consists of the porous bearing material filled with a lubricating liquid. Under load, the squeezed-out fluid contributes to the lubrication of the joint.

3.3 Systems with Increased Friction

The roughness of specialized biological surfaces and their material properties are the two main factors responsible for the frictional behavior of biological material on a variety of substrates. There are three types of friction-based attachment to the substrate: (1) permanent attachment; (2) temporary attachment and (3) dynamic attachment used for locomotion on sloped and vertical substrates or even on a ceiling. All systems usually require specialized structures on their surfaces, a defined degree of roughness of interacting surfaces, or the ability to replicate surface profile [355].

3.3.1 Surface-Replicating Mechanisms in Plants

There is an interesting example of dry friction due to replication of surface profile. An important attachment-related property of living tissues is its growth ability. Growing plant tissues can exactly replicate a surface profile by growing into surface depressions. The living tissues, after filling up the surface profile, become hard by thickening of the cell wall. Such a mechanism is hard to define as either purely adhesive or frictional. After hardening, the

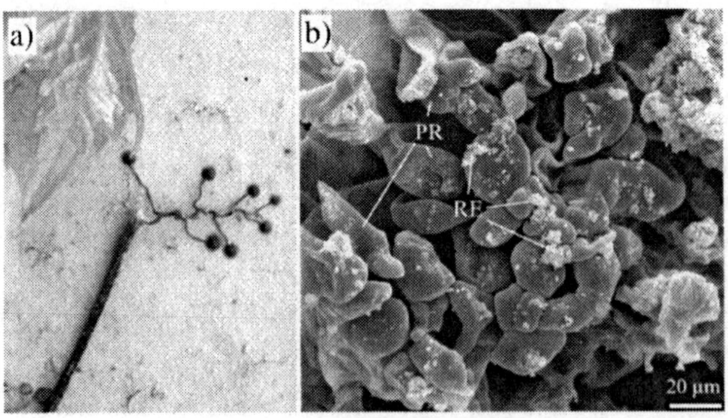

Fig. 3.11. "Suckers" of the liana *Parthenocissus tricuspidata*. (**a**) Attachment devices as they are when connected to a surface. (**b**) "Sucker" removed from the wall of a building. RE, remaining stone material. PR, plant roots grown into the irregularities of hard, stone-like material.

plant interface resembles a mechanical grip firmly attached to the surface. To some extent it is analogous to the glue principle. By the use of such a mechanism, some lianas can climb up and spread over other plants and rocks (Fig. 3.11).

3.3.2 Egg Filaments of Teleostean Fish

The eggs of teleostean fish have filaments with a length between 1.5 μm and 3.0 μm that can be distributed over the whole surface or located exclusively around the micropyle of an egg. Their origin is not clear. It is supposed that they may have originated from the egg before maturation (oocyte) or come from the material secreted by the follicular cells [356]. The function of the filaments is to interlock the eggs to substrate irregularities. However, an alternative hypothesis is that filaments attach an egg not by interlocking but rather by the adhesive properties of the filaments [357].

3.3.3 Attachment Apparatus in Hill Stream Fish

Hill stream fish often bear highly specialized attachment organs. The organs are modifications of the skin (integument) in several regions of the body. The organs are usually supplied with muscles and nerves. Such fish can be found in evolutionary diverse families, such as Echeneidae, Sisoridae, Homalopteridae, Gobiidae, Cycropteridae, and Gobiesocidae. The best-studied at-

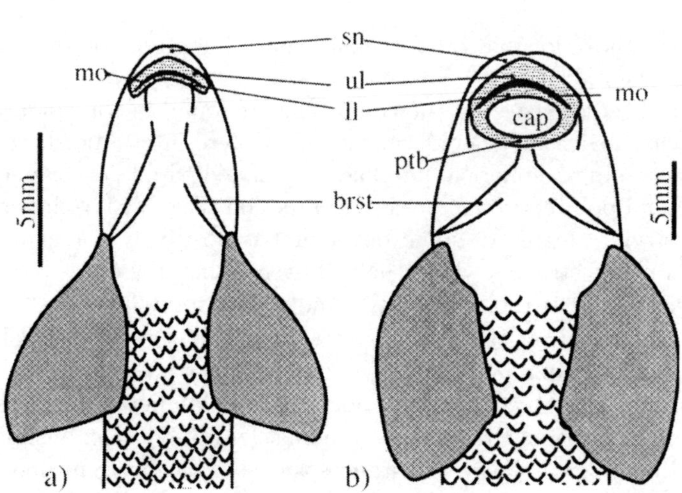

Fig. 3.12. Ventral aspect of hill stream fish, *Crossocheilus latius punjabensis* (**a**) and *Garra mullya* (**b**). brst, branchiostegal region; cap, callus portion of the menthal disc; ll, lower lip; mo, mouth; ptb, postero-lateral papillated border of the disc; sn, snout; ul, upper lip.

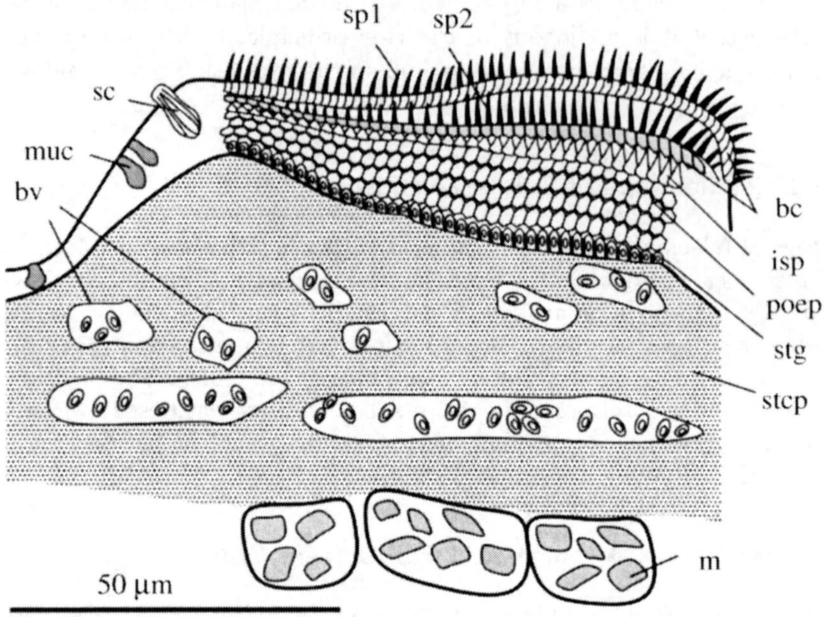

Fig. 3.13. Longitudinal section through the attachment device of the hill stream fish, *Pseudecheneis sulcatus*. bc, basal cell; bv, blood vessel; isp, intermediate spine layers; m, muscle bundle; muc, mucous cell; poep, polygonal epithelial cells; sc, sensory cell; sp1, sp2, first and second layer of spines; stcp, stratum compactum; stg, stratum germinativum.

tachment devices are those located on the mouth part (mentum) or thorax (Fig. 3.12, redrawn from [358]).

A large number of secrete-covered (mucous) cells are found in the epidermis of the attachment devices. The epidermis in the postero-lateral border of the disc has thick, prominent spherical papillae. The single papilla is covered with spines curved and pointed in one preferred direction (Fig. 3.13, redrawn from [358]). The varying degree of spine development positively correlates with the flow speed in the streams where the fish lives. When the outer layer of spines gets worn away, it is replaced by the underlying spiny layer.

It is generally accepted that the thoracic attachment devices of clingfish operate on a frictional principle [359–362]. The coarse ridges prevent the fish from slipping. In addition to mechanical interlock, the suction between ridges may also be involved in fixation. Presumably, the muscle supply is responsible for vacuum generation. Some adhesive substances, such as glycoproteins, have also been reported to be found in the sucker-like areas in some fish species [363].

Two distinct types of epidermis are present on the disc surface of the clingfish *Gobiesox maeandricus*: multilayered skin (cuboidal epithelium), in which four cell types can be distinguished, and papillate stratified columnar epithelium. The papillae are usually arranged in a characteristic pattern. Generally, the suction mechanism was experimentally proven to be responsible for attachment. Negative pressure, generated by the fish, decays in time and is periodically renewed. However, the epidermal papillae represent an additional gripping mechanism for irregular surfaces. The papillary surface apparently acts as an abrasion-resistant frictional surface. Connective tissue layers and muscle underlying the epidermis of the disc are arranged to provide structural integrity and shock absorption while allowing some flexibility to the disc and papillae [364]. Such a dual attachment system, employing both papillate epithelium, is responsible for attaching to rough surfaces, and the sucker, working on smooth surfaces, has an advantage of being highly adaptive to a variety of natural substrates with unpredicted roughness. Suckers, occurring in other groups of animals, such as monogenean flukes (Trematoda), and deuterophlebiid larvae (Insecta; Coleoptera), are often supplemented with frictional surfaces to interlock with irregularities of the surface [309,365].

3.3.4 Friction in Fish Spines

Many fish, such as sticklebacks and catfish, bear strong sharp spines in their fins, which are part of defense systems against predators. The spines are usually folded flat but can be raised at right angles to the body when danger appears. When the spine is raised, it cannot be folded down by pressing on them from the outside. Spines are locked by friction, and the muscular force is used only to raise spines, but not to hold them (Fig. 3.14, redrawn from [366]).

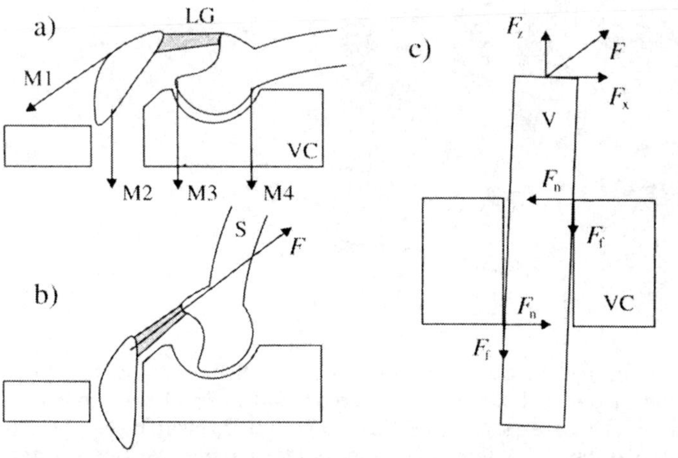

Fig. 3.14. Mechanism of spine locking in a catfish (*Synodontis*). (**a**) Folded state of the spine. (**b**) Raised state of the spine. (**c**) Model of locking mechanism. *F*, external force and components; LG, ligament; M1–M4, muscles; S, dorsal spine; V, V-shaped bone; VC, vertebral column.

The spine locking mechanism consists of several bones, which are firmly attached to the vertebral column. The spine is joined to the vertebral column by a hinge-like system composed of an inverted V-shaped bone and a strong ligament connecting the bone to the spine. When the spine is raised, the V-shaped bone slides down into a depression in the skeleton. Due to the roughness of the bone surface and the skeleton depression, frictional forces will hold the spine in a vertical position when an external force acts on it. The spine cannot be folded down by pressing its tip. To unlock this system, a vertical force, which does not jam the spine, but pulls it out of the depression, is required. Such a force is a result of two forces, the first one acting anteriorly on the V-shaped bone (M1) and the second one pushing obliquely upwards on the bone through the ligament (M2).

3.3.5 Attachment for Locomotion in Lizard Pads

Three primary groups of lizards living on trees (arboreal) exist, the family Gekkonidae, the anoline section of the family Iguanidae, and several representatives of the family Scincidae which all have specialized sub-digital pads. These animals are able to climb up vertical glass surfaces and even to hang upside down on a ceiling.

Most researchers have reported the presence of microscopic bristles on the underside of the feet [367–372], which are keratinized epidermal derivatives [373]. Similar structures are also known from the ventral side of the tail in geckos from the family Diplodactilidae [374]. The bristles arise in pairs from

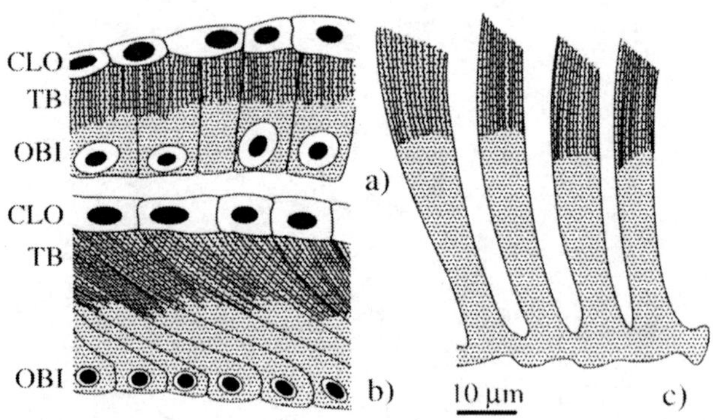

Fig. 3.15. Bristles on the feet of the lizard *Gekko gecko*. Schematic of the three stages in the formation of new bristles from the epidermal layer of the inner generation. (**a**) Early stage. (**b**) Later stage. (**c**) End stage. CLO, clear layer of outer epidermal generation; OBI, epidermal layer of inner epidermal generation; TB, terminal sub-divisions of major bristles. The cells of the epidermal layer of inner epidermal generation produce polymerized keratin in their distal parts. At a certain stage the layer of outer epidermal generation dies, and the keratinized surface become exposed.

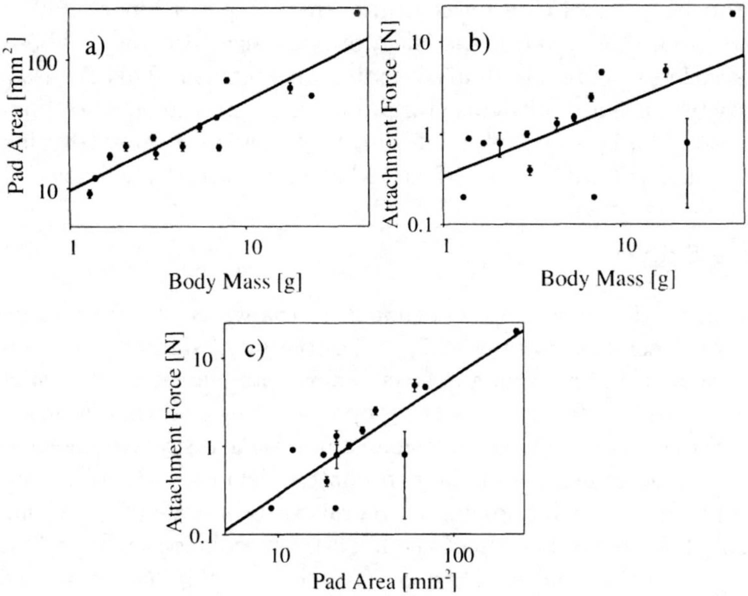

Fig. 3.16. Interrelationships between body mass, pad area and attachment force for 14 lizard species.

the surface of the keratin of the outer scale side. They are about 90 μm long and about 10 μm wide (Fig. 3.15, redrawn from [373]).

At the distal ends of the bristles, the brush-like structures called terminal divisions occur. The method of formation of the terminal divisions is complex. The hairs (setae) themselves consist of material of high tensile strength and the epidermal–dermal junction is an extremely tight bond [375]. Among lizards, at least four different structural designs are capable of attaching the animal to smooth surfaces: three types of setae and a fold ruffle structure [376].

There are several theories explaining the attachment abilities of keratinized epidermal derivatives in gekkonid lizards. Terminal sub-divisions are very small (of the order of 0.1 μm – 0.5 μm) so that they can be interlocked with irregularities even on a smooth glass surface. The flexible nature of these outgrowths lets them adapt to the surface profile. However, the frictional component of attachment is, to a large extent, excluded when the lizard walks on a smooth ceiling. Knowing that every surface, under normal conditions, has a water layer bound to the surface, such small sub-divisions may use capillary forces. An additional hypothesis about the attachment mechanism is the use of electrostatic charges between keratin and a variety of natural materials.

The pad area of lizards has been shown to be the primary causal factor influencing the clinging ability in geckos, skinks and anoles [377]. However, despite the tight correlation between pad area and attachment ability, the pad area depends on body mass to a lower degree than the attachment ability (Fig. 3.16, data from [377]). When the effect of body size is removed, about 50% of the variation in clinging ability remains unexplained. This suggests that the microsculptural and ultrastructural differences may affect the clinging ability. Geckos, skinks and anoles differ in the structure of setae covering the attachment pads [372,378]. These factors have not been studied yet.

3.3.6 Primate Skin

The frictional properties of skin are dependent on the water content and on the amount of oily secretion (sebum) [379]. The friction of skin on wood enables small clawless primates to cling to steeper slopes and larger vertical supports. In human subjects, friction is more proportional to load than in small primates. This is due to the fact that the small animals have discrete, elevated palm (volar) pads. Pad coalescence in large primates may be an adaptation for increasing the power to which a load must be raised to become proportional to friction [380]. The detailed study of the fingertip morphology among four small-bodied New World monkeys (*Saimiri sciureus*, *Leontopithecus rosalia*, *Callithrix jacchus*, and *Saguinus oedipus*) revealed that development of the epidermal ridge becomes more pronounced as the frequency of small branch foraging increases. Terminal phalanx breadth and epidermal ridge complexity are both positively correlated with apical pad size. The large, flexible apical pad increases the stability of the hand and foot on small diameter arboreal substrates, because the pad can contact the substrate on several planes which, in turn, enables the pad to resist disruptive forces from different directions by friction and interlocking [381]. Adaptations of frictional surfaces of primate skin reflect evolution of some lineages and can be, therefore, used for primate phylogeny [382].

3.3.7 Frictional Surfaces of Snake Scales

Friction-modifying nanostructures of the scale surface of snakes have recently been described [383]. These include an ordered microfibrillar array, presumably to achieve adaptable friction characteristics. Significant reduction of adhesive forces in the contact areas caused by the double-ridge microfibrillar geometry provides ideal conditions for sliding in a forward direction with minimum adhesive forces. Low surface adhesion in these local contact points may reduce local wear and skin contamination by environmental debris.

The highly asymmetric profile of the microfibrillar ending with a radius of curvature of 20 nm–40 nm may induce friction anisotropy along the longitudinal body axis and functions as a kind of stopper for backward motion, providing low friction for forward motion. The system of micropores penetrating the snake skin may serve as a delivery system for a lubrication/anti-adhesive lipid mixture that provides boundary lubrication of the skin.

3.3.8 Interlock in Parasites and Plant Diaspores

Hookworms have a specialized attachment mechanism adapted to the soft tissues of the host. Usually, they bear characteristic toothed plates to hook into the tissues. Such systems are also called multiple-hook devices. There are three main types of hooks described in cestod worms: primary, accessory and localized. Primary hooks (macrohooks) surround the mouth structure (proboscis) in typical half rings. The microhooks are located in spaces between macrohooks. Localized hooks are found only at certain places on the proboscis [309]. A very regular arrangement of hooks is characteristic for proboscis of acanthocephalan worms. If the proboscis is invaginated, the hooks point inwards. When the proboscis is everted by an increase of internal pressure or special protractor muscle, the teeth contact the soft tissues of the host and interlock. A similar principle is convergently employed by plants. Fruits and seeds of many plants possess hooking devices and use animals to disperse themselves some distance from the parent plant. The hooks contact animal hairs, interlock, and are transported by the animals (Fig. 3.17a). The frictional properties of the huge variety of plants employing the hook principle remain unknown.

The hook principle was used for an industrial application called the Velcro fastener, which comprises two surfaces. One surface is covered by relatively stiff nylon hooks or mushroom-shaped structures and the other by soft adaptive threads or loops (Fig. 3.17b). When both parts are pressed together, they become securely attached. An important property of hook systems is that detachment usually requires a large force and that the detachment action consists of a strong jerk pulling two parts apart.

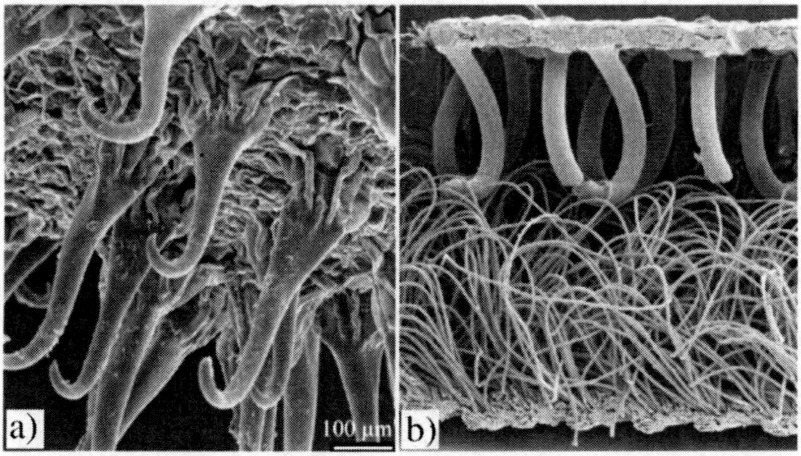

Fig. 3.17. Hook-like frictional systems. (**a**) Fruit part of the plant, *Galium aparine* (courtesy of Olena Gorb, MPI, Tübingen). (**b**) Two co-opted surfaces of a Velcro fastener (courtesy of Jürgen Berger, MPI, Tübingen).

3.3.9 Bird Feather Interlocking Device

An interesting frictional system, working at low attachment force, is found in the feathers of birds. The feather vane can be separated into two parts by pulling tangentially. The original state can be re-established easily by a light stroking through the feathers [309].

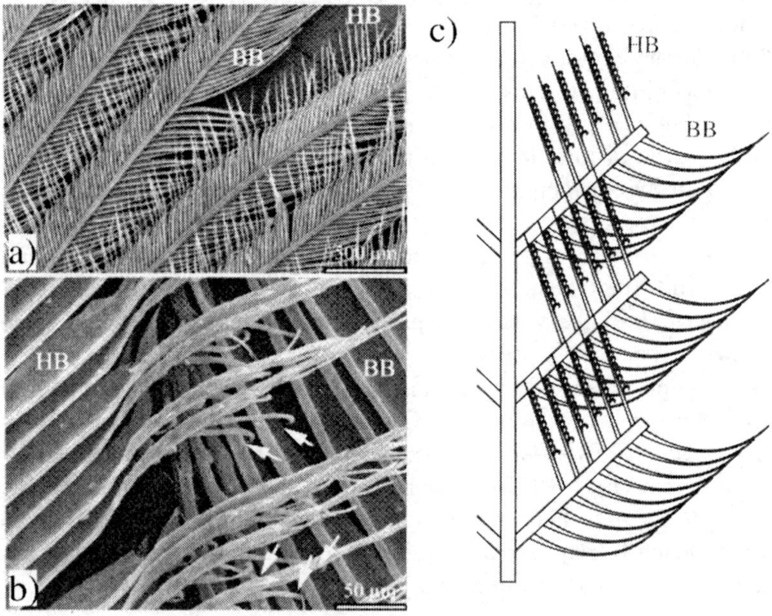

Fig. 3.18. (**a**) and (**b**) Bird feather details. (**c**) Schematic of probabilistic fasteners of the bird feather. Arrows point to hooks; BB, bow barbules; HB, hook barbules.

Such an interlocking mechanism, called a probabilistic fastener, uses small elastic hooks (Fig. 3.18). The vane of a typical contour feather consists of a number of side branches called barbs which are aligned in parallel and at an angle to the shaft. The barbs carry, on either side, a similar array of second side branches called barbules. In most birds the row of barbules on the side of the barb toward the tip of the feather bears very fine, backward facing, hooks. These barbules are called hook barbules. The barbules on the other side of the barb are sharply curved and come without teeth or with very short outgrowths. These barbules are called bow barbules. Hook barbules become automatically anchored on the bow barbules, due to the preferred orientation of the hooks and micro-irregularities on both types of barbules. The detachment of the barbules can be easily caused by an external force. This mechanism prevents real damage of the feather and provides easy repair of microdamage of the feather coverage. Although the structure and function of the system is well known, the attachment force holding two barbules together has not been measured previously, but it presumably lies in the micronewton or millinewton range.

In summary:

- There are three types of friction-based attachment to the substrate: (1) permanent attachment; (2) temporary attachment and (3) dynamic attachment used for locomotion on sloped and vertical substrates or even on a ceiling.
- Permanent attachment is the result of intimate contact due to surface replication at the micro- and nanoscale by growing cells.
- Temporary attachment of different biological systems is based on a variety of mechanisms: suction, mechanical interlock, capillarity, viscous force, molecular adhesion, and electrostatic interaction.
- Dynamic attachment is mainly mediated by adhesion.

3.4 Mediators of Adhesion

3.4.1 Design of Adhesive Polymers

An ideal adhesive can be viewed in relation to particular adherends (substrates). It should fulfill the following conditions: (1) zero contact-angle; (2) low initial viscosity; (3) joining of liquid to solid in a manner which allows displacement of contaminant (air/water) from the surface [384]. The properties of adhesives and adherends vary from one to the next. However, polymers used in adhesives demonstrate several common properties. The polarity of molecules as well as charged and hydrophilic groups, as a rule, enhance adhesion. Such molecules are able to form strong ionic or hydrogen bonds with the polar groups of surfaces, which usually have an abundance of bound water or oxides. The chain length and the number of side chains are additional important properties of adhesive polymers (Fig. 3.19). Since van der Waals dispersive interactions (see Sect. 2.4.1) increase with molecular size, a longer polymer chain will also enhance adhesion. Adhesion is also improved by the formation of ordered arrays of linear molecules with regular sequences, such as collagen or keratin. Small side chains may result in the formation of polycrystalline structures. Therefore, cross-linking may contribute directly to the elastic modulus of a material by preventing the flow of molecules relative to one another. A viscoelastic polymer with few cross-links is a stronger adhesive than a rigidly cross-linked one. This is due to better abilities of stretching in longer linear sequences prior to failure. Desirable properties for adhesive polymers are listed in Table 3.1 [384].

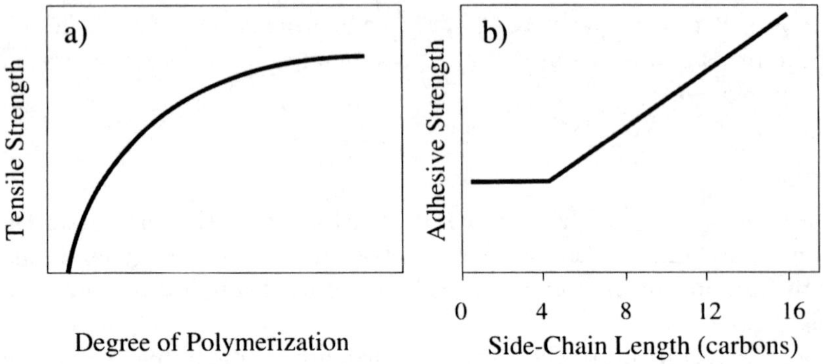

Fig. 3.19. The effect of the degree of polymerization (**a**) and side-chain length (**b**) on the adhesion of polymers.

Table 3.1. Properties of polymers enhancing adhesion

Polymer property	Function
high molecular weight (> 100 K)	better cohesion between molecules
bulky side chains	prevents crystallization, improves interfacial entanglements
polar, hydrophilic side chains	promotes ionic and hydrogen bond formation with adherend
zero contact-angle	spontaneous spreading on adherend
low initial viscosity	facilitates diffusion into depressions of surface
cross-linking	improves cohesion; retards degradation

3.4.2 Coupling Agents

Even chemically incompatible materials can be attached, when a coupling agent is used. The coupling agent is necessary when: (1) the spreading of the adhesive on the adherend is poor; (2) when severe conditions, such as high humidity or high temperature, prevent formation of adhesive joints; (3) when the surface tension of the adherend is lower than that of the adhesive. Coupling agents (detergents, surfactants) have a dual functionality, interacting well with both the adhesive and adherend [384]. Usually, they are monolayers with a preferred orientation of chains or molecular groups (Fig. 3.20, from [384]). For example, sea organisms have to displace or absorb all water, in order to make intimate contact with the surface. A similar problem with dental adhesives is solved by the use of large polar hydrocolloids (sodium potassium salts, guar gum).

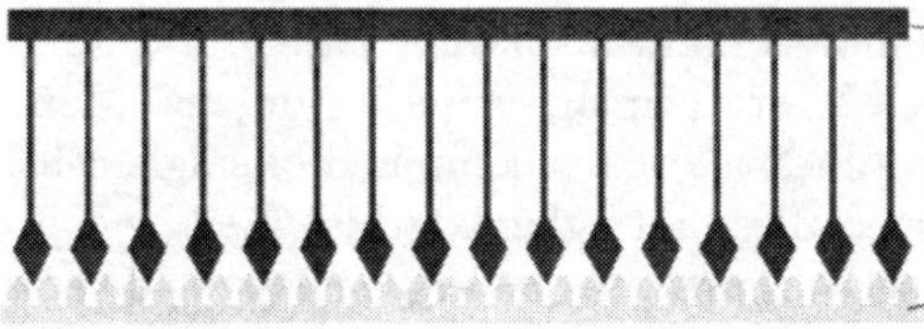

Fig. 3.20. A coupling agent promotes and strengthens adhesion between otherwise incompatible materials. It forms an orientated monolayer on the substrate over which the adhesive spreads.

3.5 Systems with Increased Adhesion

Materials preventing the separation of two surfaces may be defined as adhesives. Why use adhesives for attaching materials to surfaces? There are a variety of natural attachment devices based on mechanical principles [309]. There are at least three reasons for using adhesives: (1) they join dissimilar materials; (2) they show improved stress distribution in the joint; and (3) they increase design flexibility [384]. These reasons are relevant to the evolution of natural attachment systems and man-made joining materials. However, it is sometimes desired that materials do not stick to each other. For this purpose, nature has devised a number of systems that prevent adhesion.

In general, adhesive-bond formation consists of two phases: contact formation and generation of intrinsic adhesion forces across the joint [385]. Adhesive forces generated by biological adhesives may be of a different nature. The action of the adhesive can be supported by mechanical interlock between irregularities of contact surfaces. Increased surface roughness usually results in an increased strength of the adhesive joint. The last result, however, may be simply explained by increased contacting area between contacting surfaces and the adhesive substance.

Strong adhesion is also possible between two ideally smooth surfaces. The basic interactions (intermolecular forces) are present in all compounds. If sufficient contact between the substrate–adhesive interface is reached, forces will be set up between atoms and molecules of both contacting materials, and they will adhere. Van der Waals forces are the most common of such forces together with secondary forces, such as the hydrogen bond. The electrostatic forces, produced by electron transfer in the contact area, have been clearly demonstrated. However, for typical biological adhesion, their contribution is not significant.

Adhesive organs, used for attachment to substrates as well as those involved in catching prey, demonstrate a huge diversity among living creatures due to their structural and chemical properties. Biological adhesion underlies the organization of all living tissues. Cell contact phenomena have been extensively reviewed in the biomedical literature [386,387], and the physical bases of adhesive phenomena have been reported by many authors in physico-chemical contributions [388]. The function of attachment appeared very early in evolution; early unicellular organisms have a variety of cellular adaptations to adhesion. In fact, cells are primary adhesive organs. Many multicellular organisms often bear adhesive organs composed of single cells. In many cases, however, cells are specialized into glands, which may be composed of several cell types. This section provides a brief review of diverse biological adhesive systems intensively studied in the last two decades. The study of biological adhesion is very difficult because of the variety of cellular structures and chemical substances involved. Additional problems are the very small amount of material which can be obtained for analysis and impurities leading to artifacts.

3.5.1 Cell Adhesion

Multicellular organisms consist of many cells, which adhere to each other, forming the anatomic integrity of an organism. Cell contact phenomena may be divided into three basic processes: (1) cell contact itself, (2) cell adhesion and attachment, and (3) cell separation [389].

The forces which contribute to cell adhesion are (1) chemical bonds, which can be sub-divided into electrostatic, covalent, and hydrogen bonds; and (2) van der Waals interactions. These forces act, approximately, over a range 0.2 nm–0.4 nm in distance. Therefore, for cells to adhere to each other or to the substrate, they must come within this distance. The initial adhesion and spreading of cells is dictated by the particular surface properties of the non-living substrate or surface of another cell. Since cells are dynamic systems, the adhesive interactions can also occur not only directly between two cell surfaces but between adhesive substances produced by the cells and delivered to the surface. Such cellular products as fibers and desmosomes are often found in the contacting area of adhering cells. For example, the organic union between the tooth and adjacent epithelial cells is at least partly based on a matrix produced by epithelial cells [390]. The matrix consists of collagen and adhesive molecules.

There are several adhesive molecules contributing to the maintenance of organism integrity: proteoglycans, mucopolysaccharides, proteins and phospholipids [390]. Recently, experiments have revealed that proteoglycans are substances with extremely high binding forces. Direct measurements of the adhesive forces, by the use of AFM, show that a single pair of a proteoglycan molecule of the marine sponge can hold the weight of 1,600 cells [70].

Cell adhesion phenomena are mainly the result of intermolecular interactions between cell-produced substances and various substrates. In attempts to modify biological surface interactions, contradictory requirements arise. In the case of dental and bone prosthesis, good tissue–to–plastic interaction is desired. For the vessel implants, the surface should minimize possible thrombus formation, but simultaneously should be compatible with the complex antibody and rejection mechanisms [388]. This challenge requires more experimental measurements, which are currently in an initial phase.

3.5.2 Invertebrates

Three types of adhesion at the organism level are known: (1) Temporary adhesion allowing an organism to attach strongly to the substrate and detach quickly, when necessary; (2) Transitory adhesion permitting simultaneous attachment and movement along the substrate; (3) Permanent adhesion involving the secretion of a cement. These three types of adhesion do not have the same purpose and use different adhesive systems (Table 3.2) [391].

Table 3.2. Tenacity of some marine invertebrates on smooth glass

Taxa	Species	Structure involved in adhesion	Type of adhesion	Tenacity ($\times 10^5$ N/m^2)
Cnidaria	*Metridium senile*	foot	transitory	0.43
Anthozoa	*Actinia equina*	foot	transitory	0.19
Mollusca Gastropoda	*Patella vulgata*	foot	transitory	2.28
Mollusca Bivalvia	*Mytilus edulis*	byssus	permanent	3.16–7.50
Crustacea	*Semibalanus balanoides*			
Cirripedia	cyprid larva	antennules	temporary	0.98
	adult	base	permanent	2.30
	Balanus crenatus	base	permanent	5.20
Echinodermata Asterias vulgaris	*Asteroidea*	podia	temporary	1.70
Echinodermata Holothuroidea	*Holothuria forskali*	cuvierian tubules	permanent	6.00

Flatworms (turbellarians), annelids, and gastrotrichs (worm-like animals) use glandular organs composed of three cell types: two distinct gland types and specialized epidermal cells for secretion discharge [392]. Such a triplet of structures, usually operating with three distinctive functions of adhesion, de-adhesion, and tension-bearing, is known also in most marine organisms. Because of the dual function of such glands, they are called "duo-glands" independently of their homology. It has also been suggested that organisms adhere by secretion of poly-cationic substances, which bind to substrates, especially to anionic sites [393]. Release involves secretion of poly-cationic substances capable of competing for active sites in adhesive molecules. The

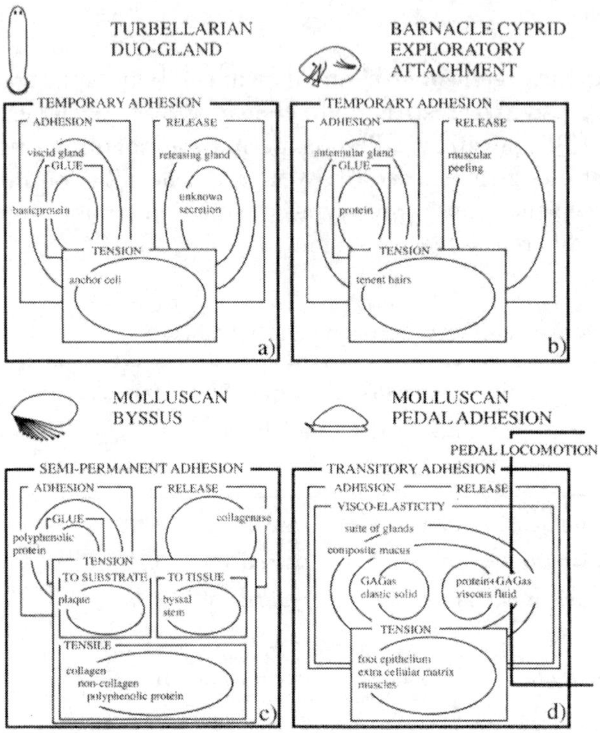

Fig. 3.21. Schematic comparison of functional elements of adhesive organs in a turbellarian (**a**), barnacle (**b**), mussel (**c**), and (**d**) gastropod. Functions are marked by boxes; structures performing them as ovals.

actual mechanism of adhesion depends on the chemistry of secretion. These compositions are different in various groups of invertebrates. Attachment systems of polychaete larvae use lectins as adhesives. Insects seem to use lipid-containing secretions. The majority of other systems bear proteins and glycans (glycoproteins, proteoglycans, polysaccharide protein complexes).

Systems adapted for permanent attachment, such as mussels and barnacles, bear proteins and polyphenols [394–398]. Polyphenols play a role in tanning proteinous elastic components of the secretion.

In systems using adhesive substances as a kind of viscoelastic lubricant for locomotion, glycans are the main component. These adhesives, reported from molluscan pedals and echinoderm tube feet, are called mucus. Due to its viscoelastic properties, mucus behaves as an elastic solid under rapidly applied forces (adhesive function), but as a flowing fluid at slow velocities (lubricating function) [399]. The variety of adhesive systems, based on two-component adhesion, shows that there is no universal duo-gland mechanism (Table 3.3 and Fig. 3.21 from [392]).

Table 3.3. Chemical compounds of proteinaceous and/or glycan nature in invertebrate adhesive systems. AB, Alcian–blue; CBB, Coomassie–brilliant–blue; GAG, Glycosaminoglycan; MBB, Mercuric–bromphenol–blue; PAS, Periodic–acid–Schiff; (+), positive reaction; (−), negative reaction

Taxon and type	Proteinaceous component	Glycan component
Plathelminthes Rhabditophoran turbellarian duo-gland, viscid gland	basic protein, with arginine, polyphenolic, resistant to pronase, trypsin and pepsin	minor vicinal–diol component (weak PAS+ detectable only by EM)
Gastrotricha *Turbanella sp.* duo–gland, viscid gl.	non-basic protein, polyphenolic	minor vicinal–diol component?
Nematoda *Theristus* duo gland, viscid gl.	non-basic protein?, with sulfhydryl groups	minor vicinal–diol component
Mollusca *Mytilus byssus*	polyphenolic–protein adhesive on byssus plaque; semipermanent attachment through tanned byssus threads; catechol–oxidase–cross-linked polyphenolic protein; collagen formed rigid region; non-collagenous elastic region	sulfated GAGs from separate gland mixed incompletely with adhesive in plaque
gastropod pedal mucus	protein in 2 of 6 pedal glands, reportedly not used in adhesion, mostly acidic amino acids	neutr. and acidic (GAG) mucopolysaccarides in 4 of 6 glands; GAGs (+ proteins?) or glycoprot. function as viscoelast. adhesive
Annelida branchiobdellid duo–gland	basic protein component of major gland; non-basic protein with tryptophan in minor gland	neutral mucosubstance (PAS+, AB+) in major gland; minor gland (AB-, PAS–)
histriobdellid duo–gland, viscid gland	protein? (basic?) (MBB+)	neutral mucosubstance (PAS+)
Arthropoda barnacle cyprid larva	protein (CBB+)	(none detected AB-, PAS-)

Table 3.3. Continued

temporary adhesion		
barnacle adult permanent attachment	glycoprotein, protein collagen-like, lysine–rich (basic) or glutamic–acid–rich (acidic); phenolic–tanned	hexosamines detected biochemically, otherwise AB-, PAS-
spider–web adhesives	basic protein, lysine–rich, also contains other polar and charged side–chains	
Echinodermata *Leptasterias*	proteoglycan? protein moiety indicted by trypsin and chymotrypsin–sensitivity	anionic (ruthenium–red positive component) possibly GAG; amylase–insensitive
other asteroid and echinoid genera	? (none–detected; MBB-)	neutral and acidic (GAG) mucosubstances and sulfated GAG's in 2 or 3 gland types; (PAS+ and AB+ moieties)
Cnidaria *Hydra* basal disk		neutral acidic (GAG) mucosubstances (PAS+, AB+)

3.5.3 Byssus Adhesion in Molluscs

Marine bivalve molluscs are probably the most-studied adhesive organisms. Their adhesive capacity is selectively advantageous, because it allows molluscs to exploit nutrient-rich turbulent zones of the sea. Oysters are permanently fixed to the substrate by secretion of cement glands, whereas byssated molluscs adhere but are not necessarily permanently attached. They are able to abandon an old attachment site and move to a new one. An additional advantage of the byssus are its mechanical properties, allowing better damping of the wave impact. The byssus is an extracellular structure consisting of a thread (0.05 mm–0.20 mm in diameter and 2 cm–4 cm long) attached to the animal at one end and to a substrate at the other (Fig. 3.22, redrawn from [384,400]). The byssus is produced by the foot, and the attachment to the animal is at the ventral base of the foot. The threads end at the stem, a root-like process embedded deeply in the base of the foot. The distal ends of the threads are attached to oval plaques (2.5 mm long and 2.0 mm wide).

The average attachment strength of the whole composition of threads of an animal ranges from 10 N–12 N in *Mytilus edulis* to 90 N in *Septifer*

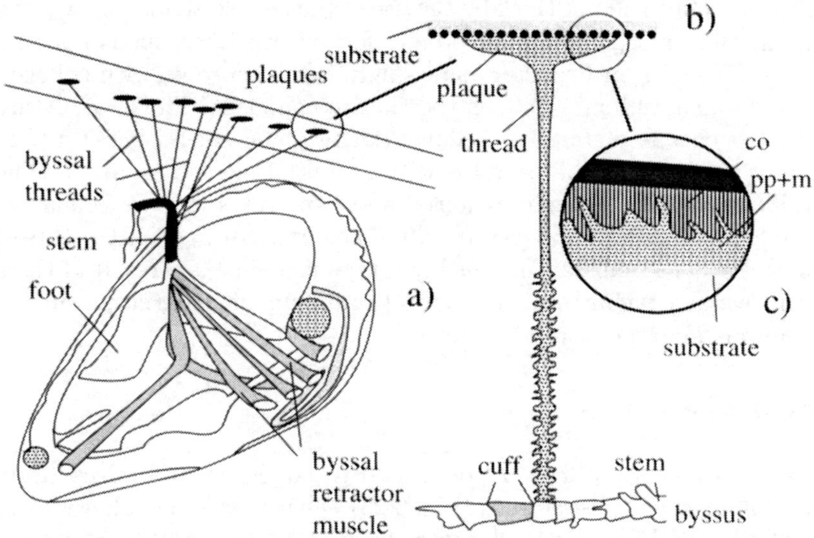

Fig. 3.22. Mussel byssus is a bundle of threads consisting of extracellular material. By drawing itself up on these tendinous threads, the mussel can control the tension of its attachment. The distal tips of the threads end in flattened adhesive pads called plaques. (**a**) Schematic view of a mussel suspended by a byssus. (**b**) Schematic of the byssus. (**c**) Distribution of mucosubstance (m), collagen (co), polyphenolic protein (pp) and accessory or b-type protein in a region of plaque adhering to the substrate.

bifurcatus [384]. However, such a force is summed over all structural elements securing animals against external forces. It includes an adhesive bond of the plaque to the substrate, the thread attachment strength to the mussel, and the tensile strength of the threads. Tensile tests revealed a lower initial modulus, a lower ultimate stress, and a higher ultimate strain in the proximal region compared to the distal one [401]. Usually, the breaking strength of threads is an order of magnitude greater than that of the attachment sites. The proximal region and adhesive plaque are common sites of structural failure [401]. Byssal thread consists of highly ordered crystalline polymers (collagen and b-type protein). Collagen fibers are packed inside a b-type protein sheath [402]. Sometimes byssal ribbons consist of striated fibers.

Five groups of cells or glands participate in the byssus production. Many chemical details of byssus secretion remain unknown. In most species, the byssus consists of at least four essential components: acid mucopolysaccharides, adhesive protein, fibrous proteins, and an oxidative enzyme, polyphenoloxidase.

Waite [384] has reviewed the sequence of events during the formation of the attachment plaque (Fig. 3.22). (1) A dispersion consisting of phenol granules and a mucoid substance with a contact-angle of about 6° is secreted onto the substrate surface. This dispersion is presumably mixed and applied to the surface in films 50 nm thick by the use of the paddle-shaped cilia. Such a mixture results in a polyphenolic protein forming the continuous bonding phase. (2) The secretion becomes increasingly heterogeneous, with collagen added via the longitudinal ducts as the third component. The architecture of this heterogeneous material is cellular (trabecular): spaces between trabeculae of phenolic material are filled with an initial mucoid secretion. The collagen interacts with phenolic material and penetrates the trabecular architecture to the depth of about 3 μm. (3) The outer coverage of the thread and plaque is a dense substance, possibly b-type protein. As a result of these events, the byssal attachment plaque joins two dissimilar materials, a surface and the fibrous material of the threads.

3.5.4 Tube Feet of Starfish

Echinodermata bear tube feet or podia used to adhere to and release from substrates by chemical interactions (Fig. 3.23). This mechanism allows them to move, feed, and burrow on substrates [403,404]. The podia are external appendages of the hydraulic (ambulacral) system and are driven partly by muscles and partly by hydraulic pressure [405]. There are six types of podia: descending, penicillate, knob ending, lamellate, ramified, and digitate [391].

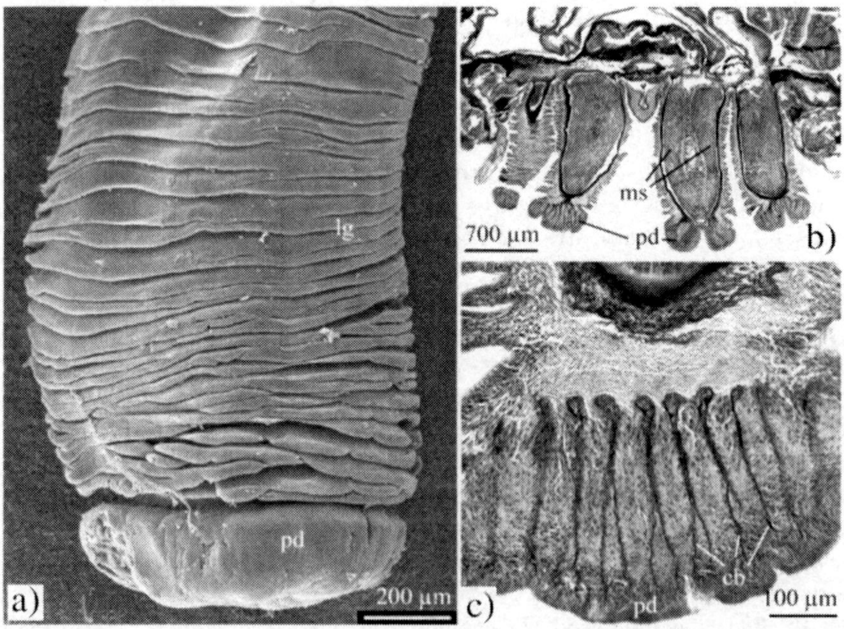

Fig. 3.23. Podium of the starfish, *Asterias rubens.* (**a**) External morphology. (**b**), (**c**) Longitudinal sections. cb, collagen bundles; lg, leg; ms, muscles; pd, podium.

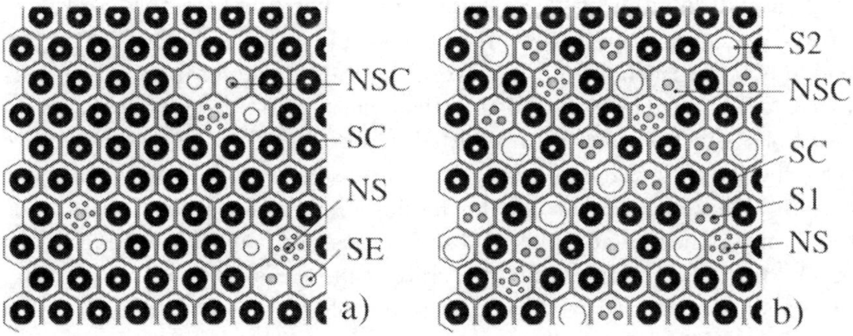

Fig. 3.24. Schematics of transverse sections through adhesive areas of echinoderm podia. (**a**) Handling podia of *Asteronyx loveni* (Ophiuroidea). (**b**) Locomotory podia of *Holothuria forskali* (Holothuroidea). NS, neurosecretory-like cell; NSC, sensory cell; S1, secretory cell of type 1; S2, secretory cell of type 2; SC, support cell; SE, secretory cell.

The histological structure of podia is relatively constant. They consist of four tissue layers: an inner and outer epithelium, with a connective tissue layer and a nerve plexus between them. The connective tissue layer is made of an amorphous substance that encloses bundles of collagen fibers and various types of cells (Fig. 3.23). The layer consists of a diffuse external layer and a more compact internal layer. With the internal layer, the collagen fibers are circular in arrangement.

In echinoids and holothuroids, the external connective layer may contain one or several kinds of embedded calcareous needles (spicules) [406, 407]. Externally, the epidermis is covered by a multi-layered cuticle [391]. The cuticle consists of fibrous and, sometimes, granular material. There are 3–5 sub-layers, the outermost of which is composed of numerous fibrils.

Only the distal parts of the tube feet (knobs, discs and digitation structures) are involved in attachment. Adhesive organs operate according to mechanical and chemical mechanisms. Podia possess a duo-gland adhesive system with two types of secretory cells: those releasing an adhesive secretion and another releasing a de-adhesive secretion. In podia epidermis, there are three main types of specialized cells: secretory, neurosecretory-like, and sensory cells. In handling podia, cells are always closely associated, forming sensory–secretory complexes in which the three cell types occur (Fig. 3.24a, redrawn from [391]). In locomotory podia, three cell types form a homogeneous cellular layer together with support cells (Fig. 3.24b). Such a cellular organization of the adhesive surface correlates well with the functions of the podia: a large adhesive area correspondingly provides a stronger attachment during locomotion, whereas discrete adhesive areas seem to be more efficient for the handling of small substrate particles.

Adhesive and de-adhesive secretions are produced by secretory and neurosecretory-like cells, respectively. In most species, the secretory cells contain acid mucopolysaccharides. In several species, proteins instead of mucopolysaccharides have been revealed in these cells. The third group of species possesses both substances. Presumably, different functions of podia require different adhesive strengths and, therefore, adhesive secretions with different composition. Thus, a coarse molecular model for adhesion in echinoderms is a protein–polysaccharide complex whose composition varies from one taxon to another [391]. Interestingly, neurosecretory-like cells, which are secreting de-adhesive substances, do not stain with any of the classical histochemical dyes. The mechanism of detachment is not fully understood. There are two hypotheses explaining podia detachment. (1) De-adhesive material interferes with the electrostatic bonds between the adhesive layer and external layer of the cuticle. (2) The material works as an enzyme, releasing an external layer of the cuticle from the underlying cuticular layers.

The mechanical mechanism of attachment is restricted to the sucker-mediated attachment. This hypothesis presumes that descending podia can deform the disc shape into a kind of sucker that is additionally supplemented by chemical adhesion. The levator muscle in the podial disc is a probable candidate for suction generation. The mechanics of the podial end may play an important role in an increase of the contact area. Unfortunately, to our knowledge, the mechanical properties of the tube feet ends have not yet been studied.

3.5.5 Transitory Adhesion of Soft-Bodied Invertebrates

Gastropod molluscs, annelids, sea anemones, and free living turbellarians attach to the substrate by a viscous film, which they leave behind as they move. These animals can move along the substrate by ciliary gliding or by means of waves of muscular contractions running along their feet. The power of locomotion in gastropod molluscs is provided by rhythmic muscular contractions on the ventral surface of the foot, and the force due to these waves is coupled to the substrate by the pedal mucus. Consequently, movement of the animal is a result of a balance of forces. In the case where the adhesive force of the stationary portion of the foot surpasses the friction force caused by the portion of the foot moving forward, an anteriorly-directed reactive force is produced [399,408]. Both portions of the foot remain in contact with the substrate. Differences in the adhesive and frictional properties of different parts of the foot are caused by the mechanical properties of pedal mucus at different velocities (Fig. 3.25, redrawn from [408]).

An interesting mechanism of intermediary type between transitory and temporary adhesion occurs in branchiobdellids (Annelida). Branchiobdellids are small annelids living as ectosymbionts on fresh-water crustaceans. These soft-bodied animals use anterior and posterior attachment sites to move in a leech-like way. The anterior attachment organ is a glandular pad on the

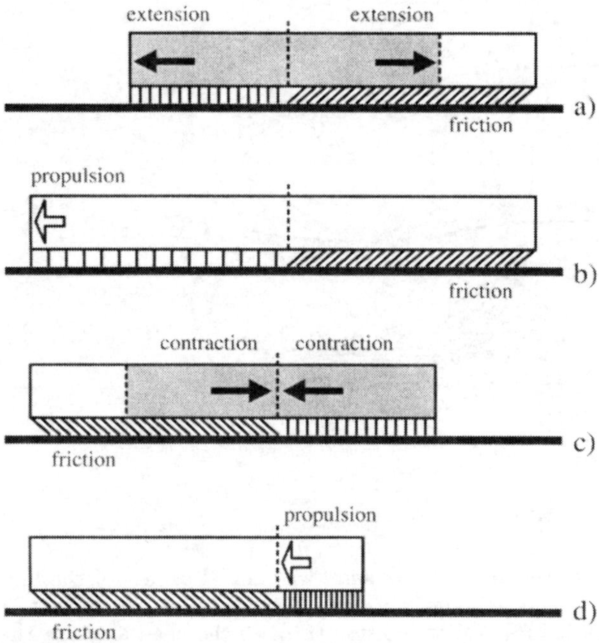

Fig. 3.25. Schematic of the propulsive movements based on differences in adhesion-mediated frictional forces of the gastropod slug.

ventral surface of the peristomium. The posterior site is an adhesive disc located on the terminal body segment (Fig. 3.26, redrawn from [409]). Two gland types have been reported: the first consists of the viscid cells with PAS+ granules, and the second mainly contains smaller granules. It has been previously suggested that two types of glands compose a kind of duo–gland type of adhesive system, where one gland type is responsible for attachment and another serves to release attachment [409]. Thus, these animals use a duo-gland system not only for temporary attachment, but also for locomotion. The pedal disc of sea anemones is also an organ adapted for temporary and transitory adhesion driven by muscles. Muscle contraction, however, is not important in the adhesion mechanism. Anemones utilize a proteinaceous secretion and a small amount of mucus for pedal adhesion. The tenacity of the pedal disc on a variety of substrates ranged from 0.16×10^5 N/m^2 to 0.85×10^5 N/m^2 in *Metridium senile* and from 0.09×10^5 N/m^2 to 0.27×10^5 N/m^2 in *Actinia equina* [410]. Lower tenacities of sea anemones compared to other marine organisms, such as mussels and barnacles, might be explained by higher the mobility of the sea anemones.

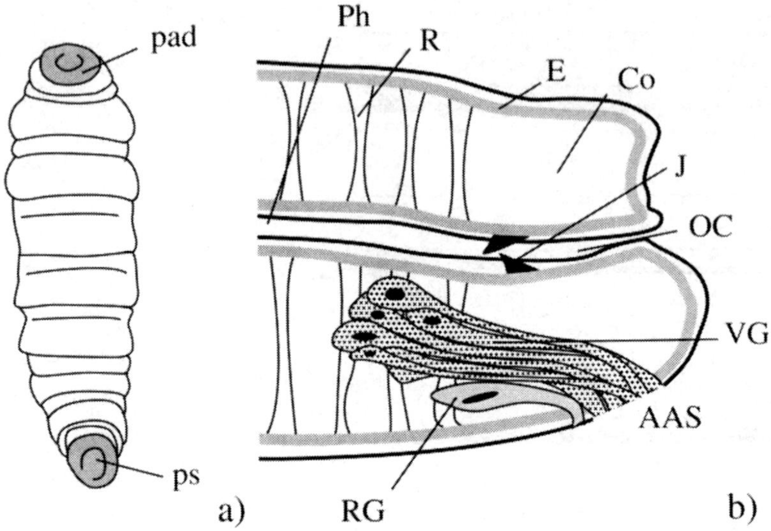

Fig. 3.26. Branchiobdellid *Xironodrilus bashaviae*. (**a**) Ventral aspect of the animal with anterior and posterior attachment sites. ps, peristomium; pad, posterior attachment disc. (**b**) Schematic of a sagittal section through the head showing the anterior attachment site (AAS), viscid (VG) and releasing (RG) glands. J, jaws; Co, coelom; E, epidermis; OC, oral cavity; Ph, pharynx; R, radial muscle of pharynx.

3.5.6 Adhesion in Barnacles

There are two types of adhesion occurring in these crustaceans. The first one is temporary, which occurs in larval barnacles. The second one is the permanent adult adhesion. The antennular attachment organ holds the larva on the substrate while it explores, prior to permanent settlement. The bond allows the larva to remain attached against moving sea waves, but still provides some movements. The so-called antennulary sucker is the attachment disc and encircling velum. However, this organ could not produce suction. Attachment is due to the presence of secretion on the attachment disc. The secretion is produced from the numerous (about 20) unicellular glands located in the second antennal segment. These gland cells open individually, in two concentric rings on the surface of the attachment disc [411]. Additionally, the attachment discs demonstrate high deformability, allowing them to adapt to the substrate profile.

In the larva, two cement glands are located just posterior to the compound eyes. From each gland, a collecting duct leads into the single cement duct, which passes down an antennule and opens out at several points on the attachment disc surface. There are two cell types responsible for permanent adhesion. The first type (a-type) gives positive reactions to histochemical tests for proteins, phenols, and polyphenoloxidase, whereas the second one (b-type) only for proteins. The cement is relatively liquid when secreted, but

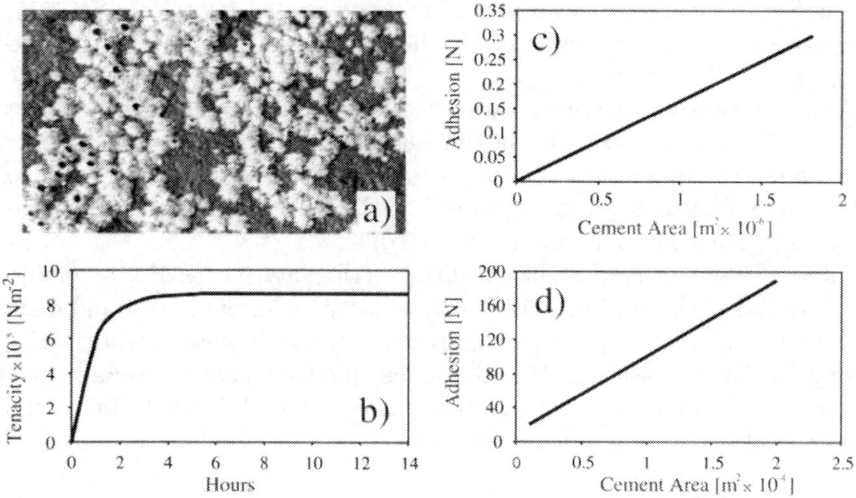

Fig. 3.27. Barnacle adhesion. (**a**) Barnacles on the sea substrate. (**b**) Tanning of the barnacle cement. Non-linear asymptotic regression of the tenacity of the barnacle cement versus time from settlement. (**c**), (**d**) Linear regressions of the relationships between the area of cement of small (**c**) and large (**d**) individuals of *Balanus balanoides* and the attachment force.

its tanning begins immediately. The liquid cement fills in the microscopic depressions of the surface. The tanning process is gradual, during which the attachment force, plotted versus time from settlement, shows an asymptotic increase (Fig. 3.27, data from [412], redrawn from [411]). This natural adhesive apparently meets the criteria for good bonding agents for further use in medicine [413].

3.5.7 Permanent Adhesion in Insects and Spiders

The coxal vesicles in representatives of Machilidae (Archaeognatha) serve to cement the old cuticle during skinning (ecdysis), while legs are used to free the body from the skin (exuvium). The cuticle of the collapsed everted coxal vesicle remains attached firmly to the substrate after ecdysis is completed [414]. The coxal vesicle is a thin sac-like structure located on the posterior margin of the abdominal coxae. The primary function of these organs is to take up water from the substrate. The cement from the vesicles has glandular origin. The cement cannot be removed from the surface by the use of water, saline or acetone. The exact composition of the cement remains unknown.

Parasitic insects such as lice or dipterans from the family Oestridae (Hypodermatinae) possess specialized glands to glue their eggs to the host hairs. Hypodermatin eggs also have specialized attachment organs consisting of a flexible petiole and a prominent clasper. The clasper, which is responsible for affixing the egg on a host hair, consists of an attachment groove filled with adhesive substance and a pair of adhesive-coated lateral flanges [415]. The host hair is primary locked within the groove, and the adhesive solidifies around it.

Adhesive substances are also used for prey capture by orb-web spiders. These spiders use the secretion of the aggregate glands to cover a so-called sticky spiral, where the prey is mostly captured. The essential components of this adhesive substance are glycoproteins [392,416], which also occur in other systems of permanent adhesion (Table 3.3).

Various types of spider silk (primary ecribellate, cribellate, secondary cribellate and sticky silk) have different tenacity under dry, moist and dusty conditions, and are adapted to a particular range of items of prey [417]. Among four species studied, the silk of the spiders *Filistata insidiatrix* and *Tegenaria atrica* demonstrated a higher tenacity (2.3 mN/cm^2–2.7 mN/cm^2). Dusty conditions usually reduce tenacity.

3.5.8 Temporary Adhesion in Cladoceran Crustaceans

Little cladoceran crustaceans, such as *Sida crystallina*, attach themselves to the substrate by adhesive anchors connected to the cuticle. *S. crystallina* bears three anchor organs at the dorsal side of the head. An animal attaches to an anterior organ first and then additionally attaches to two movable posterior organs. The attachment is presumably due to a cementing secretion because fine threadlike structures remain attached to the substrate even after the animal frees itself [418]. However, the nature of this secretion and the strength of attachment remain unknown.

3.5.9 Chordata – Larval Adhesion in Fish and Ascidians

Larvae of cichlid fish can attach themselves in the course of several days after emergence by the use of the secretion of adhesive apparatus, which is very variable among the species. As a rule, it consists of three pairs of cup-shaped glands located on the dorsal surface of the larval head [419–421] (Fig. 3.28, redrawn from [420] and [421]). The glands produce a filamentous secretion, which aids in attachment of the larvae at the location of emergence and remains under parental defense for several days. Later, the larvae leave and their attachment organs disappear. The secretion gives positive histochemical reactions on polysaccharides [422].

In larval ascidias, the anterior end of the cephalenteron bears three adhesive papillae enabling the larva to attach to the substrate just prior to metamorphosis. In early studies, they have often been referred to as suckers. However, recent studies have provided evidence about the adhesive nature of these organs [423]. Papillae bear two types of cells (A and B). The first type is a typical secretory one, the second is probably the sensory one [424]. The A-cells produce proteins and mucopolysaccharides, which often occur in marine organisms as those providing attachment (Table 3.3).

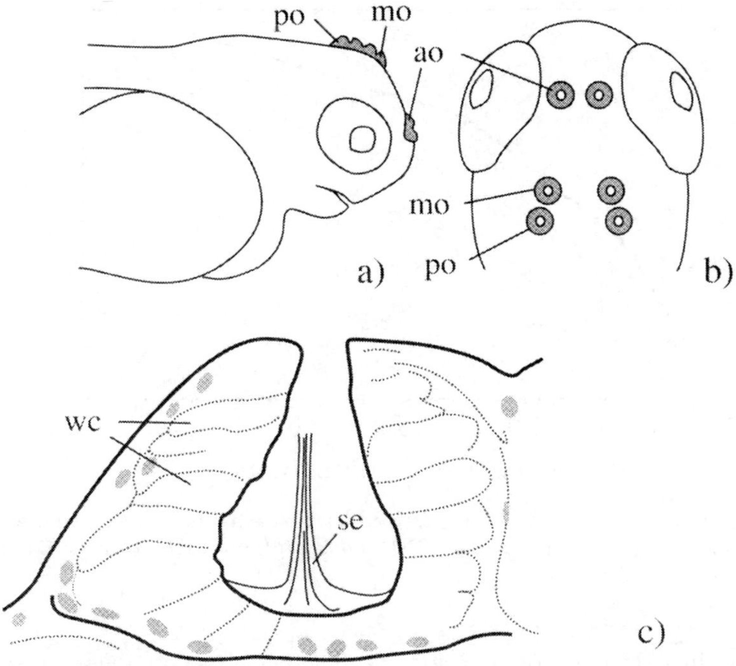

Fig. 3.28. Schematic of the larval adhesive devices in *Tilapia* fish. (**a**) Lateral aspect. (**b**) Dorsal aspect. (**c**) Longitudinal section through the posterior adhesive organ. ao, anterior organ; mo, median organ; po, posterior organ; se, threadlike secretion; wc, wall cells.

3.5.10 Sticking in Tree Frogs

Tree frogs of several families possess large, disc-like pads at the tip of each toe. These pads are responsible for attachment to the substrate during climbing. The pads have a specialized epidermal layer made of several layers of cells [425,426]. The outermost cell layer consists of flat-topped hexagonal or cuboidal columnar cells, separated from each other by canal-like spaces [427, 428]. The cell surface appears smooth in the light microscope, but is rough in the scanning electron microscope. The surface is covered by little outgrowths called hemidesmosomes. Underlying mucous glands open between cells. The mucous-containing fluid of the glands is an essential part of the adhesive mechanism of the tree frog pads [425,426].

There are several hypotheses explaining the attachment mechanism of tree frogs: friction, suction and wet adhesion [309,429]. Careful experimental studies have provided evidence that surface tension is the main component of the adhesive force [428,430]. The following evidence supporting the surface tension mechanism is provided: (1) The adhesive force is directly proportional to the toe-pad area; (2) The meniscus between the pad and substrate is present [431]; (3) The suppression of the surface tension by the use of de-

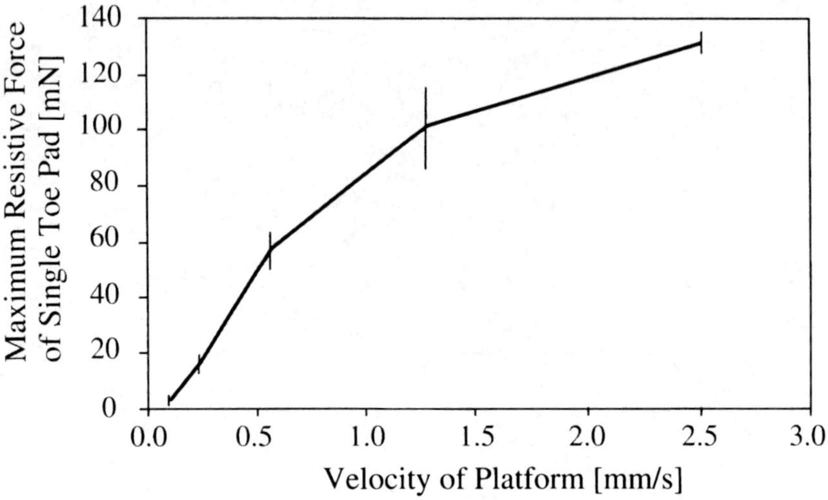

Fig. 3.29. Velocity dependence of the tree frog adhesive mechanism. Resistive force of a single toe pad of an 8.7 g heavy frog *Osteopilus septentrionalis* to movement in the shear plane.

tergents results in a decrease of adhesion; (4) The resistance of shear forces is velocity-dependent, as would be expected in any system employing a fluid as an adhesive mechanism [429] (Fig. 3.29, redrawn from [429]). The viscous forces (Stephan adhesion) are also involved in the attachment mechanism. Hemidesmosomes, covering the pad surface, may be responsible for an increase of frictional forces in the contact area by mechanical interlock with the substrate during walking on sloped and vertical substrates [430].

The mucus produced by glands in the depth of the pad and delivered to the pad surface is not the mandatory adhesion agent. It has been previously shown that even distilled water has sufficient adhesive ability as an intermediary fluid in the frog adhesion mechanism [309]. An important advantage of mucus in the frog adhesive system is a low drying rate compared to water.

In tree frogs, the regular structure of hexagonal cells results in guiding the fluid in all directions, thus wetting the whole pad in a regular manner and,

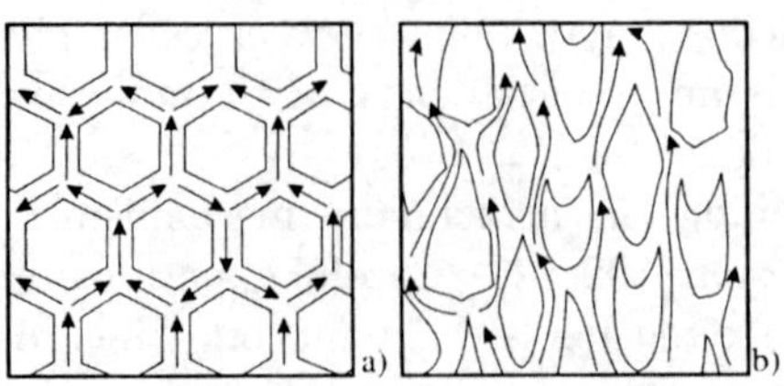

Fig. 3.30. Hypothetical schematic of fluid flow on a frog digital pad with regular hexagonal cells occurred in *Hyperolius sp.* (**a**) and with elongated cells of *Amolops sp.* (**b**).

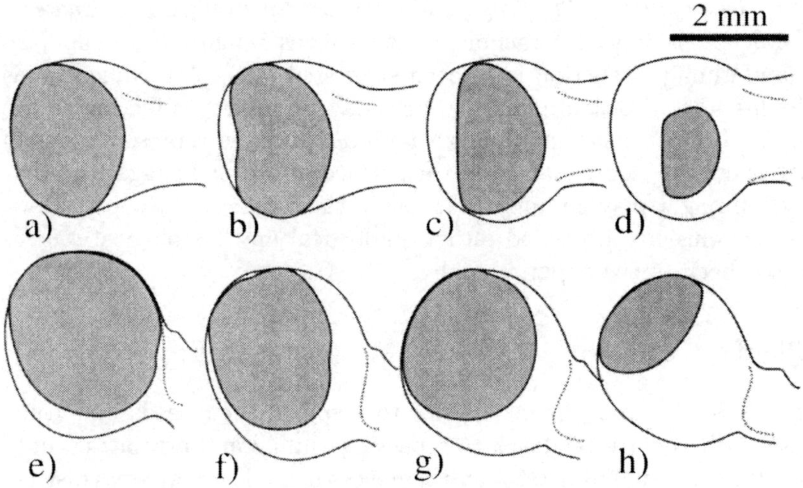

Fig. 3.31. Frame sequences (20 ms intervals) showing a detachment mechanism of the pad of the tree frog *Osteopilus septentrionalis*. During forward walking up a vertical sheet of perspex, (**a**) – (**d**), peeling began at the rear of the pad and progressed forward towards the tip, while during backward walking down the same surface, (**e**) – (**h**), peeling began at the tip of the pad and progressed rearwards.

presumably, optimizing the use of the fluid (Fig. 3.30a, redrawn from [432]). The elongated cells of *Amolops* frogs guide the fluid in the disto-proximal direction (Fig. 3.30b). In this case, the fluid will be directed out of the pad, and the distance from the pad to the substrate will be minimized. Such an action will result in an increase of the sticking force [432].

Since a toe-pad attachment system is used in locomotion, how do frogs overcome the substantial sticking force between pads and substrate when they take a step? To overcome forces directly would be energetically ineffective. Tree frogs have developed a specific mechanism of detachment by peeling. During forward locomotion up a vertical surface, the pads are removed from the rear. When the frogs walk backwards down the vertical surface, peeling occurs from the front of the pad towards the rear [429] (Fig. 3.31, redrawn from [429]). Peeling seems to be the method that minimizes the forces needed to overcome adhesion.

3.5.11 Adhesion in Bats

Most mammals living on vertical or sloped surfaces hold themselves on the substrate by the use of their claws. Specialized attachment devices rarely occur. Vespertilionoid bats utilize adhesive pads on their thumbs, wrists, or ankles to cling, especially to smooth surfaces. The detailed description of these structures has been previously done for representatives from the genera

Thyroptera, *Tylonycteris*, *Glischropus*, *Myzopoda*, *Pipistrellus*, and *Eudiscopus* [433–438]. A variety of attachment mechanisms ranging from capillary adhesion and gluing to suction have been suggested [439]. Attachment pads of *Pipistrellus nanus* contain glands, which may be used in adhering to the substrate. Gluing is a likely mechanism for *Myzopoda.* In representatives of this genus, there are many glands on the surface of the pad suggesting that their secretion might play an important role in the function of the pad. However, these conclusions are based on morphological and histological studies and have not been tested experimentally.

3.5.12 Plants

Adhesion is used by many plant species to disperse their seeds and fruits (diaspores). Adhesive fruits attach to a passing animal and are carried until they drop off or the animal detects and removes them. There are two distinct attachment mechanisms found in plants: mechanical interlock or hooking (84%) and gluing of viscid fruits (13%) [440]. Among viscid fruits, three main types can be recognized: (1) Fruits that become sticky only when wet; (2) Tarry fruits; (3) Fruits with hairs that do not need moisture to become sticky. Diaspores can be viscid in five ways: (1) sticky cuticle (calyces); (2) viscid (glandular) hairs (bracts); (3) viscid seed bodies; (4) viscid drupes/berries; and (5) viscid diaspores with drupes.

The sticky diaspores may be dispersed in three ways. First, seeds may be attached to the head or other parts of the plumage of birds consuming the fruit pulp. Second, sticky seeds may be carried on feathers after they have passed through the gut of a frugivorous bird. Birds usually defecate the seeds on branches, but some seeds can stick to the bird body and may be carried for long distances. Third, seeds remain sticky for 6–12 minutes after fruit explosion and can attach to birds foraging in the vicinity. However, there is no evidence that bird movements trigger explosion. Additionally, chemical adhesion is used by carnivorous plants to catch insects. The adhesive properties as well as chemical composition of plant glues and the mechanical properties of plant diaspores exploiting mechanical adhesion still need further experimental studies.

3.5.13 Bonding to the Dental Surface

Since dental research deals with a variety of questions related to friction, wear and adhesion of dental surfaces, tribological methods and approaches are widely used here. One of the most important problems in this research field is the bonding of dental adhesives to the surface of teeth. Enamel is the outermost layer of dental tissue. It consists of about 95% hydroxyapatite with calcium and has a crystalline structure. The crystallites are separated from each other by thin sheets of threads of a keratin-like protein. Such a composite material has extreme anti-bonding properties, which can cause difficulties in artificial tooth coatings and dental adhesives. In addition, in water and enzymatic environments, temperature fluctuations and compressive forces applied to the bonding sites result in a shorter life of artificial adhesives. Modifications of the tooth surface and the synthesis of new adhesives are two main topics in studies of dental tribology.

The surface preparation of the adherend surface is essential for optimal bonding. The preparation method should take into account the nature of the adherend and the type of impurities of surface contaminants likely to be present [441]. Since an enamel surface possesses a fully reacted, low-energy surface and is usually covered by bacterial contaminants and chemical agents derived from food and drinks, the mechanical and chemical cleaning of the surface is the first step of the surface preparation. Surface imperfections can be filled by a variety of materials and then etched by hydrochloric and phosphoric acid prior to the application of an adhesive. Etching has the following effects: (1) it cleans surface contaminants and exposes a fresh reactive surface more favorable to adhesion, (2) it effects a tremendous increase in the surface area of the enamel, (3) it exposes a surface of an adsorbed layer of highly polar phosphate groups derived from the acid [441]. The next step is to apply an adhesive to the enamel surface. The liquid basis of this adhesive is a 25% water solution of a polyacrylic acid; the solid phase consists of a finely powdered mixture of zinc oxide (90%) and magnesium oxide (10%). The compressive and tensile stress of the polyacrilate materials is comparable with that of zinc phosphate cement. These materials provide a large number of polar carboxyl groups, which are neutralized by oxide powder. The neutralization of the mixture should happen after the mixture is applied to the enamel surface, otherwise the bonding is not likely to occur [441].

Abrasive wear is an important problem in the application of dental composite resins. Studies of the friction and wear behaviors of different types of dental composites, containing various particles, is another wide field of tribological approaches in dental research [442]. Experimental results obtained by the use of scratching machines show that the size, shape and distribution of the particles are more important variables than applied load and sliding speed in controlling the wear mechanism [443].

In conclusion:

- There are at least three reasons for the use of adhesives: (1) they join dissimilar materials; (2) they show improved stress distribution in the joint; and (3) they increase design flexibility.
- Adhesion can be the result of mechanical interlock between irregularities of the contacting surfaces. Increased surface roughness usually results in an increased strength of the adhesive joint.
- An ideal adhesive can be viewed only in relation to a particular adherend (substrate). It should fulfill the following conditions: (1) zero contact-angle;

(2) low initial viscosity; (3) joining of liquid to solid in a way that allows contaminants (air/water) to leave the surface.
- Coupling agents (detergents, surfactants) have to interact well with adhesive and adherend.
- The coupling agent is necessary when: (1) the spreading is poor; (2) when severe conditions, such as high humidity or high temperature, prevent formation of adhesive joints; (3) when surface tension of the adherend is lower than that of the adhesive.
- Contact phenomena of living cells may be divided into three basic processes: (1) cell contact, (2) cell adhesion and attachment, (3) cell separation
- Forces which contribute to cell adhesion are chemical bonds, which can be sub-divided into electrostatic, covalent, and hydrogen bonds and van der Waals interactions.
- Proteoglycans, mucopolysaccharides, proteins and phospholipids are adhesive molecules contributing to the maintenance of organism integrity.
- Many invertebrates use glandular organs composed of different cell types (duo-glands). Two distinct gland types exist, equipped with three distinctive functions: adhesion, de-adhesion, and tension-bearing.
- Invertebrates adhere by secretion of polycationic substances, which bind to substrates, especially to anionic sites.
- In systems using adhesive substances as a kind of a viscoelastic lubricant for locomotion (molluscan pedals), glycans mainly predominate.
- Cementing glues are widely spread agents in systems responsible for permanent attachment.

3.6 Anti-adhesive Mechanisms

Some biological systems have developed anti-sticking mechanisms, which are interesting in several respects. Knowledge of such mechanisms may aid in understanding principles of the functioning of other biological frictional and adhesive systems. The best example is the competition between anti-wetting mechanisms of plants and attachment devices of insects walking on them [444, 445]. Furthermore, biological anti-sticking surfaces provide a rich source of ideas for development of industrial applications, such as self-cleaning surfaces based on the lotus effect [446, 447].

3.6.1 Plants

The majority of (vascular) plants are covered by a hydrophobic cuticle, which is composed of lipids embedded in a polyester matrix. The cuticle surface is often sculptured by cellular outgrowths (trichomes), folds, or covered by wax crystalloids (Fig. 3.32). It has been previously shown that the roughness of the hydrophobic plant surface decreases wettability [448–450], which is reflected in a greater contact-angle of water droplets on such surfaces, compared to smooth surfaces of the same chemical composition.

This property of structured hydrophobic plant surfaces results in their self-cleaning ability, recently called the lotus effect [446,447]. Particles contaminating plant surfaces consist, in most cases, of material that is more readily wetted than hydrophobic wax components. Contaminants usually rest on the tips of the surface structures, so that the real contact area between the particles and plant surface is minimized. Thus, these particles can be easily removed by water droplets rolling over the surface. In this case, adhesion between particles and water droplets is greater than between particles and plant surface due to the reduced contact between the particles and plant surfaces (Fig. 3.33a–c, redrawn from [447]). In the case of a smooth plant surface, the real contact area between the contaminating particles and the surface is large enough to avoid particle adherence to rolling water droplets (Fig. 3.33d–f).

In a series of experiments, it was recently revealed that surfaces of such plants as *Nelumbo*, *Colocasia* and *Brassica* are richly covered by wax crystal-

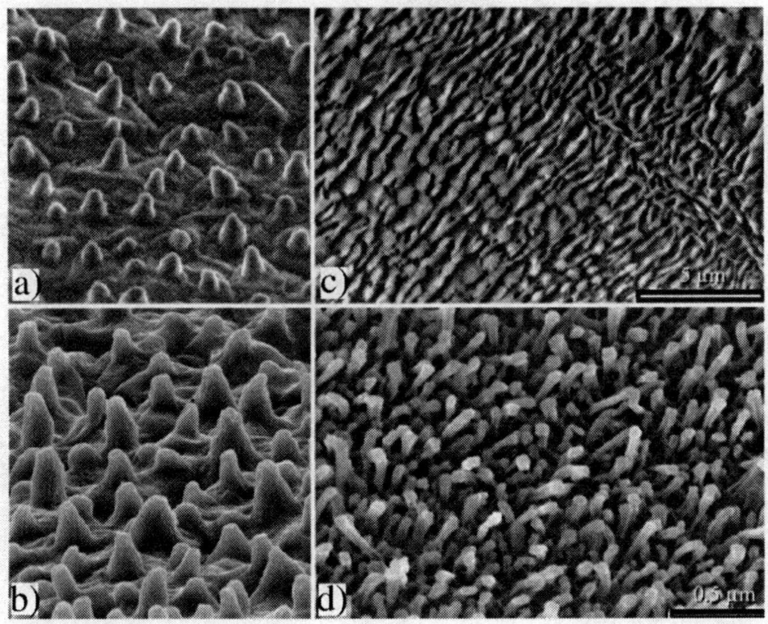

Fig. 3.32. Anti-wetting surfaces in plants and animals. (**a**) Intact surface of the plant *Nelumbo nucifera* covered with wax crystalloids. (**b**) The same surface after treatment with organic solvents. (**c**) Wax crystalloids on the surface of the plant *Iris palustris.* (**d**) Wax-coverage on the wings of the damselfly *Pyrrhosoma nymphula.* [(**a**), (**b**) courtesy of Jürgen Berger (MPI, Tübingen).]

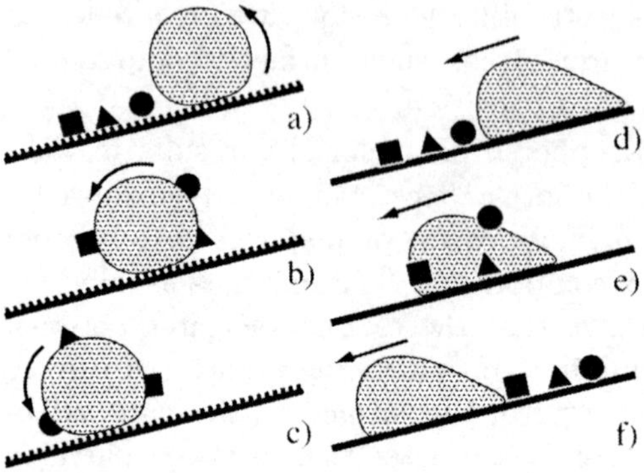

Fig. 3.33. (**a**)–(**c**) Self-cleaning effect of the hydrophobic sculptured plant surfaces. On the rough surfaces, the particles adhere to the droplets, and are removed from the surface by water droplets; (**d**)–(**f**) On the smooth surface, contaminating particles are mainly redistributed by rolling water droplets.

lites, which are responsible for the self-cleaning effect of their surfaces [447]. Similar adaptations have also been described for animal surfaces.

3.6.2 Animals

Wings of Odonata, Ephemeroptera and Neuroptera are covered by wax crystallites (Fig. 3.32d), with dimensions that are comparable to those found in plants. Thus, such surfaces have been experimentally proven to be non-wettable and self-cleaning [451]. The single eye (ommatidia) of many insects has a roughness in the range of 100 nm–200 nm. These structures have been reported to have the function of a light antireflector [452]. However, since these surfaces have a roughness similar to the self-cleaning surfaces of plants, it can be suggested that they may be involved in a self-cleaning mechanism of ommatidia.

Some animals have developed an alternative mechanism of self-cleaning [314]. Dung beetles *Geotrupes* or some tenebrionid beetles, which live in extremely dirty conditions, keep their surface clean by the continuous secretion of an oil-like hydrophobic substance, which prevents adherence of the hydrophilic contaminants, and even rinse out the contaminants by large amounts of the secretion. The epidermal cells of these insects contain huge lipid-containing vesicles and the cuticle is densely penetrated by pore canals responsible for delivering the secretion to the surface (Fig. 3.34).

In summary:

- The surfaces of many plants, insects and marine organisms are heavily covered by microstructural elements which are responsible for the self-cleaning effect (lotus-effect).

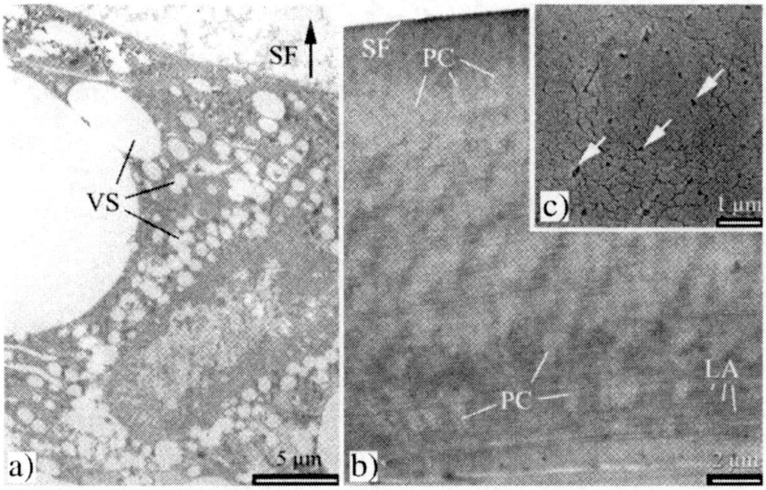

Fig. 3.34. Anti-adhesive mechanism of the dung beetle *Geotrupes stercorosus*. (**a**) Epidermal cells underlying elytra cuticle (TEM). (**b**) Elytra cuticle (TEM). (**c**) Surface of the elytra (SEM). The arrows indicate surface pores. LA, cuticle layers; PC, pore canals; SF, surface; VS, lipid vesicles.

- An alternative mechanism of self-cleaning is the continuous secretion of an oil-like hydrophobic substance, which prevents the adherence of hydrophilic contaminants, or which rinses out the contaminants by large amounts of secretion.

4. Frictional Devices of Insects

Since our case studies (Part IV) have been performed on cricket (orthopteran) attachment pads, this chapter is written to give a short survey of insect surfaces, and the material structure of the outer layer of insect skin (cuticle), as well as to provide a short overview of the frictional surfaces described for the case of insects. Frictional devices have been reported on terminal segments of legs, head-arresting systems, mouth-part (galeal) linkage in some flies (Lepidoptera) [453], and armored membranes in butterflies and moths (Diptera) [454]. Copulation organs, the egg-laying apparatus (ovipositor), and intersegmental coupling are also often supplemented with frictional surfaces. The notion that insects are "animals of the surface" reflects the general significance of the microsculpture of the insect skin (integument) in the biology of this group of animals. Functions of cuticular protuberances are listed in Fig. 4.1. Highly specialized, jointless cuticular microstructures originating from a cell complex (trichoid complex of epidermal cells) are mainly sensory organs, but in some cases the original sensory function has been lost in evolution (scales). Some skin outgrowth (setae) may also have a defensive function, as in the dermestid beetle larva or the function of pheromone release. Cuticular microstructures originating from one cell (acanthae), or as a pattern from one epidermal cell (microtrichia), serve diverse mechanical func-

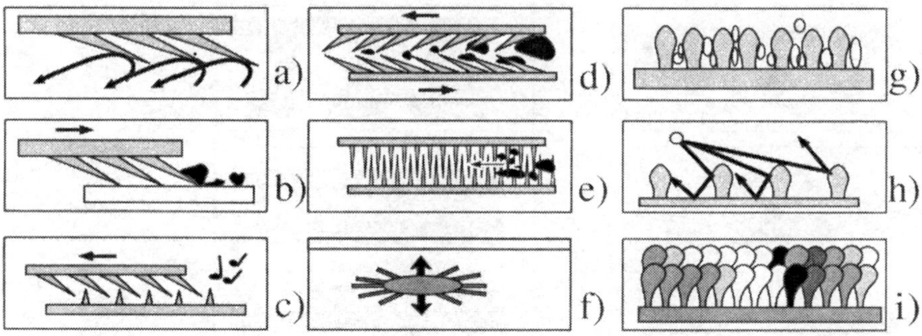

Fig. 4.1. Cuticular microstructures can function (**a**) as aerodynamically active surfaces; (**b**) for body cleaning; (**c**) for sound generation; (**d**) for food grinding; (**e**) as filtration devices; (**f**) as hydrodynamically active surfaces; (**g**) for air holding; (**h**) for thermoregulation; (**i**) as body coloration pattern.

tions or form a coloration pattern. Some functions are the following: body cleaning, air holding, thermoregulation, increasing the aerodynamic activity of the surface, sound generation, food grinding, filtering of food particles. Often cuticular protuberances are poly-functional.

4.1 Design Principles of Insect Attachment Devices

Cuticle microsculptures often promote the fixation of the surfaces. Different attachment systems of insects can be grouped together into eight fundamental classes based on the form and size of the microsculpture, presence of the muscles, cuticle secretions, and according to the general design of the system. Friction is the underlying principle for which most systems have been evolved: (a) hooks, (b) lock or snap, (c) clamp, (d) spacer, (e) sucker, (f) expansion anchor, (g) adhesive secretions, (h) friction (see Fig. 4.2, from [355]). However, real systems may be designed in accordance with any one of these principles or their combination. Actually, almost all species of insects are supplied with diverse attachment devices. The morphology, therefore, depends on the biology of each particular species.

Frictional devices usually contain two surfaces covered with surface micro-outgrowths, such as acanthae or microtrichia, which increase the frictional forces when the surfaces come into contact. Furthermore, there are frictional systems specialized for only one type of surface. The most interesting are

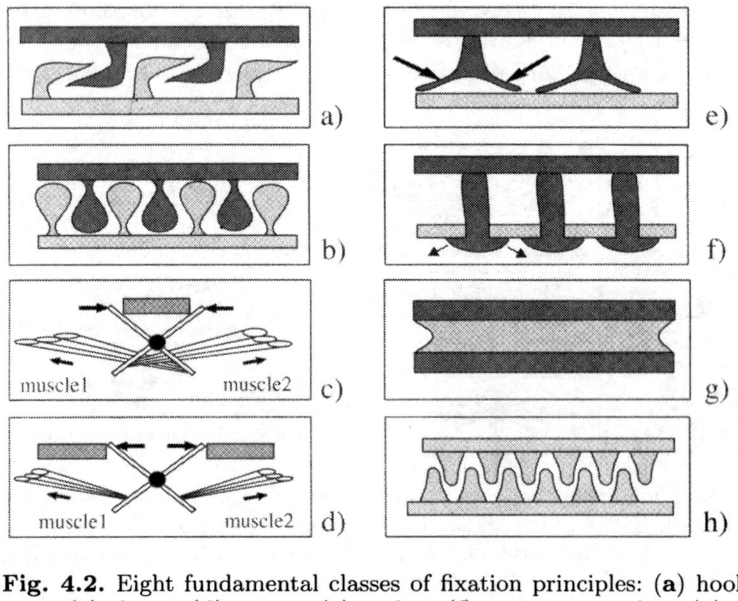

Fig. 4.2. Eight fundamental classes of fixation principles: (**a**) hooks, (**b**) lock or snap, (**c**) clamp, (**d**) spacer, (**e**) sucker, (**f**) expansion anchor, (**g**) adhesive secretions, (**h**) friction.

"smooth" surfaces with a particular material design. In both of these latter cases, the system may be adapted to a variety of surfaces. Parameters of frictional systems such as surface profile and material structure may be optimized for particular frictional properties with one co-opted surface or with many different surfaces. Attachment by these systems is usually fast, precise, and reversible. Adhesive devices occur most widely among representatives of insects with an outer skeleton (Arthropoda).

4.2 Insect Cuticle: Ultrastructural Architecture of the Material

The insect skeleton, often called the cuticle, is a fibrous composite material. It is sub-divided into two main layers: the epicuticle and the procuticle. Each layer consists of several sub-layers. Additionally, there are two layers belonging to the cell layer underneath the skin (epidermis): (1) a layer of epidermal cells producing cuticle and other epidermal secretions and (2) a basement membrane underlying epidermal cells.

This composite structure does not contain chitin and overlies the chitin-bearing deeper layer of cuticle (procuticle). The outermost layer of cuticle (epicuticle) covers the surface of an insect. The epicuticle is the very thin outermost layer of the cuticle. However, it is complex and consists of several sub-layers, namely (1) cement sub-layer, (2) wax sub-layer, (3) outer epicuticle, and (4) inner epicuticle. The epicuticle has high extending ability, especially in arthrodial membranes, for example in fluid-sucking insects [455]. The primary function of the layer is protecting the lipid monolayer and outer epicuticle from possible abrasive or impact damage. The chemical composition of the wax layer varies even between individuals of the same species, but usually the waxes are composed of long-chain saturated alcohols and fatty acids. The outer epicuticle (cuticulin) is a trilaminar layer about 12 nm–18 nm thick that is believed to consist of polythene-like polymers and quinone-tanned proteins [456]. The surface patterns of insect-serving functions, named above, are formed at the same time as the outer epicuticle [457]. The inner epicuticle is an optically isotropic layer consisting of polymerized lipoprotein stabilized by quinones [455]. It serves as the reservoir for extracellular substances, for example, extracellular enzymes.

The procuticle is the chitin-bearing part of the cuticle and it represents the secretory product of epidermal cells. It is generally known that it consists of two principal components, chitin and protein. Chitin is a structural polymer, which occurs in insects in crystalline form. Structures of chitin molecules and their arrangement are well studied; however, the structure of chitin–protein assemblages still remains unknown. Chitin-bearing cuticular microfibrils have a complex pattern of orientation in the three-dimensional space of the procuticle [458,459]. Within the cuticle, the chitin microfibrils occur in layers

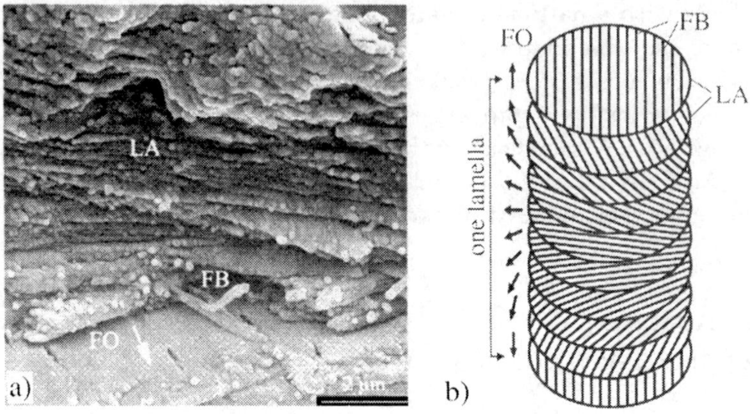

Fig. 4.3. Ultrastructural architecture of the procuticle. (**a**) Fractured cuticle of the rear surface of the head of the dragonfly *Sympetrum sanguineum*. (**b**) Schematic of the cuticle lamella. FB, fibers; FO, preferred orientation of the fibers; LA, layers of fibers.

(sheets), and within any sheet all of the microfibrils are orientated parallel to the surface plane (Fig. 4.3a). Neighboring sheets of microfibrils are rotated in relation to each other in the same direction. A set of sheets that is rotated about 180° is usually called lamella (Fig. 4.3b). The thickness of a lamella depends on the thickness of the microfibrils and the rotation angle. The interplay of different cuticle compounds results in a wide spectrum of properties. Cuticles can be functionally grouped into (1) the so-called "solid" cuticle typical for sclerites, skin ingrowth (apodems), infolding of a portion of the outer layer (phragmata), insect mouth parts (mandibles) and claws; (2) thin arthrodial membranous cuticle as well as the cuticle of caterpillars; and (3) resilin-containing cuticle from tendons of the flight muscles, springs of mouth-parts and wing joints [460,461]. The solid cuticle is usually rigid and the membranous one is flexible, whereas the resilin-containing cuticle has elastic properties. Frictional devices can be composed of different types of cuticle. The two complementary surfaces in the friction process are usually composed of one hard and one soft surface. Such systems are mainly used for the fixation of two body parts or for preventing sliding between two mechanical parts of the body. Frictional devices adapted to a variety of surfaces have elastic or viscoelastic properties to adapt to the surface profile. An overview of both types of systems is given below.

4.3 Systems of Two Complementary Surfaces

4.3.1 Dragonfly Head-Arresting System

The arrester system (Fig. 4.4, from [462]) serves as an adaptation to save the head from violent mechanical disturbances and to stabilize gaze in a variety of behavioral situations [462]. The area of the joint of the dragonfly head with the neck is very small compared to the head dimension; it is virtually a single point providing high head mobility on the roll, pitch and yaw planes [463]. Such mobility of the head compensates for the absence of a specialized gravity organ. The movements of the head are monitored by the head, neck and prothoracic fields of the hair-sensory receptor (sensilla). At the same time, such a design of the head–neck joint has weak mechanical strength. The head, however, must be stable with the rest of the body, for example during feeding to prevent disturbances caused by big mandibular muscles and to stabilize gaze while perching or holding prey in flight.

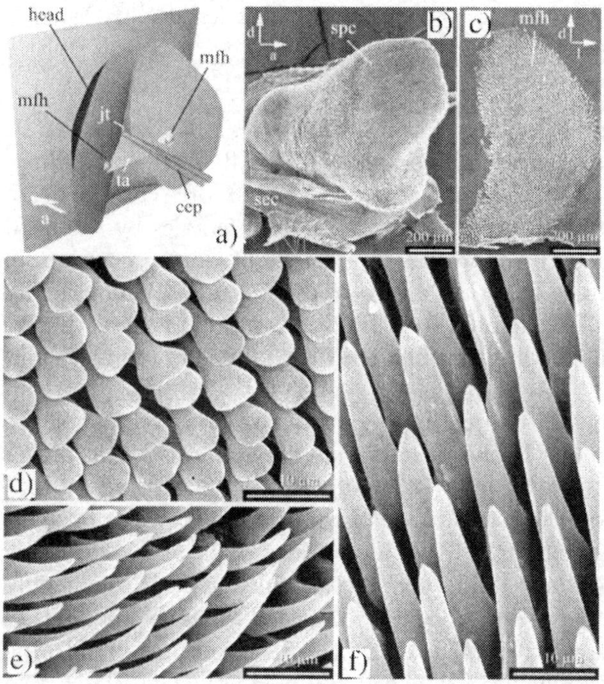

Fig. 4.4. Head-arrester system in dragonflies. (**a**) Schematic of the head–neck joint. (**b**) Postcervical sclerite, lateral aspect. (**c**) Microtrichia field on the right-hand side of the head. (**d**), (**e**) Microtrichia from the dorsal (**d**) and ventral part (**e**) of the postcervical sclerite. (**f**) Microtrichia from the field on the rear surface of the head. cep, cephaliger; a; anterior direction; d, dorsal direction; head, head; jt, joint; l, lateral direction; mfh, microtrichia field on the rear surface of the head; ta, supporting triangle.

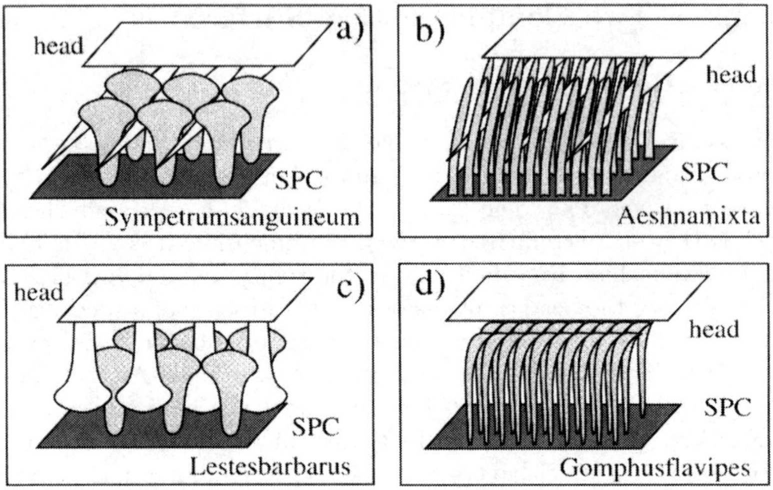

Fig. 4.5. Schematics of corresponding frictional surfaces occurring in odonate head-arrester system. (**a**) spines – plates system of libellulids; (**b**) spines – filaments of aeshnids; (**c**) clavate microtrichia – clavate microtrichia of lestids; (**d**) leaves – smooth surface of gomphids. SPC = postcervical sclerite.

During the initial stage of tandem formation, the male holds the female insect by the head or prothorax with its appendages [464]. This usually happens in flight and at high speed, and one may expect high forces on the head. Additionally, when two animals are mating, males hold females by the head (Anisoptera) or prothorax (Zygoptera) in tandem flight during copulation. In such a position, the female head can easily be damaged if it is not rigidly connected to the thorax. The arrester is a frictional mechanism serving to immobilize the head during feeding or when the dragonfly is in tandem flight [465].

The arrester system includes adjusting organs of two body segments: the head and neck. It consists of: (1) the skeleton–muscle apparatus that moves the head and neck sclerites [466,467]; (2) co-opted microsculptures, which are fields of microtrichia on the rear surface of the head and on the postcervical sclerites of the neck [468,469]; (3) the secretory apparatus, consisting of epidermal cells that produce lipid substances, which pass through pore canals in the cuticle into the region of contact between microtrichia fields of the head and postcervical sclerites [470–472]; (4) sensory organs monitoring contact between microtrichia fields of the head and postcervical sclerites, and the position of the sclerites relative to the other neck sclerites [473].

The microtrichia field on the rear surface of the head and the postcervical sclerites covered with microtrichia are functionally corresponding structures. The shape, size and density of microtrichia of the head and neck parts of the arrester are different in various taxa of dragonflies, but they correlate in the same species (Fig. 4.5, from [462]). When the muscle driving the postcervical sclerite contracts, the sclerite moves medially to the longitudinal body axis and loses contact with the microtrichia fields (the head is free). This movement enables the disengagement of the arrester system when the animal needs to move its head. The backward movement of the sclerite is passive. Cuticular "springs" return the sclerites to their lateral positions (the head is fixed).

4.3.2 Wing-Locking Devices

Some insects have convergently developed an ability to attach their wings to the body when resting. Such wing-to-body locking devices have very different designs in different lines of descent (phyla) of insects. They include snap mechanisms in some Hemiptera [474], microtrichia fields in some Hymenoptera [475] or the thoracic brush consisting of hairs (chetae) in Dermaptera [476].

In beetles (Coleoptera), the system responsible for an attachment of forewings (elytra) to the thorax is composed of several interlocking mesoscale structures located between the thorax and the body and between the left and right forewings. These structures, studied in detail in representatives of some Carabidae [477], are designed as co-opted lock-and-key devices.

In the majority of beetles with flying ability, these structures are supplemented with microtrichia fields of the inner surface of the forewing and adjacent parts of the pterothorax (Fig. 4.6, [478], Fig. 4.7 [478]). Such fields differ in density, design, and directionality of microtrichia [479–481]. Quanti-

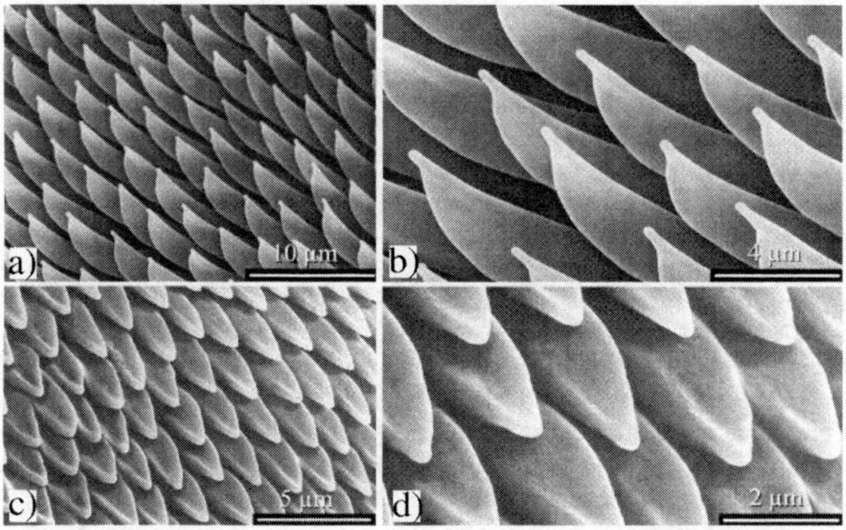

Fig. 4.6. Wing-locking devices in beetles. Microtrichia of the postero-lateral thoracic field (TPL) in *Tenebrio molitor* (**a**), (**b**) and *Tribolium castaneum* (**c**), (**d**).

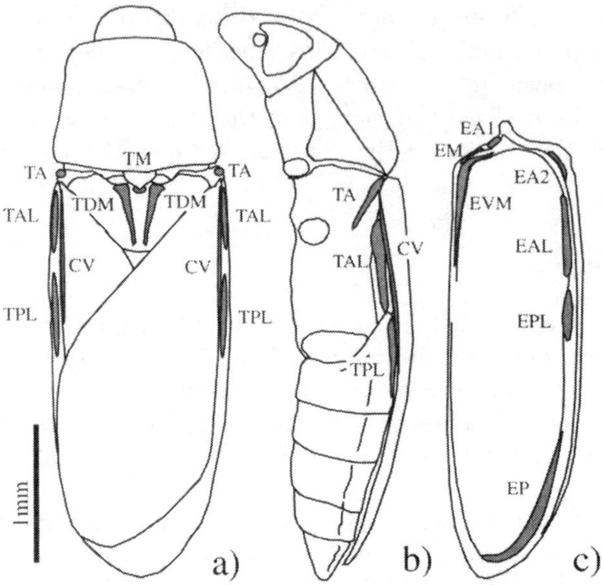

Fig. 4.7. Locations of microtrichia fields of the wing–locking devices (grey) in the beetle *Tribolium castaneum.* (**a**) Dorsal aspect, elytra removed. (**b**) Lateral aspect, elytra removed. (**c**) Left elytra, ventral aspect. CV, costal vein; EA1, first anterior field of the elytra; EA2, second anterior field of the elytra; EAL, anterio-lateral field of the elytra; EM, medial field of the elytra; EP, posterior field of the elytra; EPL, postero-lateral field of the elytra; EVM, ventro-medial field of the elytra; TA, anterior field of the thorax; TAL, anterio-lateral field of the thorax; TDM, dorso-medial field of the thorax; TM, medial field of the thorax; TPL, postero-lateral field of the thorax.

tative analysis of microtrichia of all corresponding structures involved in the locking of the forewing to the body shows the scale effect on microtrichia length, width and density [478].

The forewings provide complete cover for the fragile hindwings and the last part of the insect body (abdomen). When they are interlocked with thorax and abdomen, the body forms a stable unit. The forewings protect the flight-wings from becoming wet, dirty, or damaged. In many species of Coleoptera, elytral coverage minimizes the loss of water. As previously suggested [479], these strengthening and protective functions provided the selection pressure that resulted in successive changes of the forewing design and interlocking mechanisms. M. Karner, in his recent study on the arrangement of bifunctional muscles in Coleoptera, has hypothesized that the effective attachment of both forewing and hindwings to the body may be important in enabling the bifunctional muscles to work effectively during walking [482].

4.3.3 Insect Unguitractor Apparatus

An insect pretarsus (last part of the leg) has only a flexor muscle; an extensor muscle is absent [483, 484]. The pretarsus is connected to the terminal segment of the tarsus by a sclerotized plate called an unguitractor plate (or simply an unguitractor) [485]. The unguitractor is an insertion point of the claw flexor muscle, which is responsible for claw bending (Fig. 4.8). The return of the claw to an extended position is passive due to resilin-filled cuticle areas at the base of the claw [486].

Scanning electron microscopy of the anterior surface of the unguitractor has revealed the presence of denticle-like structures which have been interpreted as a friction system providing anchorage of the unguitractor [487]. The anterior part of the terminal sub-unit of the leg (tarsomere) bears a wall invagination that forms an additional plate-like structure facing the unguitractor plate. This structure, comprising the unguitractor from the lateral and anterior side, is covered by microtrichia in its distal part. Its microtrichia are elongated and usually directed inwards (proximally).

The unguitractor system has been previously interpreted as an energy saving device anchoring claws in the grasping position [488, 489]. The analysis of pretarsal structures suggests the functioning of the anchorage mechanism

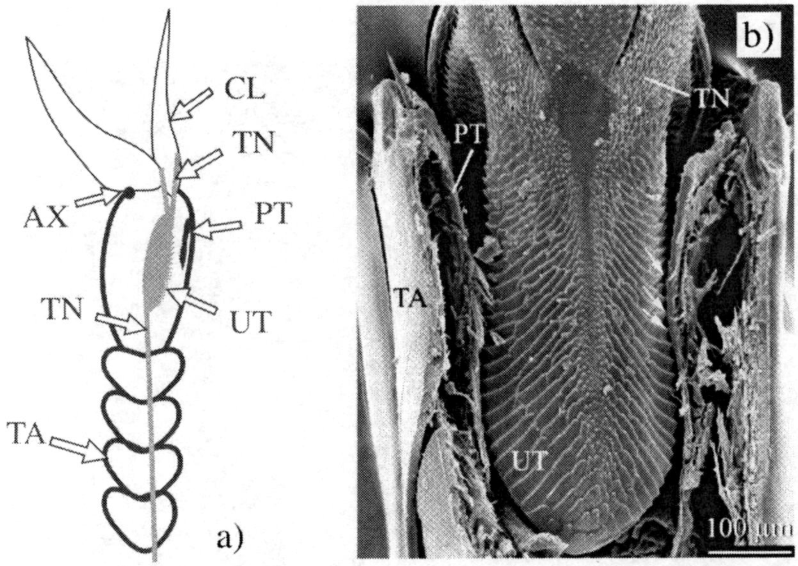

Fig. 4.8. Insect unguitractor apparatus. (**a**) General schematic of insect pretarsus. (**b**) Unguitractor plate of the beetle *Melolontha melolontha*, the ventral wall of the terminal tarsomere is removed. AX, joint between claw and tarsus; CL, claw; PT, anterior plate of the tarsal wall; TA, tarsomere; TN, tendons; UT, unguitractor plate; white arrows indicate contact between the unguitractor sculpture and the microtrichia of the corresponding surface.

as follows. During contraction of the flexor muscle, the claw is moving proximally from an extended position. Substrate resistance returns the claws to their initial position. Therefore, the resulting force is directed to the anterior wall of the segment where the corresponding field of microtrichia is located. Due to the frictional surfaces on the contacting structures, the unguitractor system locks the claws in the grasping position.

4.3.4 Coxal Interlocking Mechanism in Lacewings

Lacewings possess so-called bifunctional thoracic muscles. These muscles are used for walking when an insect is on land and for wing-beat generation while an insect is flying. In flight, lacewings immobilize the coxae of the middle legs by the use of frictional fields located on the medial surface of coxae covered by long microtrichia (Fig. 4.9). Contacting fields prevent sliding of the coxae and stabilize the origins of the bifunctional thoracic muscles.

4.3.5 Coxal Interlocking Mechanism in Jumping Cicadas

Jumping cercopid cicadas bear a similar mechanism to immobilize their hind coxae while jumping. Both the synchronization of the movement of a pair of jumping legs and their stabilization during a jump seem to be important

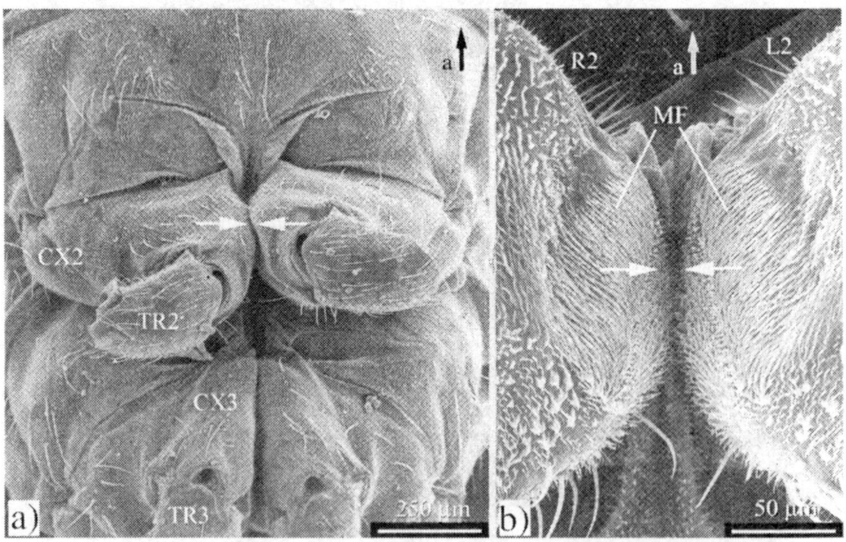

Fig. 4.9. Coxa interlocking mechanism in lacewings; (**a**) ventral view of the meso- and metathorax in *Chrysopa flava*, (**b**) a, anterior direction. CX2, coxa of the middle leg; CX3, coxa of the hind leg; L2, left middle leg; MF, fields of microtrichia; R2, right middle leg; TR2, trochanter of the middle leg; TR3, trochanter of the hind leg; white arrows show where both coxae are interlocked.

problems for small jumping insects. Interestingly, for this purpose different lineages of this insect group have developed different attachment devices.

Some leafhoppers (Homoptera) have a remarkable mechanism of leg synchronization. This mechanism involves cogwheel sectors of the cuticle of the medial surface of the hind leg coxae. Previous authors have presumed that these attachment devices aid attaching left and right coxa during jump performance [490,491]. Other Cicadina, such as representatives of the Cicadellidae family, interlock their coxae using the snap system in which two asymmetrical structures (outgrowth and depression) on the left and right coxae are involved [492].

In cercopid cicadas, the median surfaces of the hind coxae bear frictional surfaces, which interlock the coxae during a jump [493]. In the cicada *Cercopis vulnerata* (Cercopidae), the hind coxae are rounded in their anterior and lateral parts, whereas the medial parts are planar and contact each other over a rather large area. Microtrichia fields (270 μm× 320 μm) are located at the medial surface of each hind coxa (Figures 4.8 and 4.9). Single microtrichia are a kind of microplate in the central and ventral regions of the field or comb-shaped crests in the dorsal region. The microplates are 3 μm–4 μm high, 3 μm–10 μm wide, 2 μm–5 μm long, and lengthened along the longitudinal body axis.

4.4 Systems of One Adaptable Surface

Flies holding onto the surface of smooth substrates fascinated scientists already in the 19th century [494–496]. To date, both the morphological and ultrastructural bases of the ability of many insects to walk on ver-

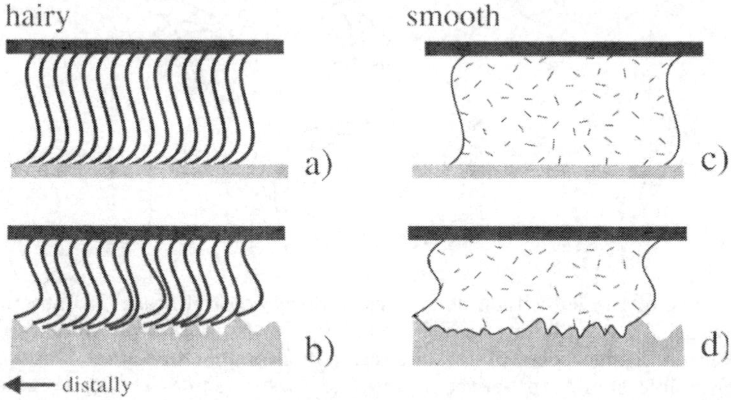

Fig. 4.10. Schematic of action of the "hairy" (**a**), (**b**) and "smooth" (**c**), (**d**) pad attachment systems on smooth (**a**), (**c**) and structured (**b**), (**d**) substrates. Both systems are able to adapt to the surface profile.

tical surfaces have been studied only in selected insect groups, including Diptera [497–500], Coleoptera [501–503], Homoptera [504,505], Heteroptera [506,507], Orthoptera [508–510], Blattaria [511,512], Hymenoptera [513,514] and Thysanoptera [515,516]. The fine structure and functional morphology of leg-attachment structures of representatives of other insect orders still need attention.

In their evolution insects have developed two distinctly different mechanisms to attach themselves to a variety of substrates: setose or hairy surfaces and smooth flexible pads. Both mechanisms can maximize the possible contact area with the substrate, regardless of the substrate microsculpture (Fig. 4.10, from [517]), because of the flexible material of pads. It is interesting that these highly specialized structures are not restricted to one particular area of the leg. They may be located on different parts such as claws, derivatives of the pretarsus, tarsal apex, tarsomeres, or tibia (Fig. 4.11). Therefore, it appears rather obvious that structures adapted for attachment to the substrate have evolved several times independently [436,518]. The diversity of leg-attachment structures of insects is generally based only on two main principles: hairy surfaces or smooth flexible pads (see Fig. 4.11 [517]). Possibly, natural fibrous composite materials, such as insect cuticle [519], are a kind

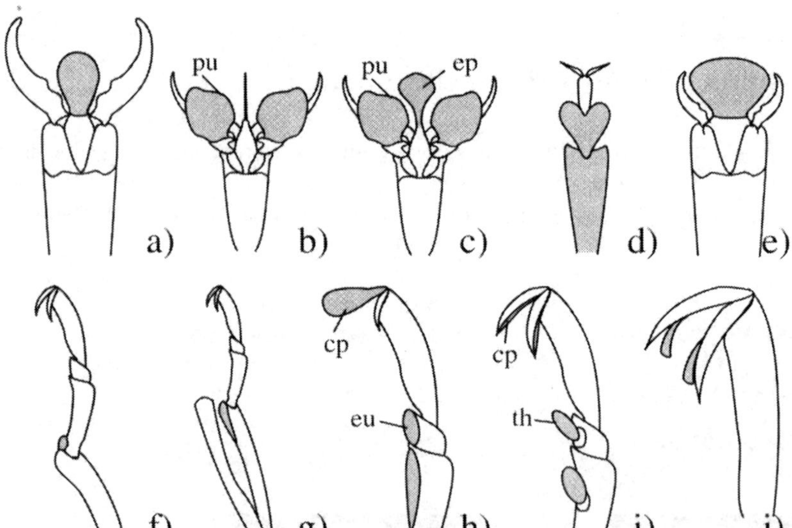

Fig. 4.11. Diversity of the leg-attachment devices (grey colored areas) in insects. (**a**) Arolium (smooth). (**b**) Pulvilli (smooth or hairy). (**c**) Empodial pulvillus (ep) (hairy). (**d**) Hairy adhesive soles of tarsomeres. (**e**) Eversible pretarsal bladder (smooth). (**f**) Eversible structure between tibia and tarsus (smooth). (**g**) Fossula spongiosa (hairy). (**h**) Euplantulae (eu) and claw pad (cp) (both smooth). (**i**) Tarsal thorns transformed into adhesive structures (th), claw pad (cp) (both smooth). (**j**) Adhesive claw setae.

of pre-adaptation, which may provide only limited possibilities for the design of attachment systems.

4.4.1 Surface Characteristics

Hairy systems always contain cuticle protuberances on their surfaces. The protuberances occurring in the hairy pad of Coleoptera, Dermaptera, and Diptera belong to different types. Representatives of the first two lineages have hairs (setae) with sockets providing additional mobility of the hairs. Setae range in their length from a few micrometers to several millimeters.

Dipteran protuberances are acanthae, single sclerotized protuberances originating from a single cell. Acanthae range in length from a few microm-

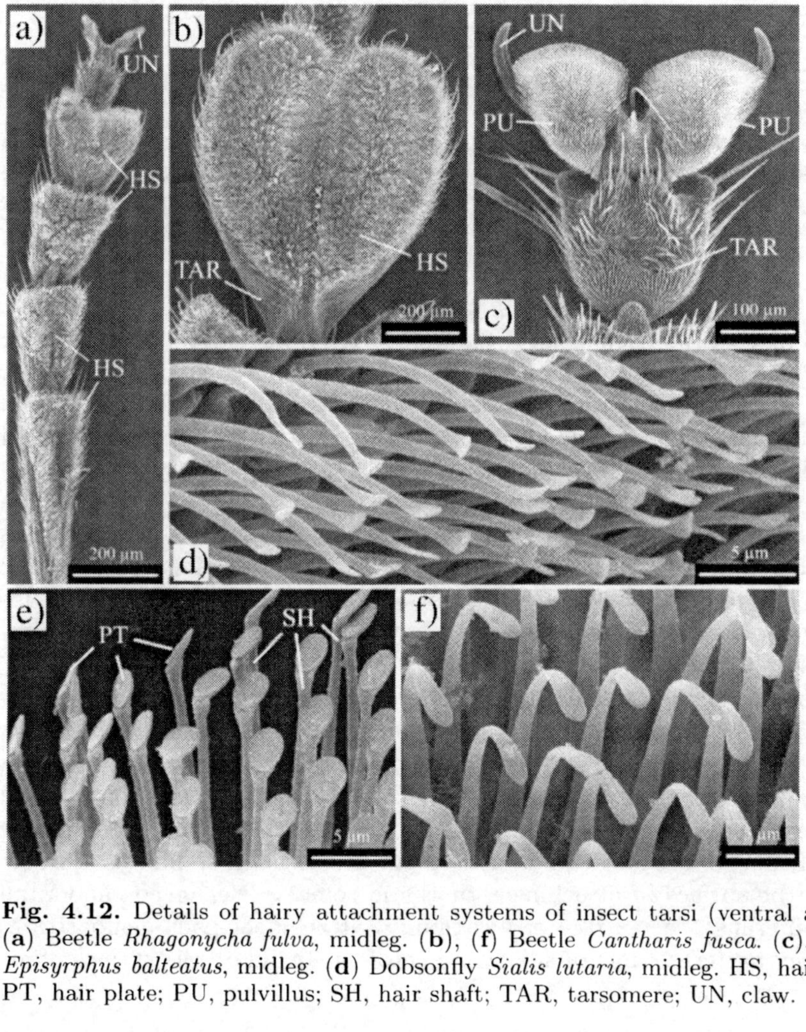

Fig. 4.12. Details of hairy attachment systems of insect tarsi (ventral aspects). (**a**) Beetle *Rhagonycha fulva*, midleg. (**b**), (**f**) Beetle *Cantharis fusca*. (**c**), (**e**) Fly *Episyrphus balteatus*, midleg. (**d**) Dobsonfly *Sialis lutaria*, midleg. HS, hairy soles; PT, hair plate; PU, pulvillus; SH, hair shaft; TAR, tarsomere; UN, claw.

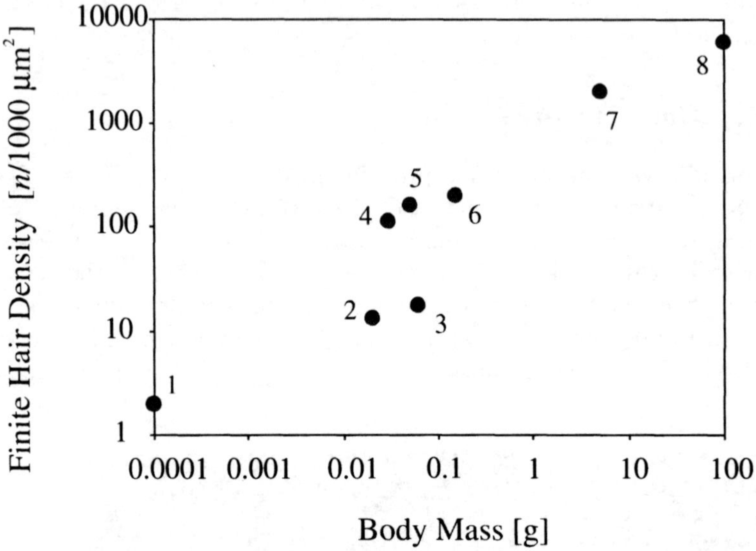

Fig. 4.13. Dependence of the finite hair density of the attachment pads on the body mass in hairy pad systems of representatives from diverse animal groups. 1–fly *Drosophilamelanogaster*, 2–fly *Luciliacaesar*, 3–beetle *Cantharis fusca*, 4–fly *Episyrphus balteatus*, 5–fly *Calliphora erythrocephala*, 6–bug *Rhodnius prolixus*, 7–spider *Cupienniussalei*, 8–*Gekkonidlizards.*

eters to at least 0.5 mm (Fig. 4.12, from [517]). Some hairs bear discs or widened compressions at their tips. The important morphological characteristics of acanthae is that they lack a socket [520,521]. Both types of structure, setae and acanthae, can be additionally covered with tiny outgrowths called microtrichia.

The density of hairs strongly increases with increased body weight. Because of the different scaling behavior of mass and surface area, animals cannot increase the area of the attachment devices proportional to the body weight. Therefore, the increase of the attachment abilities is realized by increasing the number of single contact points, i.e., by increasing the hair density as shown in Fig. 4.13.

A surface of the "smooth" systems appears smooth in the light microscope and at low magnification in the scanning electron microscope (Fig. 4.14, from [517]). In reality, the surface may be patterned by hexagonal structures, as in representatives of Orthoptera (see Case Studies). Hymenopteran and mecopteran attachment pads (arolium) may be patterned in lines running perpendicular to the longitudinal axis of the pretarsus. A similar pattern occurs on the surface of attachment thorns in some species of non-apocritan hymenopterans. The surface of fly (tipulid) attachment pads demonstrates a complex pattern of microfolds. The microfolds found on the surface of attachment thorns in some species of non-apocritan hymenopterans are much larger.

4.4.2 Ultrastructural Architecture of the Pad Material

The structure of the fly adhesive hairs has been previously reported for the syrphid fly *Episyrphus balteatus* [500]. Its acanthae are hollow inside, containing pores under the end plate. These pores probably provide a delivery of the adhesive secretion directly to the contact area. Also, pore canals located at the base of the shaft can contribute to secretion transport to the surface [498]. The membranous cuticle of hairy pads is designed as a fibrous composite material in which the fibers are not densely distributed. In Coleoptera

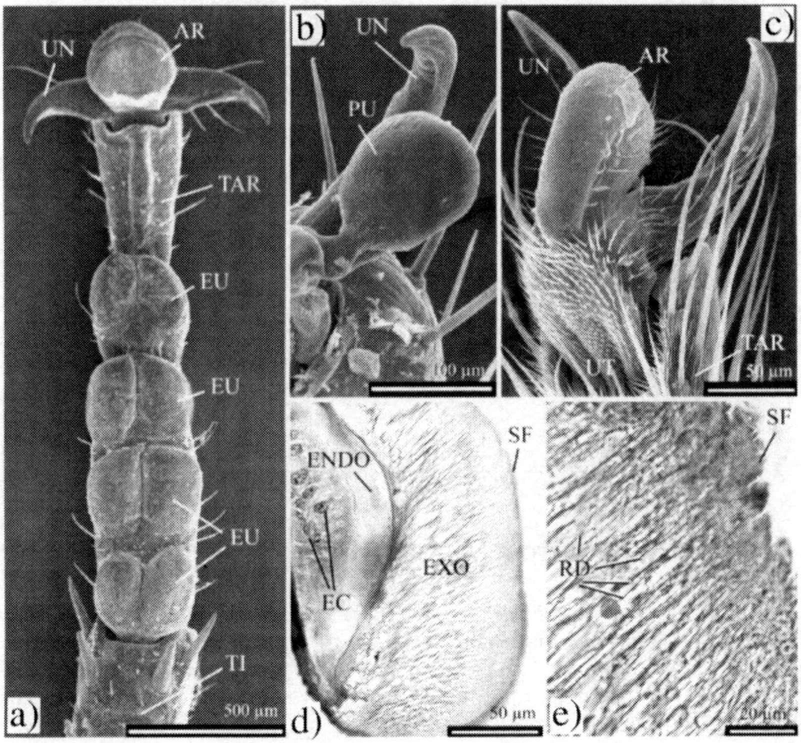

Fig. 4.14. Details of smooth attachment systems of insect tarsi (ventral aspects). (**a**) Arolium and euplantulae in the foreleg of the grasshopper *Omocestus viridulus*. (**b**) Left pulvillus in the bug *Coreusmarginatus*. (**c**) Arolium of the dipteran *Tipula sp.* (**d**), (**e**) Arolium of the orthopteran *Schistocerca gregaria*, foreleg, sagittal section, triple stain after Cason. AR, arolium; EC, epidermal cells; ENDO, endocuticle; EU, euplantulae; EXO, exocuticle; PU, pulvillus; RD, chitin rods; SF, surface; TAR, tarsus; TI, tibia; UN, claw; UT, unguitractor.

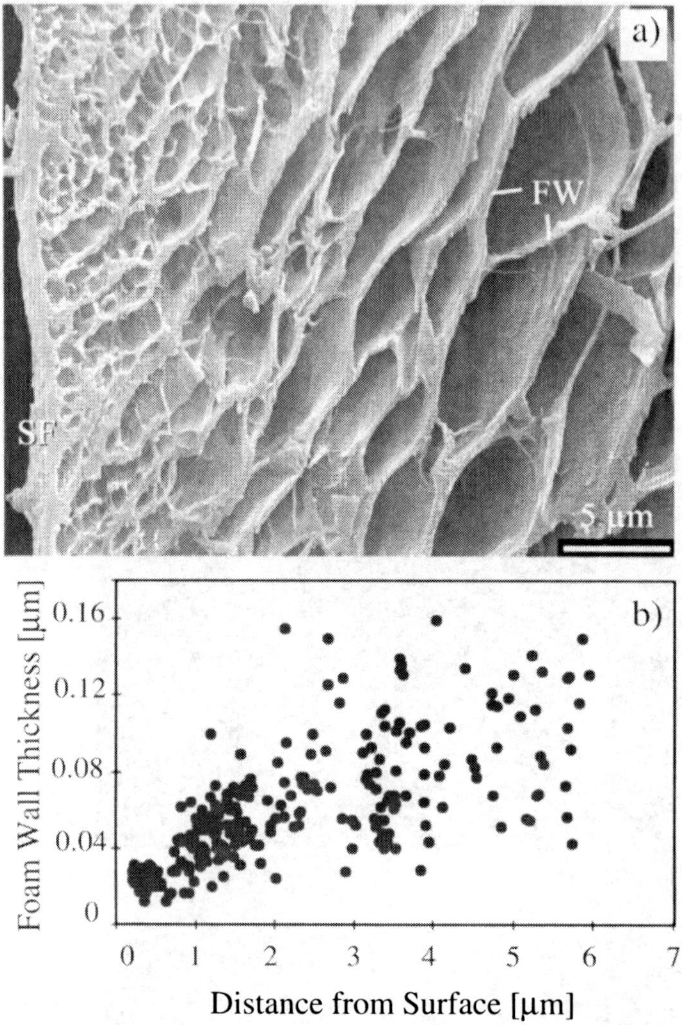

Fig. 4.15. Foam-like inner architecture of the pad material in the cicada *Cercopis vulnerata*. (**a**) Fractured material viewed in a scanning electron microscope. (**b**) Dependence of the foam wall thickness on the distance from the surface. The measurements are based on results obtained from ultra-thin sections taken in the transmission electron microscope. SF surface; FW foam wall.

the hair bases are embedded in this material, which provides high mobility of hairs and their adaptability on a variety of surface profiles.

Ultrastructural studies show that the cuticle of smooth pads are natural friction-active materials with a specific inner structure. The attachment pads of the bee and most orthopterans contain chitin rods, which may provide higher flexibility of the material (Fig. 4.14d, e). The inner architecture of the attachment pads of the cicada resembles foam, consisting of fluid-filled cells surrounded by a flexible cuticle. The diameter of the cells and the thickness of the walls in the foam-like structures are smaller closer to the surface of the material (Fig. 4.15).

4.4.3 Two Alternative Designs of Adhesive Pads

Data on the hairy pad system of the adult reduviid bug *Rhodnius prolixus* led previous authors to suggest that the mechanical seizure between adhesive setae and substrate irregularities is responsible for attachment to the substrate [522]. In the *Chrysolina polita* beetle, the attachment force increases with the total number of adhesive setae [501]. In the hairy system of the beetle, two forces mainly contribute to the overall attachment force: (1) the molecular adhesion between setae and substrate and (2) the surface tension of a thin layer of epidermal secretion. Molecular adhesion requires very close proximity between contacting surfaces. This means that the setae or setal ends have to be composed of extremely flexible material. The flexibility of beetle setae has been supported by Mallory's single stain [503]. For hairy pad systems there are no data on the dependence of the attachment of pad material to the substrate with increasing loading force.

The deformable cuticle of the pad has a different ultrastructure compared with the surrounding cuticle. It consists mainly of the endocuticle [511,523], which is organized into characteristic rods or a foam-like network, orientated at some angle to the cuticle surface [508]. Attachment of smooth systems to a surface is strongly dependent on the deformation of pads. Pads can replicate the surface to obtain maximum contact. The mechanical properties of the pad material are determined by its composite structure. Sectioned and fractured pad material provided evidence that, when the surface is compressed, the pad cuticle decreases in thickness and the rods flatten down close to one another against the inner layer. When the pressure was released, the rods returned to their previous condition [508]. Additionally, the euplantulae contain air sacs branching from the tracheae trunk [523]. Spaces within the tarsus are filled with insect blood (haemolymph). The air sacs embedded in the fluid may provide an additional mechanism responsible for the flexibility of the pad material. For smooth systems, the viscoelastic properties of the pad material have been suggested previously [524] and recently demonstrated experimentally (see Case Study). It was shown that pads deform elastically under rapidly applied forces and behave viscoelastically under slowly increasing forces.

Leg attachment devices are adaptive structures. Their construction and properties may correlate with the preferred substrates mainly used by particular insect species. Insect attachment pads appeared in their evolution mainly as adaptations for walking on plant surfaces. However, plant surfaces have a wide range of textures. They may be smooth, hairy, and covered with waxes

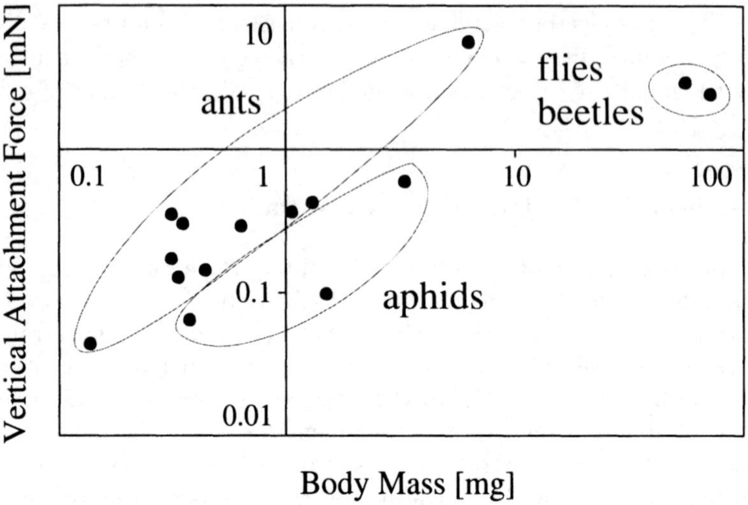

Fig. 4.16. Scale effects on the attachment forces of insect pads. Only forces measured perpendicular to the substrate were taken into account.

or moist secretions [525–527]. As for any integument, plant cuticle with related structures is a functional organ reflecting a number of environmental pressures resulting in the particular design, ultrastructure and chemistry of the plant surface.

During the long period of co-evolution between flowering plants and insects, plants have developed not only structures attracting fertilizing insects (pollinators) but also a wide variety of structural and chemical attributes of their surface related to defense against herbivorous insects. Actually, the evolution of plant surfaces and insect pads is a competition between insect attachment systems and plant anti-attachment surfaces. However, the aspects of mechanisms of insect attachment to plant surfaces remain somewhat understudied. There are only very few reports showing pad specializations to particular plant surfaces.

In oak lice (aphids), the correspondence of attachment structures on legs to the host plant surface is an important characteristic of their specialization to the plant substrates. The aphid *Tuberculoides annulatus*, feeding mainly on leaves of *Quercus robur*, has difficulties walking on the leaves of *Q. ilex*. The aphid *Myzocallis schreiberi*, living on *Q. ilex*, walks better on the surface of this plant [528]. W. Federle, with co-workers, carried out a series of experiments on ants, which are able to walk on slippery waxy blooms in *Macaranga* ant-plants [529]. Many insects, including other ant species, slip off these plant surfaces. Only the symbiotic partner ants of *Macaranga* are capable of climbing up these plants [*Crematogaster* (*Decacrema*)].

The attachment forces by which insect systems of two complementary surfaces operate are unknown. The design of many complementary surfaces often suggests the dependence of frictional forces in the direction of movement. The attachment forces of pad systems have been measured by a variety of methods ranging from strain gauges [499] to a centrifugal device [514] (Fig. 4.16, data from [499, 504, 505, 514, 530]). The forces range from 0.05 mN to 10 mN, depending on the size of the insect.

A plausible explanation for the existence of only two principal designs is that leg pads are primarily adapted to locomotion. This means that, for such systems, the ability for fast detachment from the substrate is of equal importance to fast attachment performance for the variety of substrate roughness and material properties. In this case, other principles, such as suckers, hooks or glue, cannot work well because of difficulties during rapid detachment.

4.5 Epidermal Secretions

Sometimes traces of secretory substances occur on the attachment pad surface. Footprints on a glass surface were previously reported for both hairy and smooth systems as consisting of round droplets (hairy systems: Coleoptera [530], Diptera [494]; smooth systems: Blattaria [511], Homoptera [504]). This secretion is an essential component of attachment in both hairy and smooth pad systems. For example, the bug *Rhodnius prolixus*, with pads treated with lipid solvents, had poorer attachment performance compared with untreated pads [531]. Aphids (*Aphis fabae*), after some period of walking on silica gel, lost their ability to attach to smooth surfaces [505].

The pad secretion of diverse insects contains a non-volatile lipid-like substance, which can be observed in footprints under water [498, 532] or in footprints stained by Sudan Black [504]. This fluid adheres well to hydrophilic as well as to hydrophobic surfaces, such as wax [511] and salinized glass [505]. In beetles, the chloroform-soluble part of the pad secretion is composed mainly of hydro-carbons, fatty acids and alcohols [530, 533]. However, the water-soluble fraction of the fluid remains unknown. It has been previously suggested that the fluid contains a kind of surfactant, such as Triton X–100, which would make adhesion less sensitive to the nature of the substrate [505].

Presumably, adhesive forces generated by the secretion are of capillary and/or viscous origin. To separate two surfaces with a viscous solution between them takes some time. However, the capillary force does not have such a time dependence. Hence, the two mechanisms of adhesion can be distinguished by measuring the velocity dependence of surface separation. The adhesive force of the fly pad secretion on a glass surface was probed by AFM in the force–distance mode at vertical scan speeds ranging from 0.1 μm/s to 10 μm/s. The adhesion force between the glass and the AFM tip is commonly attributed to capillary force due to the water layer. In a reference experiment, the measurements showed that the capillary adhesion between the tip and the clean glass surface was a single breakage of the water layer. This adhesion did not exhibit any scan-velocity dependence. In case of secretion, the maximum

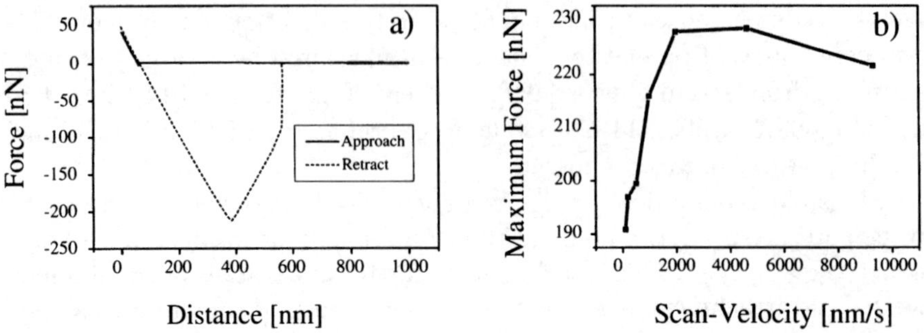

Fig. 4.17. Properties of fly pad secretion. (**a**) Typical force–distance cycle measured by AFM in a droplet of the secretion on a glass surface. The pull-off portion shows a linear force increase, probably due to the viscoelasticity of the secretion, which is followed by a sharp jump-off indicating the breakage of the liquid bridge. (**b**) Dependence of the pull-off force (minimum of the force–distance cycle) on the scan speed.

force, which occurred immediately after the separation of the tip from the glass surface, increased with increasing scan-velocity reaching a plateau at high velocities (Fig. 4.17, data from [534]). The force was related to the real velocity of the tip. It was concluded that the viscosity of the pad secretion contributes to the measured adhesion [534].

Systems of two complementary surfaces usually demonstrate high secretory activity of epidermal cells, underlying fields of cuticular outgrowths at least at one of the co-opted structures [475,535]. In freshly killed air-dried damselflies, microtrichia are viewed as those covered with a thin layer of an amorphous substance. Thus, it can be hypothesized that the secretion builds up a regular layer on the surface. In the systems of two complementary surfaces, the secretion may be responsible for a lubricating function during release of interlocking.

4.5.1 Pore Canals

Secretions are delivered to the contact area by the use of a system of pore canals. Large surface pores have been found only in cockroach pads [512] and in the microtrichia of the mesothoracic part of the wing-lock device in water bugs (Gorb, unpublished data). Tenent hairs of the proximal part of blowfly pulvilli contain large numbers of terminal canals, which have been previously suggested to be transporting devices of lipid-containing secretions [498]. In the leg pads of aphids, the cavity of the adhesive organ is connected to the external surface through fine epicuticular filaments, which are probably sites of elaboration of the final secretory product and are also loci for the extrusion of the adhesive secretion [504]. The cuticle of the postcervical sclerite in damselflies is densely penetrated by pore canals and resembles a sponge-like tissue that can easily free its contents under pressure (Fig. 4.18, from

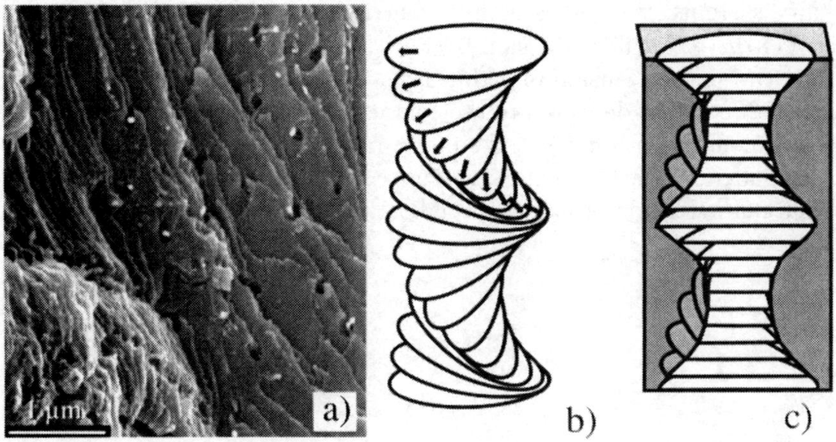

Fig. 4.18. Pore canals of the insect cuticle. (**a**) Cuticle fracture of the postcervical sclerite in the damselfly *Lestes barbarus.* (**b**), (**c**) Model of the design of pore canals. (**b**) Three-dimensional ribbon-like organization of the canal. (**c**) Longitudinal section through model shown in (**b**).

[471]). However, only a small number of large pore canals reach the surface of a microtrichium. Most of them terminate in tiny terminal canals. Terminal canals are quite different structures than pore canals. They do not penetrate the epicuticular layer. It seems that secretion reaches the epicuticle through terminal canals and comes through the epicuticle without any canal structures. It can be suggested here that they may transport substances not only physically (i.e., due to the capillarity forces), but chemically (due to active enzymatic transport by some substances).

In summary:

- Real attachment systems of insects may be designed in accordance with any one of following principles or their composition: hooks, lock or snap, clamp, spacer, sucker, expansion anchor, adhesive secretions, or friction.
- Friction is the principle for which most biological attachment systems have been evolved.
- The insect skeleton, often called cuticle, is a fibrous composite material. Frictional devices can be composed of different types of cuticle.
- Frictional devices of insects, which employ two complementary surfaces, are usually composed of one hard and one soft surface.
- To attach themselves to a variety of substrates, insects have developed two clearly different mechanisms: hairy (setose) surfaces and smooth flexible pads.
- Smooth pads are natural friction-active materials with a specific inner structure.

- In hairy systems, two forces mainly contribute to the overall attachment force: (1) the molecular adhesion between the hairs (setae) and substrate, and (2) the surface tension of a thin layer of epidermal secretion.
- Epidermal secretion delivered to the contact area is an essential component of attachment in both hairy and smooth pad systems.
- Correspondence of attachment structures and host plant surface is an important characteristic of insect specialization to the plant substrates.

Part III

Test Equipment

5. Microscale Test Equipment

5.1 Microtribometer

5.1.1 Force Detection

Forces from newtons to millinewtons are commonly measured using strain gauge load cells. Load cells come with a Wheatstone bridge configuration, featuring a mV/V output or with an integrated amplifier providing common process output signals. Upon bending a deflection beam, areas with mechanical strain and compression develop that can be measured with mechanical-electric transformers, e.g., foil-strain measurement stripes. These metallic stripes have an isolating carrier foil, which is glued to the backside of the beam. The force-dependent changes of the resistance are measured in a Wheatstone bridge, and in an amplifier the changes are transformed to a voltage or current signal for further processing. The use of load cells is limited to forces not smaller than a few millinewtons. If the forces become smaller, the signal-to-noise ratio does not allow a precise measurement.

Low-mass force sensors based on silicon micromechanics overcome limitations of conventional force measurement technology, such as complicated adhesive assembly, a limited overload capacity and a low-level signal output. Although such sensors can be applied easily, the data processing is complicated, since internal parameters like the spring constant of the micromachined part are hard to determine. Very often it is impossible to measure the spring constant. To overcome this problem, calculations are performed by applying procedures like the Finite Element Method (FEM). The FEM simulation is run with material parameters of bulk materials. However, the systems under consideration are mainly controlled by thin film material properties leading to large errors.

Another way to measure low forces is the optical detection of a spring deflection as used in the Atomic Force Microscope (AFM). AFMs have been successfully applied to measure friction, adhesion and microwear [188, 260, 536]. By means of a sharp tip, the molecular interaction inside the adsorbate layer can be detected [537].

In the following sections microtribometers are described that optically detect the deflection of a double-leaf spring allowing the measurement of forces in the range from mN to nN for a sample size that can range from micrometers to several millimeters.

The following technical requirements are met by the microtribometer:

- The forces are obtained in a direct manner; that means by directly measuring the deflection of the spring with nanometer resolution.
- The stiffness of the employed bending beam is exactly known.
- The instrument can be calibrated in both the normal and the tangential direction.
- The force measurement has a resolution of nanonewtons covering a range from nN to mN.

5.1.2 Springs for the Micronewton Range

Photo-structurable glass was used as the base material for high-precision bending elements. Glass was chosen because of its supreme mechanical properties with respect to the elastic constants [538]. By means of photolithography 1 mm thick glass wafers (Li_2O–Al_2O_3–SiO_2) were structured [539]. When the glass is exposed to ultraviolet radiation, its amorphous structure crystallizes. Using buffered hydrofluoric acid the exposed areas can then be removed after annealing. This techniques ensures a high aspect ratio. Two 100 μm wide and 1 cm long glass beams serve as displacement elements. The etching process is controlled to preset the spring constant by realizing the appropriate ratio of length and thickness of the bending beams. Mechanical stressing of the springs showed no degradation of the elastic properties, even after 10^8 deflections over a distance of 400 μm at 5 Hz. Figure 5.1 shows a double-leaf spring. Based on construction, the spring is stiffer in the z-direction (k_z=100 N/m). The stiffness in the x-direction (k_x) is 50 N/m. In macrotesters the spring constants are usually determined by measuring the deflection after the attachment of weights. Due to the small dimensions and forces this procedure can hardly be applied in the micronewton range. Therefore, to determine k_x and k_z the spring is excited into oscillation. Taking the period time T_p and the amplitude of two periods in succession (u_n and $u_{\mathrm{n}+1}$), k_x can be calculated from

$$k_x = m(\omega^2 + \kappa^2)\,, \tag{5.1}$$

where ω is the frequency of the oscillation and κ is the damping constant as given by

$$\omega = \frac{2\pi}{T_\mathrm{p}}, \qquad \kappa = \frac{1}{T_\mathrm{p}} \ln \frac{u_\mathrm{n}}{u_{\mathrm{n}+1}}\,. \tag{5.2}$$

The same procedure can also be applied for the determination of the spring constant in the z-direction. The mechanical design decouples the oscillation in the x- and z-directions from each other.

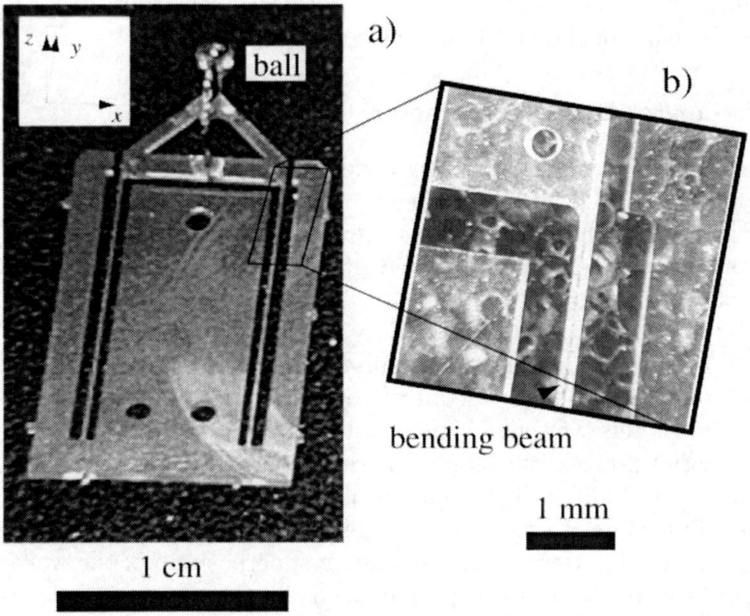

Fig. 5.1. (**a**) Double-leaf springs made of photo-structurable glass. The thickness of the spring is about 1 mm. Each spring has two bending beams with a width of 100 μm. (**b**) Detail. The spring constant in the x-direction is 50 N/m and 100 N/m in the z-direction.

5.1.3 Ball Selection

As mentioned in Sect. 2.2 a polished silicon ball with a diameter of 1 mm was used to probe the flat counter samples. Figure 5.2 shows a photo of the silicon ball. Since the ball and flat are made of the same material, identical surface preparations can be performed. This ensures comparable test results.

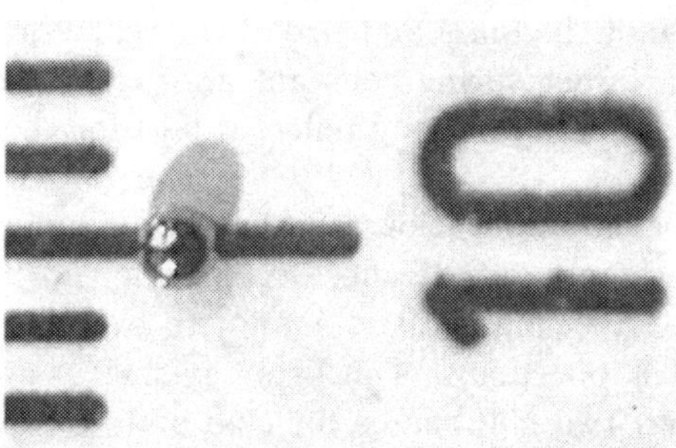

Fig. 5.2. Silicon ball used in the experiments. The flat sample and ball have approximately identical physical properties, since both materials are silicon used for integrated chip manufacturing.

In the experiments the ball was glued to the spring as shown in Fig. 5.1. Consequently, the mass of the ball has to be considered in (5.1).

5.1.4 Interferometer-Based Tribometers

In order to precisely determine the spring deflection, interferometers can be used. The most widely used interferometers in tribology are incremental interferometers [540–542]. The signal periods correspond to the distance of half the wavelength and are added or subtracted by a forward-backward counter as a function of the direction of motion of the reflector. To achieve a higher resolution, less than the signal period, a variety of interpolation routines have been developed leading to nanometer and sub-nanometer resolutions [543].

Each interferometer produces an interferogram that can be evaluated photo-electrically. In this way, two signals with a 90° phase shift are generated. The essential precondition for high resolution and accuracy is very good signal quality (that is, a high signal-to-noise ratio), and in particular a constant 90° phase shift. To meet these requirements, the interferogram is not allowed to change. Interferometers generate a superposition of the wavefronts. The geometry of the interferogram depends on the wavefront geometry of measured and the reference beam as well as on the angle between the wavefronts. If there is a change in direction or geometry of both wavefronts, then a deviation of the initial interferogram occurs. These interferometers are, therefore, designed to generate an unchanging interference pattern and retroreflectors have to be applied. If this condition is not fulfilled, the operation of the known incremental interferometers is impossible. Almost all commercially available incremental interferometers achieve the reflection of the laser beam by a retroreflector with the reflector directly attached to the object. Other types of interferometer use planar mirrors to reflect the laser. The mirrors have a lateral dimension of 20 mm, corresponding to the distance of the two necessary laser beams.

Both setups are not suitable for microtribological investigations. In the case of a retroreflector the attachment of a triple-prism directly to the object (the glass spring) is impossible because of the mass and size of the reflector. Each additional mass would lead to a larger spring constant and, thus, to lower force resolution. The application of a 20 mm long reflector fails for the same reasons.

The ideal solution for microtribological applications consists of a single-beam interferometer and a spring with integrated reflector (see Fig. 5.3). The setup involves a circular aperture [(9), diameter < 1 mm] to evaluate the passing interference pattern at just one point. The beam is then sent through a beam splitter (7b) to obtain two sub-beams for sine and cosine signal generation. First, both sub-beams are equal in phase. Therefore, the 90° phase shift has to be introduced by polarization (6). Out of a variety of possible methods the superposition of orthogonal circular polarized light was chosen, leading to a phase-stable sine and cosine signal independent of the shape of the interferogram. Since the interferogram is evaluated at just one point, a laser beam of small diameter can be used and small reflecting areas are sufficient to serve as a mirror.

The efficiency of the aperture can be ensured over just a small range. If the mirror is positioned perpendicular to the laser beam, then a fringeless interferogram appears. With increasing tilt angle (between mirror and beam) the fringes move together, thereby reducing the photo-electric contrast. The aperture works for

$$D > 1.22\frac{\lambda}{\sin 2\alpha}\,, \tag{5.3}$$

with D the diameter of the aperture, λ the wavelength of the laser and α the tilt angle. Measurements under real conditions can be carried out to the

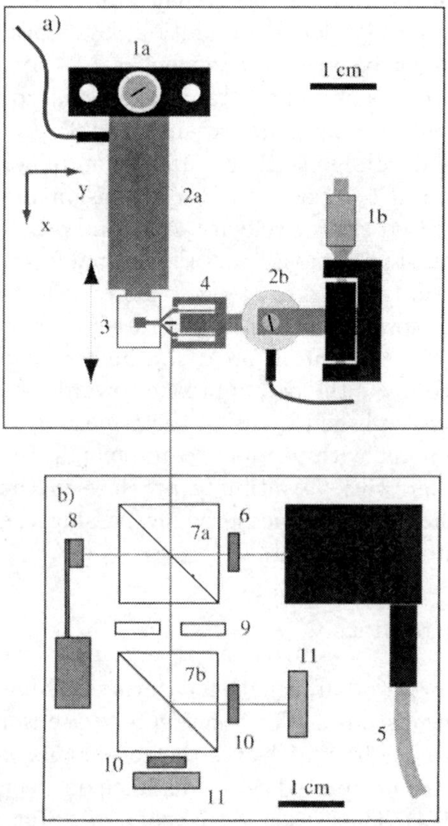

Fig. 5.3. Schematic of an interferometer-based microtribometer (top view). (**a**) Mechanical part. (1a, b) micropositioners; (2a, b) piezo-transducers; (3) sample holder; (4) double-leaf spring with ball; (**b**) Optical part. (5) laser input; (6) polarizer; (7a, b) beam splitters; (8) reference mirror; (9) aperture; (10) analyzer; (11) photo-diodes.

limit of $D = 0.75\frac{\lambda}{\sin(2\alpha)}$. Larger tilt angles require smaller aperture diameters. Since the diameter limits the beam intensity, the aperture cannot be reduced arbitrarily. All tribological tests were carried out with a low deflection of the spring. The deflection was always smaller than the highest tolerable tilt angle of $\pm 10'$. The zero position always refers to the orthogonality of laser and mirror. Deviations from the orthogonality position are expressed by the measurement error

$$\Delta L = L\left(\frac{1}{\cos 2\alpha} - 1\right) , \quad (5.4)$$

where L is the distance between the aperture and mirror. A measurement with a small error is achieved if the mirror is in the orthogonal position before and after the motion of the mirror, even when the mirror is not in orthogonal position during the motion. Since planar mirrors are neutral to polarization at orthogonally impinging beams, the sine and cosine interference signals have very good quality leading to a high length resolution. By digital ($\lambda/512$) interpolation the length can be resolved to an increment of 1.24 nm.

The mechanical part of the tester consists of a piezo-transducer (x-piezo) for dynamic application up to 400 Hz with a maximum expansion of 35 µm. The piezo-transducer is powered by a low-voltage (−10 V – 150 V) amplifier which is controlled by an input voltage-ramp to achieve a linear expansion and retraction of the piezo and by this a constant sliding velocity. A second piezo-transducer (z-piezo) is used for engagement and serves to set the normal force. After the contact is established a further expansion of the z-piezo deflects the spring in the vertical direction. The spring is stiffer in this direction and higher normal forces (with respect to the tangential force) can be set. The micropositioners (1a, b) are used first to make the ball approach towards the flat sample. The final engagement is accomplished by using the z-piezo.

Since all mechanical parts are compliant with vacuum requirements, the tester can be placed in a UHV chamber with an ultimate pressure in the 10^{-9} mbar range. The deflection of the spring is measured by passing the laser beam through a glass window onto the reflector.

5.1.5 Fiber-Optics-Based Microtribometer

Fiber-optical systems (Tetra GmbH, Germany) are length measuring systems that are easy to apply in microtribological studies. The function of the sensor is based on the fact that the intensity of light that leaves the sensor and is reflected back into the sensor head can be interpreted as the distance between the sensor head and reflector (see Fig. 5.4). The sensor consists of many fibers chosen to either emit or receive light. Infrared light, generated by a Light Emitting Diode (LED), is sent through the output fibers and returns to the input fibers after reflection. Thus, changing distance between the sensor head and the reflector modulates the intensity. After factory calibration a distance–intensity look-up table is stored in an EPROM housed in the sensor plug. The

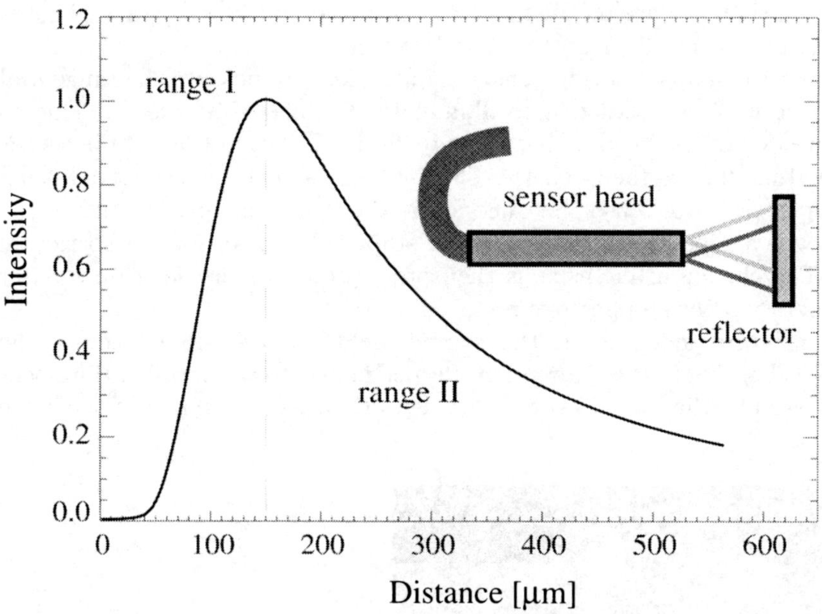

Fig. 5.4. Fiber-optical sensor and sensor signal. The sensor output voltage/distance calibration curve is stored in the plug connecting the sensor to the control unit. The distance is measured between the sensor head and the reflector.

intensity shows a peak at 150 μm dividing the curve into two ranges. Range I allows precision measurement for a reflector position between 0 μm to 150 μm. The resolution is better than 50 nm. In fact, the sensor output voltage can be amplified. After feeding the data into a computer using a 12-bit A/D card, 1 V corresponds to 1 μm. This interpolation leads to a theoretical length resolution of about 2.5 nm (10 V sampling range at 12 bit ≡ 4096 points). Assuming a spring constant of 100 N/m in the vertical direction, forces can be measured with a resolution of 0.25 μN. In the second range (range II), the reflector can be moved up to a distance of 500 μm–800 μm depending on the diameter of the sensor head.

Whereas the peak position with respect to the sensor head always remains constant, the whole curve may shift to higher or lower intensities according to the setting of the LED. By moving the reflector towards the sensor head the peak can be electronically identified and the intensity signal is transformed into a length signal by using the look-up table.

Figure 5.5a shows a photograph of the complete tester. The lower part of the tester consists of a sample holder on a vertical drive. The drive is equipped with a tilting mechanism allowing the levelling of the sample with a resolution of 10 nm. The ball (or any other probe) is attached to the double-leaf spring as shown in Fig. 5.1. The spring is placed on a metal stage that carries also the fiber-optical sensor for tangential force detection. An electromagnetic drive with a maximum traveling distance of 150 μm oscillates the stage. This is detected by another fiber-optical length sensor.

Figure 5.6 shows a spring/sensor arrangement to measure the tangential force. Sensor S1 is positioned to illuminate the edge of a reflector. Due to the deflection of the bending beams the reflector moves in front of the sensor head and modulates the intensity. This intensity signal was calibrated using the length signal of sensor S2. Therefore, the triangular part of the spring was moved against a manually inserted stopper by means of the horizontal drive. The change in intensity is then proportional to the deflection. After the calibration the stopper was removed.

To measure indentation, the sample holder moves up deflecting the double-leaf spring in the z-direction. Perpendicular to the spring a third sensor S3 was attached to measure the z-deflection. After reversal of the drive

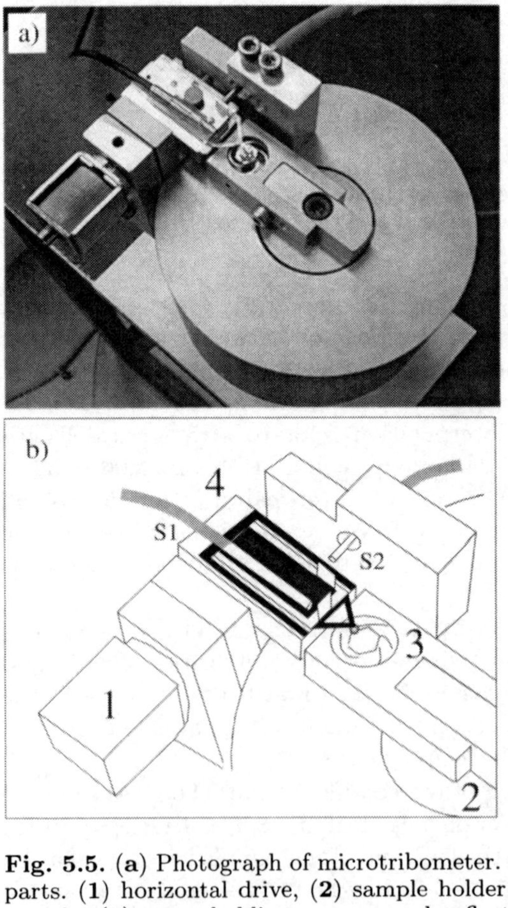

Fig. 5.5. (**a**) Photograph of microtribometer. (**b**) Schematic of the force-detecting parts. (**1**) horizontal drive, (**2**) sample holder on tilting mechanism, (**3**) ball and sample, (**4**) stage holding sensors and reflectors. S1, sensor for tangential force detection; S2, sensor for monitoring of horizontal drive motion. The sensor detecting the normal force is not shown.

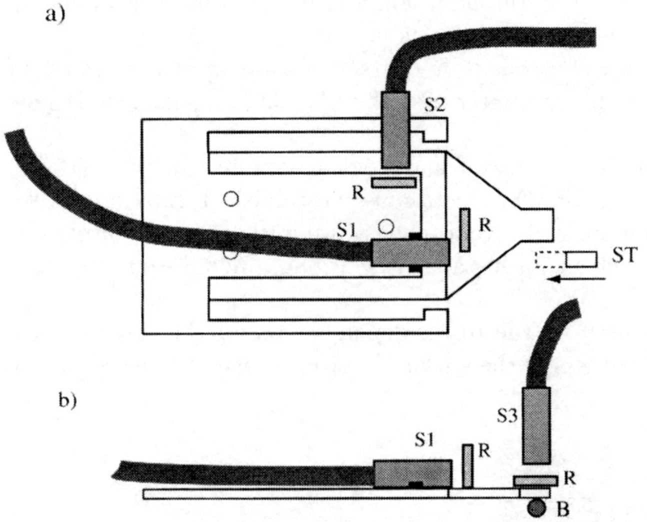

Fig. 5.6. Double-leaf spring with attached optical sensors for tangential force detection. (**a**) Top view. S1, fiber-optical sensor for tangential force detection; S2, fiber-optical sensor for the detection of the drive motion; S3, fiber-optical sensor for normal force detection; R, reflector; B, ball; ST, stopper. (**b**) Side view. The sensor for normal force detection is not shown in the top view.

motion adhesion can be measured. The contact breaks when the spring force equals the force that holds the ball and flat sample together.

The maximum deflection times the spring constant in the z-direction equals the pull-off force.

5.1.6 Bio-Microtribometer

Since biological samples are never exactly planar, new concepts had to be developed to measure the tribological properties in various configurations. In order to access every single part of a biological system, a tribometer based on a seven-axis high-precision positioning system was developed [15, 16, 544], shown in Fig. 5.7. The sample holder carries the biological sample. This sample holder can be tilted about three axis and is also able to rotate the sample around a vertical axis. In addition, the whole rotation and tilting mechanism can be moved up and down over a range of 15 mm. All positioning can be performed with sub-micron resolution. The counter sample (ball, diamond or second biological sample) is attached to the double-leaf spring as introduced in Sect. 5.1.5. The spring is part of a precision drive that is based on a piezo-electric bimorph element. This drive covers a distance of up to 1 mm with nanometer resolution. For positioning, the drive is attached to the main lever that can be expanded by 15 mm and tilted to the horizontal position in both directions. The positioning tasks are performed by the seven-axis system, whereas the actual friction, adhesion and indentation measurements are accomplished with the bimorph precision drive.

All axis are computer-controlled. A coarse positioning is performed in incremental steps using the computer and the ultimate engagement is performed automatically.

Since all points on the surface of a biological system can be accessed, friction and adhesion maps of 3D objects can be established. In addition, sequential 3D indentations provide comprehensive information on anisotropic mechanical properties such as, for instance, the Young's modulus or viscoelasticity.

The force range covered by the tester depends – as for the testers introduced before – on the stiffness of the spring. In order to measure forces in the

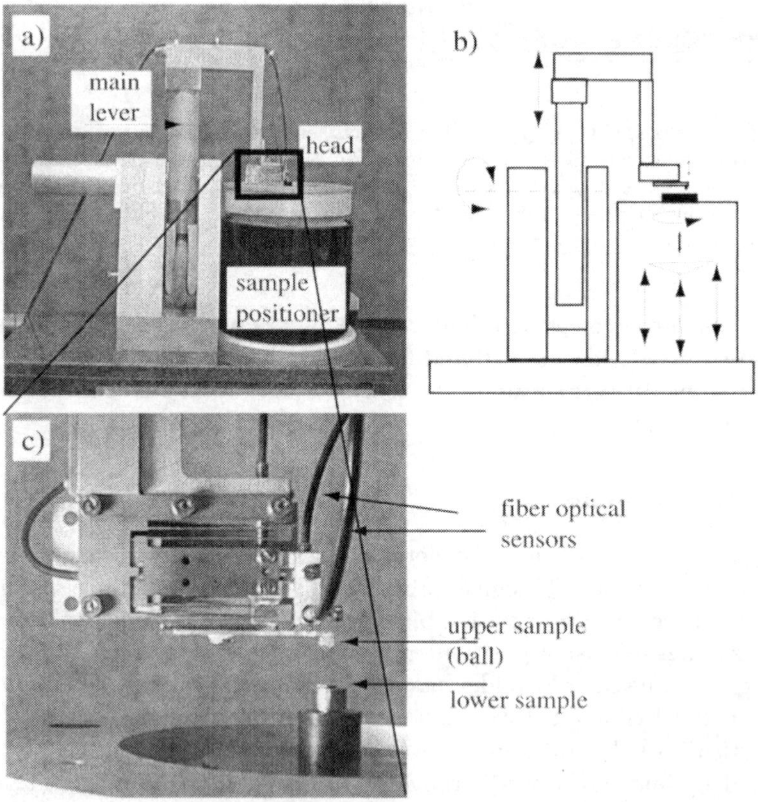

Fig. 5.7. Microtribometer BASALT (Tetra GmbH, Germany) for biological applications. (**a**) Photograph of complete tester comprising the sample positioning and tilting mechanism and the precision drive for friction, adhesion and indentation measurement. (**b**) Schematic of the tester showing details of the positioning system. (**c**) Precision drive with spring set up for adhesion and indentation measurement.

mN range double-leaf springs made of metal (with the same outer dimensions as the glass springs) can be used.

5.1.7 Data Evaluation

To measure friction, one single-length signal needs to be measured. This is the deflection of the spring. However, it is advantageous to monitor also the motion of the horizontal drive (e.g., detected by sensor S2 in Fig. 5.6). Since the drive is excited by a sawtooth signal, the measured length also has the characteristic of a sawtooth as shown in Fig. 5.8. Due to friction the ball and sample initially stick until the back driving force ($F_t = -k_x \Delta x$) generated by the spring equals the friction force (range I in Fig. 5.8). Up to this point the relative velocity is zero. After exceeding the static friction force F_{fs} the ball starts to slide (range II). When the direction of the horizontal drive is reversed, the ball and flat sample stick again until the spring force equals the friction force (range III). The peak-to-peak value is equal to twice the static friction force F_{fs}.

The force in the vertical direction can be determined by taking the deflection of the spring in the z-direction. Indentation is measured when after contact-formation part of the distance in the z-direction is absorbed by the sample. For a hard sample such as silicon the indentation is zero. Thus, the difference of this curve (a straight line for Hookean behavior) and the non-linear curve of the soft material is equal to the indentation. Upon retraction,

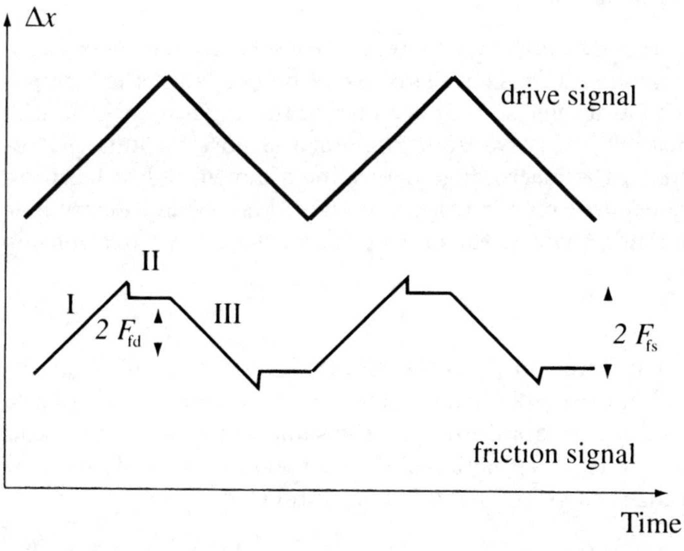

Fig. 5.8. Drive and tangential force signal. In ranges I and III the ball and flat sample stick, i.e., do not move relative to each other. The static friction force F_{fs} is defined as the force where the first slip to lower values occurs. The dynamic friction force F_{fd} is reflected by range II.

the ball and flat sample stick as long as the adhesion is stronger than the spring force. When the spring force equals the force that holds both samples together, the contact breaks. This was shown in Fig. 2.11.

Since microtribology is sensitive to structure-induced (roughness) effects as well as to subtle influences of molecular ordering, it is not reasonable to evaluate the data by determining the coefficient of friction. The coefficient of friction is the time average hiding most of the intrinsic effects. By taking the standard deviation of all dynamic friction values into consideration, fluctuations of the friction signal can also be analyzed. However, it is impossible to tell whether a strong fluctuation is caused by a rough surface or by a rapidly degrading surface.

A way to extract the information contained in the friction signal is by spectral analysis of the data in range II. Depending on the sampling rate, post-processing of the data can be performed to analyze the dynamics of sliding. By means of a Fast Fourier Transformation (FFT) [545] it can be investigated whether the friction signal contains periodic contributions. The FFT can be applied to friction values in range II (see Fig. 5.8). If there are periodic contributions, the spectra should show peaks at corresponding frequencies. The FFT can be used to determine the frequency of stick/slip, as shown in Sect. 2.6.4.

5.2 Mechanical Microanalysis

5.2.1 Hardness Measurement

Hardness is a tribologically important material property. By hardness (technical hardness) one understands the resistance of a body against the indentation of a body from a harder material. It is a measurable quantity, which until recently was not possible to derive from fundamental physical units. Hardness testing methods in the macrorange determine a dimensionless hardness number, which is dependent on the testing method. Macro- and microrange hardness testing methods quantify the hardness from the general relationship

$$H = \frac{F_{\mathrm{n}}}{A_{\mathrm{g}}}\,, \tag{5.5}$$

where F_{n} is the normal force and A_{g} is the geometric contact area calculated from the geometrical parameters of the indenter and the indentation depth. Usually only hardness values obtained with the same tester under the same conditions are comparable. As a function of the normal force one differentiates between different ranges as given in the following table.

Vickers Microhardness

In the microhardness test after Vickers a square-shaped diamond pyramid with a surface angle of 136° is used as the indenter (see Fig. 5.9). This is

Table 5.1. Hardness and force range

Method	Force range
macrohardness	50 N < F_n < 1000 N
small load hardness	2 N < F_n < 50 N
microhardness	0.01 N < F_n < 2 N
nanohardness	F_n < 0.01 N

the same setup as used in the macrorange; however, the applied normal force is lower (see Table 5.1). The indenter is pressed with the force F_n into the material and creates a micrometer-sized indent depending on the hardness of the material [546] and on the applied force [547]. After the test the permanent indent is measured by determination of the diagonals d_1 and d_2. From the median diagonal diameter the indented surface area is calculated and with this the Vickers hardness from

$$HV = 0.102\frac{F_n}{A_g} = 0.1891\frac{F_n}{d^2} , \tag{5.6}$$

with $d = (d_1 + d_2)/2$. The indentation depth δ of the pyramid should be $\approx 1/7$ of the diagonal length. This value is particularly important during the microhardness measurement of thin layers. Within this range the hardness is no longer independent of the loading force. Therefore, the test load has to be specified in each case.

Technically, the standard objective of a light microscope can be replaced by a spring-supported objective that is equipped with a Vickers diamond at the center of the objective lens. The springs serve to detect the loading force. Via an auxiliary objective and a mirror the force display is projected onto the eyepiece. A calibration of the force indicator scale is necessary. The test load has to be applied for usually 10 s – 15 s.

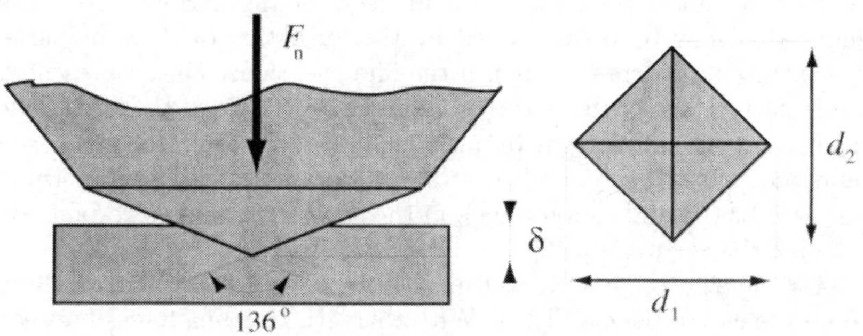

Fig. 5.9. Vickers microhardness. A diamond pyramid is pressed into the material.

Universal Microhardness

With the universal hardness testing method the hardness value is determined during indentation. The area of the indent is determined while applying the normal force by measuring the indentation depth (see Fig 5.9). This is necessary because upon release of the normal force the indentation relaxes leading to an erroneous area and unrealistic hardness values. The following definition for the universal hardness applies:

$$HU = \frac{F_{\mathrm{n}}}{26.43\delta^2} \,. \tag{5.7}$$

The unit of HU is $\mathrm{N/mm^2}$. This hardness testing method takes account of both the plastic and the elastic deformation. Thus, the hardness definition is strictly fulfilled and physically meaningful hardness values can be obtained. The procedure permits the comparison of all materials on the same hardness scale.

A further advantage of the procedure consists of the fact that from the recording of the entire force–indentation curve additional information about the plastic and elastic characteristics of the materials can be deduced (see Fig. 6.8). The Young's modulus can be obtained from the tangent to the unloading curve. The mechanical energy of plastic deformation corresponds to the area between the loading and unloading curve. A modification of the universal hardness testing method permits the investigation of the creep behavior of materials (in particular of thin layers). As a rule of thumb, for thin layers the indentation depth should be less than 1/10 of the total thickness of the thin film in order to obtain reliable hardness results.

5.2.2 Scratch Testing

Scratch testing is used to measure scratch hardness and to evaluate the adhesional strength of a thin film to the substrate. In this method a hard scratching element (indenter and stylus) is moved across the specimen surface to measure the mechanical resistance as a function of the normal force. The scratching action may be accompanied by the formation of chips or particles. With increasing indentation depth the film eventually fails, represented by a significant increase of the detected acoustic signal. With this technique thin films can be evaluated [548]. It can be seen directly from the generated groove and the emission signal whether the material is ductile and permits ploughing or is brittle and shows bursts in the emission signal corresponding to abrupt material removals [546].

The AFM version of scratch testing utilizes a very sharp diamond tip mounted on the end of the cantilever. With this setup loading forces between 0.3 μN and 18.9 μN were sufficient to wear an aluminum thin film [549].

5.2.3 Contact-Angle Measurement

Interfacial phenomena like wetting, cell adhesion and adsorption are based on thermodynamic quantities, which can be determined by contact-angle measurements [64, 550–552]. For instance, the degree of hydrophobization can easily be derived from contact-angle measurements [553]. A simple setup is shown in Fig. 5.10. By means of a pipette a microdrop of water is applied to the surface under investigation. The tangent to the cross-section of the drop tells whether the surface can be wetted or not. High angles between the tangent and the surface indicate a hydrophobic surface. When the angle is small, the surface is hydrophilic.

It was shown for micrometer-sized droplets that the contact-angle decreases with decreasing size of the droplet. This behavior is caused by long-range forces between the substrate and the fluid. These forces become unimportant for larger drops, resulting in a constant contact-angle at the macroscopic scale [554].

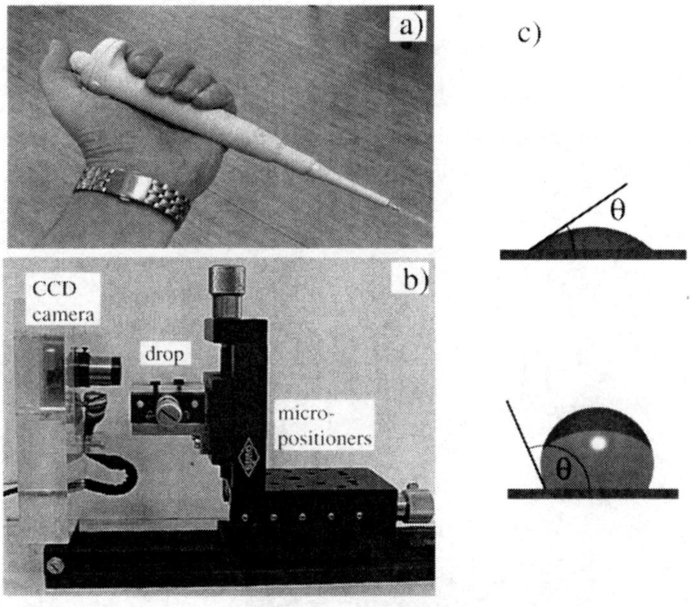

Fig. 5.10. (**a**) Pipette for microdosage (5 µl). (**b**) Simple setup to measure contact-angles. By means of the pipette a drop of water is put on the surface under investigation. A CCD camera images the profile of the drop. The contact-angle can be determined by image-processing software after screen calibration. (**c**) Contact-angles of hydrophobic and hydrophilic substrates. The higher the contact angle is, the lower is the wettability of the surface.

5.2.4 Profilometry

In previous sections it was shown how roughness and surface profile influence friction and adhesion. Profilometry is indispensable for characterizing a surface on a wider scale (up to several centimeters) with high vertical resolution. Several kinds of profilometers are introduced in the following paragraphs.

Stylus Profilometer

Surface roughness and surface profile are traditionally measured by moving a diamond needle over the surface (see Fig. 5.11). The results are comparable on a worldwide scale due to established standards. However, tactile profilometers have some limitations like the destruction of soft surfaces, long measuring times and critical calibration procedures. To obtain reliable values many scans are necessary. Furthermore, the form of the diamond needle influences the measurement results. For a fine-structured surface the diamond works like a filter, because the depth of the profile cannot be replicated truthfully. The result is always a convolution of the needle and the surface profile (see Sect. 2.3.1).

Laser Profilometer

Optical profilometers eliminate some of these limitations. With the help of a microfocussed laser beam the surface is scanned without contact. The filter

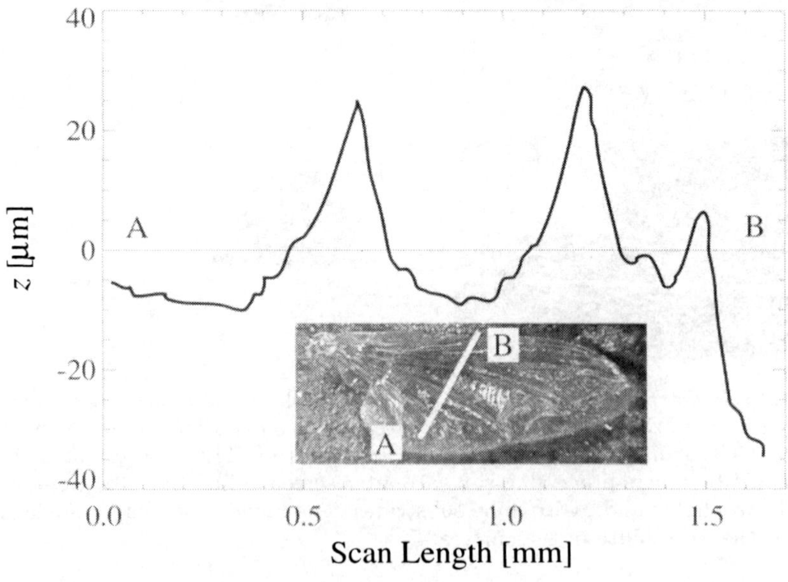

Fig. 5.11. Stylus profilometer plot of a wing of a fly. A sharp diamond needle scans the surface profile between points A and B.

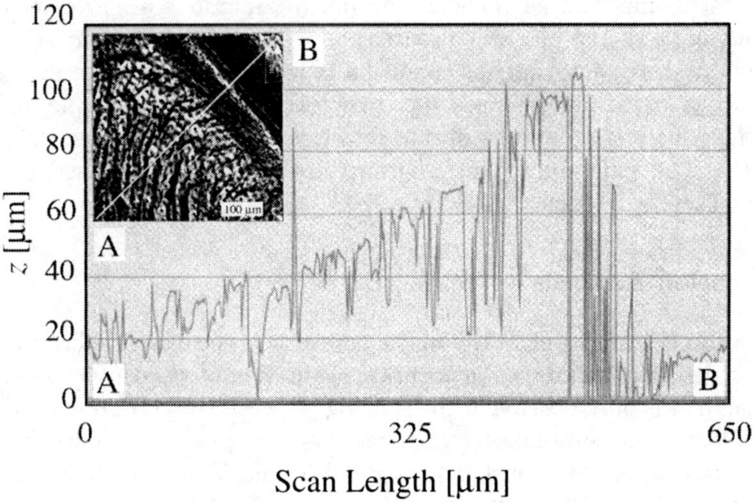

Fig. 5.12. Profile between points A and B. The inset shows a laser profilometer image of a wing of a fly. The laser scans the entire region.

effect, such as experienced with the diamond needle, is practically negligible. However, because of the radiation there can be an interaction between the laser and the material. Basically, a laser scanning microscope functions like a normal laboratory microscope. The usual lamp used for lighting is replaced by a laser. By means of a beam deflector the laser is sent line by line over the sample leading to an image of the surface as shown in Fig. 5.12. The objective projects the beam to a focus in the material. Due to the transmission of light, substances in the material (e.g., pigments) become activated and can emit radiation (fluorescence). The emitted light, thereby, has a larger wavelength than the activating light. Using two apertures the additional light arising from the laser activation is filtered out. This kind of optical arrangement is called confocal laser scanning.

Interference Optical Profilometer

A second group of contact-free profilometers are interference-optical profilometers. These types of optical profilometer generate a three-dimensional image of the surface up to a depth of 500 µm. The feature depth is determined by scanning the laser focus through a calibrated optical distance and simultaneously recording the position at which interference fringes appear on the features. The image contains depth information as well as the feature location in the x- and y-direction. There are two modes of operation available. Features with a vertical spacing below 1 µm are measured in the phase-scan imaging mode to evaluate smooth surfaces and small structures. The reconstruction of the measured surface is made with the help of phase shift interferometry. The vertical scan imaging mode is used for measuring rougher surfaces and larger structures above 1 μm. Since the surface is strongly structured, the interference condition is not met everywhere in the image field. However, if one changes the distance between the sample and the interference objective, one can obtain interference fringes at each point on the surface. The individual interferograms are computer-processed and the surface is, thereby, reconstructed.

5.2.5 Lubrication Analysis

When lubrication is tested, one of the major concerns is the exact determination of the liquid-layer thickness. An accurate evaluation of viscosity changes due to confinement is only possible if the spacing between the confining solids can be controlled in the sub-nanometer range. For this purpose, the Surface Forces Apparatus (SFA) was constructed by Tabor and Winterton [555] and later improved by Israelachvili [107, 199, 556]. The miniature SFA Mark III was introduced in 1990. The manual control of surface separation of 1 nm allowed a sensitivity of better than 0.1 nm.

Basically, two transparent samples (mica sheets) are glued on top of cylindrical sample holders made of glass. The cylinders are arranged as a cross to achieve a single point of contact. One of the cylinders is fixed and the other is allowed to perform an oscillatory motion to shear the liquid film confined between the cylinders. By means of interferometry the thickness of the film can be determined with sub-nanometer resolution [556]. To shield the whole assembly, the optical parts are located in a chamber filled with the liquid. The liquid also serves as a reservoir to maintain equilibrium conditions.

By means of this tester van der Waals forces at a surface and in its vicinity can be measured in air as well as in vacuum. In addition, the SFA is also capable of measuring DLVO forces (see also Sect. 2.4.2).

5.3 Accompanying Surface Science Techniques

The use of surface science techniques has expanded rapidly during the past 35 years. Many of the developed tools, which also can be found in macrotribological labs [557–561] are now also of growing interest in biological surface science. Surface science is necessary to fully understand and interpret results obtained in microtribology [562]. Of special importance is the interface where both solids meet. In addition, the liquid/solid interface can also be probed, disclosing the chemical and physical interactions within the electrochemical double-layer (see Sect. 2.4.2) [563]. Another important field of operation of surface science techniques is the *in situ* monitoring of tribo-chemical reactions [564].

5.3.1 Photoelectron Spectroscopy

X-ray Photoelectron Spectroscopy (XPS) provides qualitative and quantitative information regarding elements and compounds. The interaction of radiation impinging on the atoms of the topmost layers is analyzed in an Ultra High Vacuum (UHV) chamber to keep the influence of the residual gas (scattering, adsorbate formation) as low as possible. Excitation sources such as ultraviolet light, monochromatic x-rays or synchrotron radiation are typically used. The relation between the energy W and the frequency of light f_l is given by

$$W = hf_\mathrm{l} \,, \tag{5.8}$$

where h is the Planck constant (6.62×10^{-34} J·s). The excitation by ultraviolet light leads to the ionization of the valence electrons, whereas x-rays mainly ionize the core electrons. Core levels are also excited, however, with a much lower probability. In both cases the electrons leave the solid carrying chemical information about the emitting atom. The information consists of a characteristic kinetic energy W_kin and the electron momentum. The binding energy W_b can be calculated from

$$|W_\mathrm{b}| = \hbar\omega - \Phi - W_\mathrm{kin} \,, \tag{5.9}$$

where $\hbar = \frac{h}{2\pi}$ and $\omega = 2\pi f_\mathrm{l}$. For the case of impinging photons generated by an x-ray tube with an aluminum anode (Al k_α radiation), $\hbar\omega$ has a value of 1486.6 eV (or 1253.6 eV for a magnesium anode). Φ, in practice, is the work function of the analyzer and has to be obtained experimentally. The spectrum displays the number of detected photoelectrons with respect to the vacuum level as a function of their kinetic energy (see Fig. 5.13). Due to inelastic scattering mechanisms between electrons and atoms only electrons of the first atomic layers are able to leave the solid. Thus, XPS is a surface-sensitive method. Each element is characterized by its binding energy represented by a characteristic set of peaks in the spectrum. The intensity of the peaks is related to the concentration of the elements in the area under investigation ($\approx 1\ \mathrm{mm}^2$), yielding quantitative analysis.

Another important feature of XPS is the binding energy shift caused by the formation of bonds. For example, carbon can occur not only as single atom but also as a molecule such as C–C, C=C, etc. In the molecular state the binding energy is shifted to higher or lower binding energies depending on charge transfer inside the molecule. Commonly, biological specimen have a variety of different components. In this case the measured intensity peak consists of several sub-peaks. For example, the C1s peak can have contributions from (a) carbon bound only to carbon and hydrogen, (b) carbon making a single bond with oxygen or nitrogen (alcohol, ether, amine, amide), and (c) carbon making two single bonds with oxygen, among many others [565]. To deconvolute the peak, fitting routines are used to approximate the main peak by individual peaks of known molecules [566]. Therefore, the analysis of model

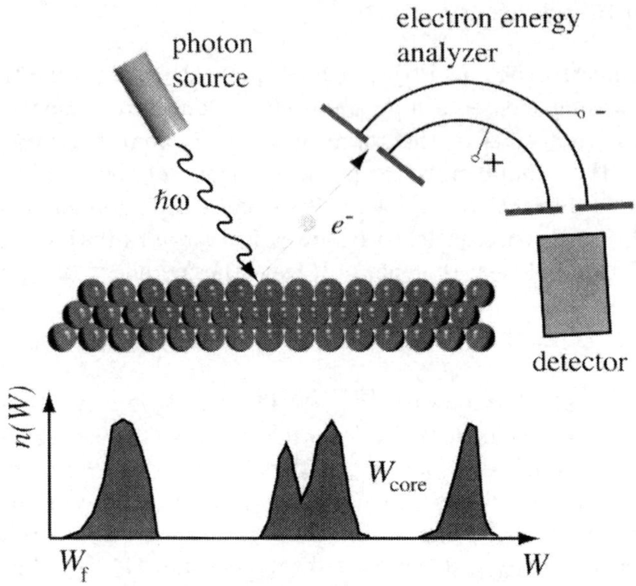

Fig. 5.13. Principle of XPS. By means of impinging x-rays photo-electrons are excited and leave the surface. The electrons carry chemical information about the topmost atomic layers. The lower part of the figure shows the electron density distribution as function of the energy states of an atom.

compounds (homopeptides, glucose-related compounds) is a necessary preliminary step of an exact quantitative measurement. In addition, computer simulation (semi-empirical quantum mechanical molecular modeling) is also used to aid the analysis of spectra [567]. Since photoelectron spectroscopy is a UHV method, biological samples have to be dehydrated in order to be measured (see Sect. 5.3.1 for details). This is necessary to prevent the samples from damage caused by the low pressure and to maintain the vacuum in the chamber (outgassing). Dengis et al. [568] reported on XPS analysis of bio-surfaces (yeast cell and model compounds) and found that sample freezing and drying are the main sources of variation. Surface alterations can also be induced by the x-rays and photoelectrons leaving the surface. The authors measured a degradation rate of 0% to 10% per hour for the intensity of the C1s peak attributed to carbon bound to oxygen or nitrogen.

Figure 5.14 shows spectra of oxidized and hydrogen-terminated silicon used in the SILICON model setup. Silicon was etched with hydrofluoric acid to remove the oxide and to terminate all dangling bonds [two for Si(100)] with hydrogen. Accordingly, the silicon contributions (Si 2s, Si 2p) showed increased intensity. At the same time the concentration of oxygen is decreased. The energy shift of the Si peaks due to oxidation cannot be seen in the spectra,

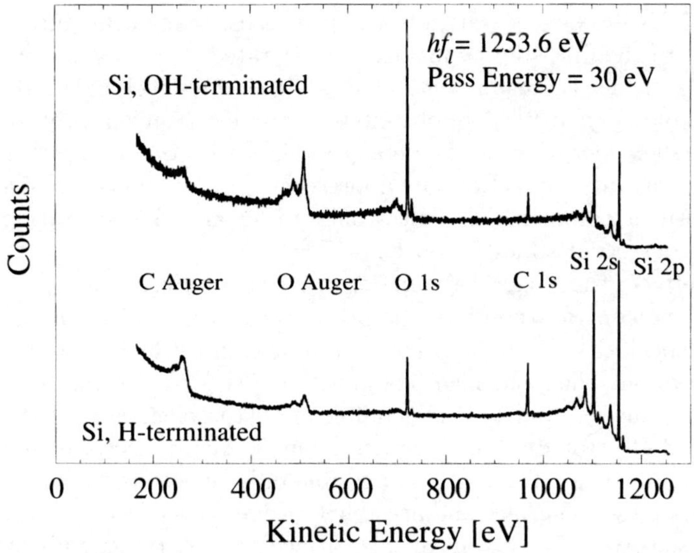

Fig. 5.14. XPS spectra (global scan) of oxidized and hydrogen-terminated silicon. The oxidized sample shows a higher content of oxygen and, accordingly, lower silicon contributions.

since Fig. 5.14 shows only a global scan. To measure the peak shift one has to zoom into the particular energy range.

5.3.2 Auger Electron Spectroscopy

Auger Electron Spectroscopy (AES) has been used since the end of the 1960s as method of surface analysis [569]. The underlying physical process was first observed by Pierre Auger in 1925 when exposing photographic plates to x-ray

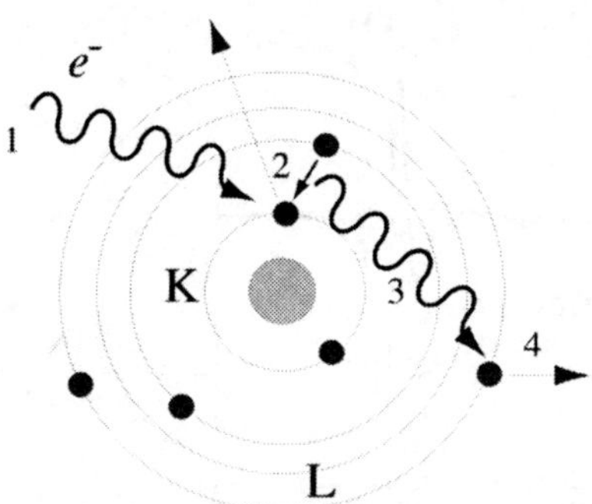

Fig. 5.15. Principle of AES. (**1**) excitation by impinging electrons results in a hole in the K shell, (**2**) jump into hole, (**3**) energy transmission to electron in L shell, (**4**) Auger electron emission.

radiation. Figure 5.15 shows schematically the processes, that occur during the Auger process. In the first step by means of accelerated electrons (energy of 2 keV to 5 keV) surface atoms are activated, so that core-level ionization takes place. The hole is then filled by an inter-atomic electron jump of an electron from a higher energy level. The energy balance is equalized either by emission of an x-ray quantum (Röntgen fluorescence) or by ionization of a further electron (Auger emission). The electron transfer and the generation of the Auger electrons occur simultaneously.

The kinetic energy of the Auger electron depends only on the energy levels involved, however, not on the energy of the primary excitation. The energy levels are element specific, so that the Auger electrons emitted by the sample carry information about their chemical composition. Auger electrons have kinetic energies of some eV and keV depending upon the element and the type of the involved electronic levels. The detection of Auger electrons and their assignment to a specific transition is possible only for electrons which are generated close the solid surface and are able to leave the surface. AES is, therefore, surface sensitive. The extent of this surface sensitivity depends on the energy of the Auger electron, which at the same time determines its mean free path in the solid. For an energy of about 50 eV one finds a minimum of the mean free path and thus a pronounced surface sensitivity of about a tenth of a nanometer.

AES was used as a measure of the oxide coverage of the samples in the SILICON model system, thus being an indirect indication for the degree of hydrophobicity or hydrophilicity. From the spectra (Fig. 5.16) it is deduced

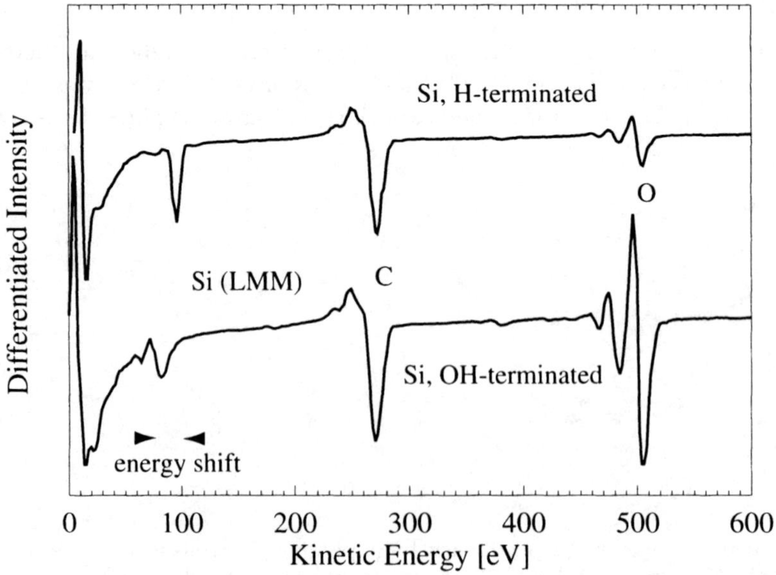

Fig. 5.16. AES spectrum (global scan) of hydrophilic and hydrophobic silicon.

that the oxygen intensity of the hydrophilic sample is about three times larger than of the hydrophobic sample, indicating the presence of an oxide layer. The small silicon peak in the spectrum of hydrophilic silicon is characteristic of the replacement of every second silicon atom by oxygen. The Auger line shape can be used to confirm that hydrophilic silicon is oxidized. The line shapes for the core–valence–valence transition represent the weighted self-convolution of the valence-band density of states. This can be shown in the LMM transition at about 90 eV. The shift to lower energies (from 92 eV for silicon to 78 eV for oxidized silicon) and the line shape are typical for oxidized silicon.

5.3.3 Infrared Spectroscopy

Infrared (IR) spectroscopy enables the chemical characterization of organic materials. For detailed technical information, refer to [570,571]. In IR spectroscopy the interaction of infrared light with a sample is recorded to measure the characteristic frequencies at which significant absorption occurs. These frequencies are used to identify the chemistry of the sample by assigning specific frequencies to chemical functional groups. As in XPS the intensity is a measure of the concentration of the component. However, since the radiation passes through the sample, the absorbance or transmittance are evaluated. The key issue of IR spectroscopy is the interpretation of the vibrational modes like stretching, rocking and wagging modes. For instance, hydro-carbon chains are stretched when excited. With the help of the IR spectrum conformational ordering can be identified as shown in [572].

In a spectrometer consisting of a light source, interferometer and detector, all optical frequencies are measured simultaneously. To identify frequency-resolved information, the data set has to be Fourier transformed (FTIR).

Infrared spectroscopy with its high information content is one of the most important methods for analytic chemistry. An unknown sample can be identified by direct or empirically derived relations existing between the position of absorption bands in an IR spectrum. The presence or absence of certain functional groups or of double bonds and other structural components can easily be detected with high probability. A more exact investigation of the position and intensity of these bands allows a closer allocation of the detected groups.

One of the greatest advantages of the IR spectroscopy is the variability of the sample pretreatments. It is possible to measure the IR spectra of almost all substances in their most diverse physical states. Additionally, the influences of the molecular environment of the sample become apparent by subtleties in the spectrum, and intermolecular interactions become transparent. It is important that the samples are pure and that the water content is less than 1% in order to prevent disturbing absorption bands. IR spectroscopy covers the spectral range of electromagnetic radiation from

400 cm^{-1} to 4000 cm^{-1}. This is equivalent to 50 MeV – 500 MeV. A necessary precondition of IR spectroscopy is that the dynamic dipole moment of the induced oscillation must change.

Electromagnetic radiation interacts with an electron as a moving electrical charge and emits energy. Strictly taken, oscillations always concern a molecule as a whole. To a good approximation, however, many oscillations of a molecule can be treated separately. The oscillation amplitude of individual molecules is indirectly proportional to their mass. Heavy atoms oscillate more weakly than light ones. At the same time, heavy atoms oscillate more slowly than light atoms, which is justified by the action of inertia. Electromagnetic radiation of appropriate frequency can, therefore, interact with a molecule if the oscillation of the atomic groups leads to a change of the dipole transition moment. It is important to note the vectorial character of the dipole transition moment. An excitation takes place only if the incident electric field has a component parallel to the dipole transition moment. If both vectors are perpendicular to each other, no excitation can take place.

The frequency of the oscillations in, for example, water can be divided into different areas: the first oscillation concerns a symmetrical valence oscillation. Because of the bent structure, a modification of the centers of charge is connected with this oscillation. The oscillation is, therefore, IR-active. Here, not only does the hydrogen atom oscillate, but also the oxygen in order to preserve the center of mass. Only a single vibration is conceivable, since an appropriate movement of the atoms perpendicularly to the molecule level would be equivalent to a rotation of the molecule. Contrary to linear molecules, here only three normal modes are possible.

Adsorbed water shows OH stretching and bending modes and it is commonly known that the bending vibrations are weaker [573]. The formation of thicker layers can be monitored by vibration modes arising from intermolecular hydrogen bonds.

5.3.4 Low-Energy Electron Diffraction

Low-Energy Electron Diffraction (LEED) is applied to determine the lattice structure of single-crystal surfaces and the structure of ordered monolayers or adsorbates [574–576]. Figure 5.17 illustrates the basic principle of LEED. As part of a mono-energetic beam, electrons with an energy of 10 eV–200 eV impinge on a solid surface. The interaction results in an elastic scattering of the electrons in all directions. The electrons can be considered as a wave with a wavelength

$$\lambda = \frac{h}{\sqrt{2m_\mathrm{e}W_\mathrm{kin}}} , \tag{5.10}$$

where m_e is the mass of the electron and W_kin is the kinetic energy of the electron.

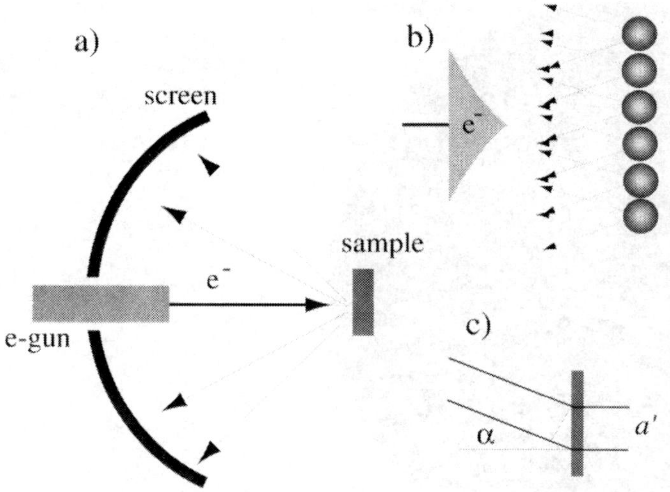

Fig. 5.17. Principle of LEED. (**a**) Electrons coming from an electron gun hit the surface. (**b**) According to the diameter of the electron beam a number of surface atoms participate in the scattering process. (**c**) Bragg condition. The LEED pattern is an image of a crystalline surface in the reciprocal lattice. This means that the LEED pattern reflects the periodicity of the crystal lattice.

The scattered electrons form an interference pattern with constructive interference in directions with

$$\sin\alpha = \frac{n\lambda}{a'} = \frac{nh}{a'\sqrt{2m_e W_{\text{kin}}}} , \tag{5.11}$$

where α is the angle between the scattered electrons and the surface normal, n is the order of diffraction and a' is the lattice constant (or more precisely, the distance between two atoms in the surface). The scattered electrons are reflected back to a fluorescent screen, thus showing a pattern of spots corresponding to the directions in which constructive interference takes place. Since the interatomic distances and the directions of constructive interference are inversely related, the separation between LEED spots is large for small interatomic distances and small for large distances. The LEED pattern develops in the so-called reciprocal lattice.

In the case of ordered organic layers the unit cell of the layer can readily be determined with LEED. LEED analysis of hydrogen bonded self assembled adenine monolayers was performed in combination with real-space scanning tunneling microscopy to determine the unit cell [577]. However, the exact position of the atoms with respect to the substrate remains undetermined, since the LEED pattern only reflects the periodicity in the layer.

LEED is a surface-sensitive technique. However, electrons scattered from deeper layers also contribute to the spectrum. When the Bragg condition in the vertical direction is satisfied – that means for a certain combination of

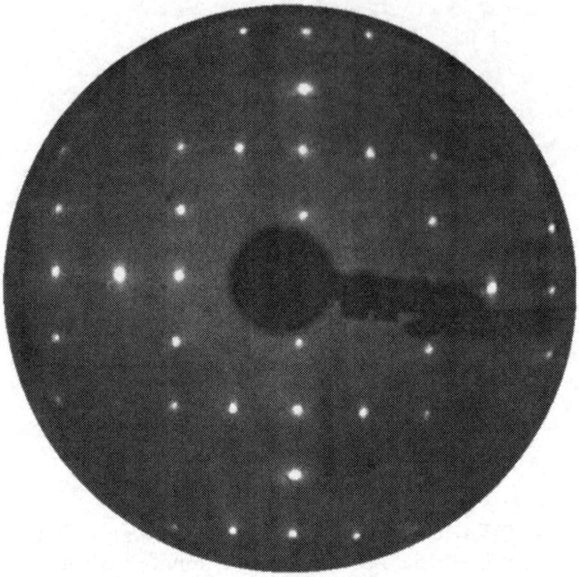

Fig. 5.18. LEED pattern of a Si(100) surface. Clearly visible is the 2×1 reconstruction with two domains.

electron energies and angles of diffraction – very intense spots are observed as shown in Fig. 5.18 for the SILICON model system. In combination with the LEED pattern the spot intensity as a function of the electron energy contains three-dimensional structure information. By this means it is possible to determine the exact location of the monolayer atoms.

In summary:

- XPS, AES, LEED and FTIR are surface-sensitive methods that allow the precise determination of surface chemistry, chemical bond formation, crystallinity and adsorbate coverage.
- In addition, mass spectrometry can be used to analyze chemical degradation mechanisms [578, 579]. (Not discussed here.)
- Whereas XPS, AES and LEED require UHV conditions, FTIR operates in air.
- For quantitative analysis the methods require comparison with model calculations.

6. Nanoscale Probe Techniques

The family of instruments which measure physical and chemical properties on the nanometer scale are called Scanning Probe Microscopes (SPM). In this section scanning probe microscopy is discussed in the context of its use in micro- and nanotribology [580]. Although the instruments – STM, AFM and their modifications – are commonly used on both length scales, distinctions between micro and nano have to be made, since the range of application differs significantly. Whereas on the microscale the instrument is mainly used to measure topography [581], friction [582], adhesion and mechanical properties, SPMs provide detailed information on structural properties [583–588], chemistry [589–592] and motion on the nanoscale [593–596]. The use of scanning probe techniques for biological applications [597,598] was promoted by the simplification of preparation techniques and the handling of different sample types. Furthermore, a new interesting feature is the option to analyze living substrates in their native environment [599,600]. The samples can be imaged in their hydrated state, eliminating shrinkage and thus degradation.

In SPMs, the sample surface is scanned with a sharp probe at a distance of less than a few nanometers or in mechanical contact. For scanning, either the tip moves against a fixed sample or the sample moves against the fixed tip. The common parts of SPMs are the piezoceramic scanner, on which the moving part is mounted, and the coarse mechanism by which tip and sample are brought close together so that the probing interactions can be measured with an appropriate detector. With the advent of new technologies for the detection of small forces, the application of microstructured tips allows the analysis of nanoscale phenomena. SPMs are often used as a complementary method in addition to Scanning Electron Microscopy (SEM) and optical microscopy, as the techniques have several major advantages [601]. In contrast to SEM no metallic coating or staining of the sample is required, preserving the specimen in its original shape and constitution. In addition, the resolution may be higher than that of the SEM. The following sections describe various types of SPMs with a focus on the Scanning Tunneling Microscope (STM) and the Atomic Force Microscope (AFM) as the most mature instruments of this family [602].

6.1 Scanning Tunneling Microscope

6.1.1 Operating Principle

The STM, developed in 1981 by Binnig [603], is an indispensable tool for obtaining information on the molecular level. Since the STM is operated without contacting the substrate, it is not possible to measure tribological properties. However, the STM can be used to elucidate the action of atoms or molecules in the dissipation process. Thus, fundamental friction mechanisms can be investigated. A comprehensive introduction to STM imaging of biological samples can be found in [604], which covers the period up to 1994. In [605] the reader finds detailed information on STM and other SPM techniques covering biological applications and experimental techniques that can be reproduced in the laboratories. The book contains extended bibliographies that guide the reader to detailed source publications.

In STM a sharp metal tip prepared from wire (tungsten, gold, or Pt/Ir, Rh/Ir) is used to inject electrons into the sample by tunneling through the potential barrier between the two solids (see Fig. 6.1). Tunneling is a quantum mechanical effect [606], allowing electrons to pass through a barrier in which electrons cannot exist. The tunneling current is a function of the width of the potential barrier, the adsorbate coverage [607] and the distribution of electronic states in the sample. It has to be pointed out that an STM image always reflects the local density of electronic states at the Fermi level. Hence, bright features may indicate locations of high electron wave function probability. When the surface under investigation has unsaturated (dangling)

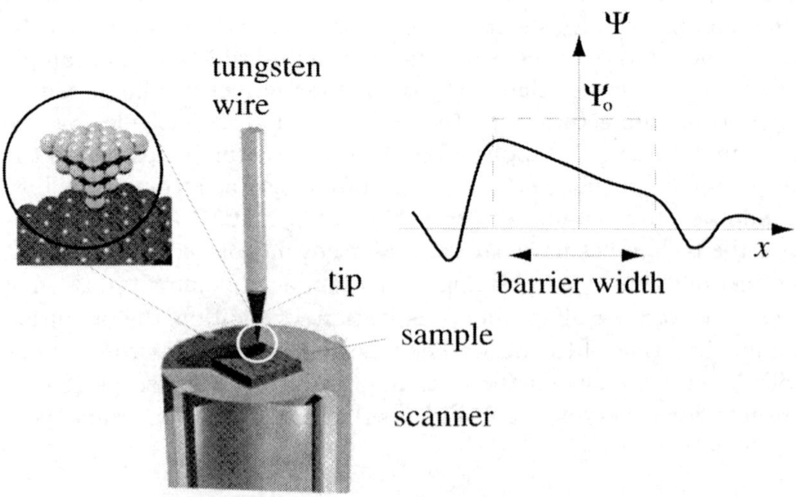

Fig. 6.1. Schematic of an STM and tunneling process. A sharp tungsten tip is moved close to the surface to induce a tunneling current. By means of a piezo-tube scanner the surface can be raster scanned.

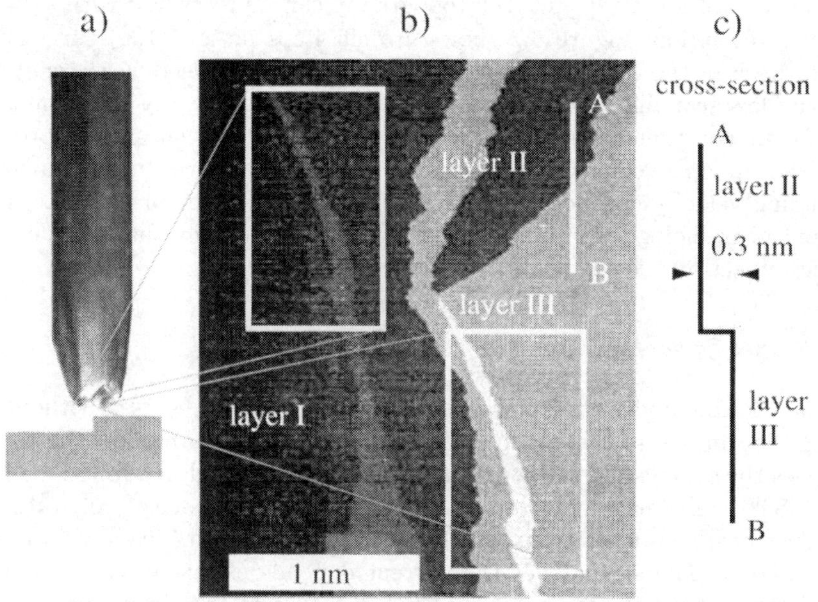

Fig. 6.2. STM image of steps on a Si(100) surface. (**a**) Damaged tip. The main tip has two sub-tips. (**b**) Image and artifact generation by the damaged tip. (**c**) Cross-section of two layers.

bonds, then those bonds rather than the atoms constitute the image. Moreover, the tunneling mechanism in the case of a clean surface in UHV differs considerably from the case when atoms are adsorbed onto that surface. It is possible that the local density of states results from the interactions between the electronic states of the substrate with those of the adsorbate.

The ideal case for the tunneling process would be a single atom at the top of the wire with a single electron orbit facing the sample. Very often, however, this is not the case, since atomically sharp tips cannot always be prepared. Thus, the image results from a multi-atom interaction decreasing the quality of the image, especially the resolution of small features. Drastic artifacts can easily be obtained when the tungsten wire has more than one asperity (group of atoms) that stick out. This is illustrated in Fig. 6.2. Two or more asperities simultaneously scan the surface leading to a tunneling current that integrates over the entire surface.

6.1.2 Constant Current/Height Mode

The most widely used mode of operation is the constant current mode. The sample is moved by piezo-electric transducers to maintain a constant tunneling current. As a result, the voltage applied to the piezo is a measure of the distance between tip and surface. After calibration, for instance by using an atomic step of known height, the topography can be measured with subnanometer resolution. Figure 6.2 shows atomic steps on a Si(100) surface. One drawback of the method is the finite response of the feedback circuit. Thus, only low scanning velocities can be achieved. Another way to image a surface is the constant height mode where – an atomically flat surface provided – the tip is moved at constant distance from the surface. In this mode the scanning velocity can be increased considerably. A common problem of STM applied to biological samples is that the tip is extremely sharp, having a drastic impact on the sample.

6.1.3 Tunnel Spectroscopy

In addition to the modes explained above the STM can also be used without scanning the sample. In this mode – the tunnel spectroscopy mode – the tip approaches the sample and the tunneling current is measured as a function of distance [608]. The image of a sample is determined by the density of states of the electrons. These electrons can be detected by the tip, leading to an electric current. The strength of the current depends exponentially on the distance between the tip and sample. Thus, the current is a direct measure of the distance between the probe and sample.

The current/distance spectroscopy can be used to measure the water film thickness on the solid by interpreting the onset of an ion current as the first

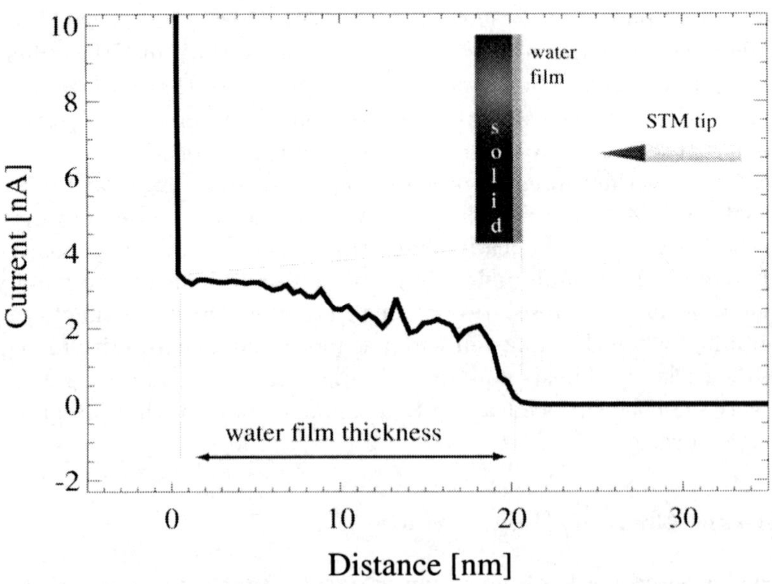

Fig. 6.3. Determination of the water film thickness by STM. Upon contact with the water film the tip induces an ionization current in the liquid. When the tip is in close proximity to the solid surface, the current increases exponentially.

contact with the water film [192] (see Fig. 6.3). When the tip contacts the solid surface, the tunneling current experiences an exponential increase. The distance between the onset of the ionization current and the exponential increase is interpreted as the water film thickness. After retraction of the tip the current drops beyond the onset point due to capillarity. Both the sample as well as the tungsten tip are covered with water. To remove the water from the tungsten tip, it can be heated before the experiment. However, this procedure yields limited success when the experiments are performed exclusively in air.

Upon direct contact with the solid and at sufficiently high voltages, material can also be removed by the tip [609,610]. This leads to an altered electronic configuration of the tip and, thereby, to unreproducible results.

6.1.4 Hydration Scanning Tunneling Microscopy

Another interesting technique with respect to living samples is Hydration Scanning Tunneling Microscopy (HSTM), where adsorbed water is used to achieve a tunneling current. Heim et al. [179] reported STM imaging of a bacterial surface protein layer. In air every surface (except a completely hydrophobic surface) is covered by several layers of water with a thickness of up to 1 nm according to [185] (see Sect. 2.5.4). After establishing contact, a thin capillary bridge connecting the surface and the tip is formed. Two different mechanisms of imaging were observed: (a) imaging with a tunneling gap and (b) imaging with a water bridge between the tip and the hydrated surface. It was shown that mode (a) is more stable because instabilities due to the jump-to-contact of the tip into the water layer and fluctuations of the capillary bridge can be ruled out. The effects depend strongly on the physical condition of the actual tip (adsorbates) and on the sample preparation. Due to repeated making and breaking of the water bridge some tips did not allow stable imaging at any voltage, while other tips showed stable imaging without current oscillations. Hydrophobic samples, for instance, lipid monolayers, make scanning difficult due to the lack of water [185].

Another technique that conserves the original (hydrated) state is accompanied by freeze-fracturing commonly used in electron microscopy (see Sect. 7.2.9). In 1997, STM imaging of fully hydrated bio-membranes was reported [611]. By using a metal replica, artifacts due to quantum electronic influences can be neglected, since the local density of states is isotropic. The resolution of this technique is about 1.0 nm laterally and 0.3 nm vertically.

STM imaging can be summarized as follows:

- STM works without contact with the material. It cannot be used to measure friction.
- Biological samples can be imaged fully hydrated.

- Freeze-substitution together with critical-point drying followed by metallization ensures isotropic tunneling conditions.
- Despite all these examples, STM has not found extensive application in biology [612].

6.2 AFM

6.2.1 Operating Principle

Based on the STM design, Binnig et al. [613] developed the first AFM to measure ultra-small forces (less than 1 nN) present between the tip and the sample surface. These small forces are measured by detecting the motion of a very flexible cantilever beam with a tip. The AFM combines the principles of the STM and the stylus profilometer. AFMs are the number one choice for biology [602,614–616]. The tip (generally made of silicon nitride or silicon) is utilized to measure topography [581,617–620], crystallography [621–625], chemistry and friction [626–630]. In addition, the elastic properties of soft samples can be determined [631]. The tips are several micrometers long and

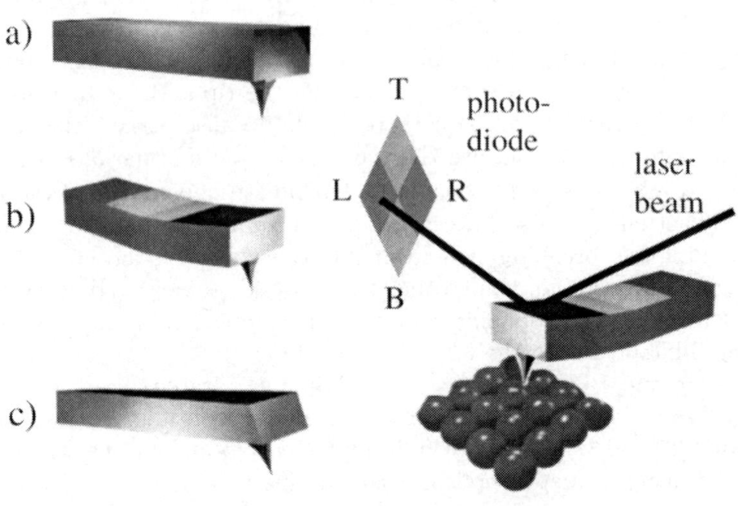

Fig. 6.4. Principle of AFM in which the detection of the vertical and torsional deflections of the cantilever are measured by different signals on the photodetector. **(a)** Undeflected cantilever. **(b)** Deflected cantilever when measuring topography. **(c)** Torsion of cantilever when measuring friction. The laser detects deflection and torsion as a sum-signal monitored by the photodiode. Reflection to the left (L) or to the right (R) correspond to friction, whereas up (T) and down (B) reflection describe the topography.

have a radius of 10 nm or larger. The tips of commercially available cantilevers are often sharp enough for high-resolution imaging of biological samples [632]. However, a significant improvement can be achieved with robust carbon nanotubes [633–635]. The force on the probing tip due to its interaction with the sample is sensed by detecting the deflection of the compliant cantilever with a known spring constant. This deflection can be measured by different detection systems: the electron tunneling method [613], optical interferometry, capacitance detection [636], the piezoresistive method and laser beam deflection [637]. The latter is the most commonly used detection mode in commercial AFMs. In this method, a beam from a laser diode is reflected from the cantilever surface to a four-quadrant photodetector. A reflection in the horizontal direction is interpreted as a friction signal, while the vertical reflection corresponds to the topography (see Fig. 6.4). It is obvious that the friction signal results from the torsion of the cantilever. Thus, a crucial part of force microscopy is the calibration of the cantilever [638]. A detailed description of the calibration can be found in [234,274]. It has to be emphasized that a proper calibration requires an exact knowledge of the cantilever geometry and of the mechanical properties, for instance the Young's and shear moduli (E and G), provided that the bulk values also apply to the properties and dimensions of the cantilever. Here, the authors only give the equations to calculate the spring constant for the vertical and horizontal directions:

$$k_z = \frac{E w_c t_c^3}{4 l_c^3}, \qquad k_t = \frac{G w_c t_c^3}{3 h_c^2 l_c}, \tag{6.1}$$

where w_c, t_c, l_c are the width, thickness and length of the cantilever, respectively. The value h_c is the height of the cantilever tip.

6.2.2 Contact mode

In direct contact the measured friction and topography signals are always a superposition of deflection and torsion (which also may introduce artifacts [639–641]), and there is no mathematical way to extract the friction or topography signals [602]. However, as in [642,643], by using special double-leaf spring cantilevers the torsion is suppressed and the friction signal arises solely from the lateral deflection.

In addition to the mechanical properties of the cantilever attention has to be paid also to the shape of the tips. Artifacts are introduced by the convolution of the tip and the sample. This effect becomes increasingly dramatic when the asperities have dimensions similar to that of the tip [644,645]. For quantitative AFM analysis a well-defined spherical apex is necessary to determine the correct values of, for example, the shear stress [646–648].

Figure 6.5 shows friction imaging of a bump in a silicon sample. While the topography image results alone from the up-and-down motion of the cantilever, the friction image can have several contributions. First of all mechanical interlock leads to torsion. This becomes obvious when the tip has

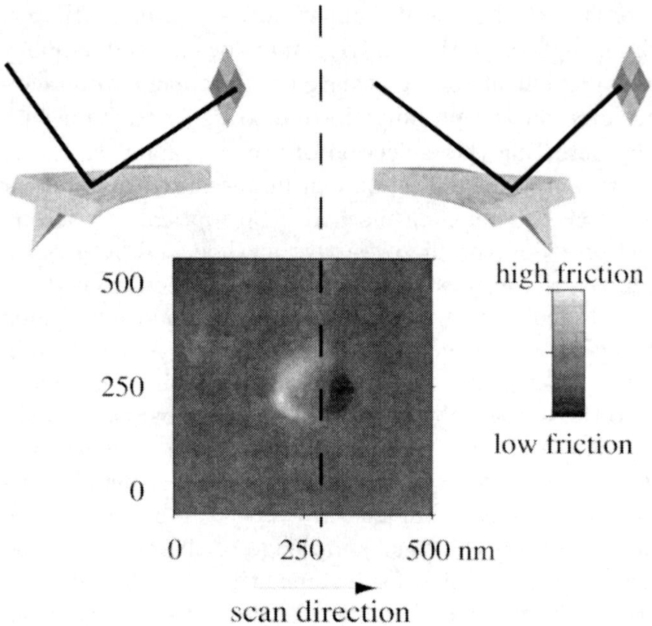

Fig. 6.5. Lateral force microscopy is based on the detection of torsion. Due to profile changes the tip is tilted (e.g., during the climb up the hill). In addition, a change in surface chemistry accompanied by different adhesive properties can lead to cantilever torsion.

to climb up an asperity [649]. Large friction is generated during motion towards the summit, while friction becomes lower during the down motion. The measured forces are typically in the range of nano- and piconewtons. In this range, besides macroscopic electrostatic and capillary forces, influences due to van der Waals or chemical interactions (bond formation) also contribute to the friction. Therefore, the friction image has always to be evaluated in conjunction with the topography image.

6.2.3 Tapping Mode

For topography imaging several types of SPM have been introduced over the last few years, allowing the imaging of a variety of different materials in vacuum, air and fluid for hydrated imaging. A first approach to reduce the large contact pressure (1 GPa for a force of 1 nN acting on an area of 1 nm^2 [601]) was undertaken by Digital Instruments. In the so-called Tapping Mode^{TM} the cantilever is operated in resonance. The distance to the sample is chosen in such a way that the tip slightly touches (taps) the atoms, leading to a shift in resonance frequency. In this mode the shear forces are decreased and the resolution increased. Despite the operation in resonance, the phase shift itself can also be evaluated [650,651]. In [652,653] the basics of the phase-imaging tapping mode AFM are described. The phase shift is quantitatively described by analytical expressions based on the harmonic approximation. The use of higher eigenmodes of the vibrating cantilever increases the sensitivity [654,655]. One of the main disadvantages of the tapping mode is that the whole mechanical base holding the cantilever must be set in vibration. Thus, when the sample is measured in a fluid cell, the liquid is also set into oscillation, affecting the quality of the images. A way to reduce unwanted oscillations is MAC-Mode SPM (MAC = magnetically-driven cantilever). Live cell surfaces, sub-cellular structures and lipids can be analyzed in their native environment or in different biological buffers. The cantilever, coated with a magnetic layer, is excited by an ac magnet underneath the sample. Similar results were obtained by photothermal modulation of the cantilever. A 2 mW laser is used to periodically heat the cantilever in solution. An amplitude of 8 nm to 20 nm can be used for imaging, as shown for high-quality images of DNA on mica [656,657]. It was shown that by oscillating a charged tip over a specimen the observation of the local charge distribution is possible by evaluating the phase differences due to charged regions [658,659]. This mode of operation can be applied to study the atomic origins of adhesion, since the spatial resolution is larger than with any other mapping method [660].

Experiments showed that non-destructive imaging was possible when the specimens were imaged with normal forces lower than 1 nN [661]. Higher forces result in the degradation of the material [662]. Thus, the ideal solution for soft biological samples would be a characterization without any contact. Morita et al. [663] formulated the conditions to achieve true atomic resolution with non-contact AFM. Equations for vertical and lateral resolution were derived and included in guidelines to achieve atomic resolution. A resolution of 0.6 nm laterally and 0.1 nm vertically was demonstrated for Si(111)–(7×7) [664]. With the help of mathematical data post-processing the resolution can be enhanced further by separation of attractive and repulsive force contributions [665].

6.2.4 Non-contact Mode

When extremely soft materials are involved, topography and morphology information can be obtained by cryo-AFM [666]. The AFM is operated in liquid nitrogen vapor under ambient pressure. The samples become more rigid, allowing the use of sharper tips. From a theoretical standpoint forces should be in the piconewton range to prevent damage [207]. Greater probe forces – increasing the resolution by decreasing the contrast, provided that the deformation remains elastic – can be applied for closely packed or cross-linked specimens [667]. In loading experiments the loss of biological activity due to mechanical interaction was demonstrated [668]. To strengthen the coherence of soft materials cross-linking agents can be applied [666] (see Sect. 7.2).

In order to image liquids with the AFM, electrically biased metallized tips are used. The feedback control maintains the electrostatic force with a tip-surface distance of tens of nanometers. This mode of operation is called Scanning Polarization Force Microscopy (SPFM) [89, 163, 177, 669]. The detected contrasts result from various contributions to the dielectric constant (e.g., solvated ions).

6.2.5 Force Modulation

In addition to topography and friction the micromechanical properties of soft materials can also be determined with the AFM. For this purpose the normal force has to be modulated to measure the storage and the loss modulus of a sample. The resulting image reflects the viscoelastic properties of the material [670]. The technique is very similar to the tapping mode. A frequency generator controls the modulation of the sample holder leading to periodic indentations of the tip. By means of the viscoelastic theory (see Sect. 2.3.3) the mechanical properties can be retrieved from the phase shift and amplitude of the cantilever response.

In summary:

- Due to its versatility AFMs are the most applicable tools in micro- and nanotribology.
- The AFM can be used to measure friction, adhesion and intermolecular forces as well as viscoelastic properties.
- Nearly all types of biological material can be probed [602].
- Most imaging problems arise from sample preparation.

6.3 Selected Results

6.3.1 Forces Between Tip and Molecule

As was shown in preliminary sections, SPM techniques are sufficiently sensitive to image molecular interactions [70, 614, 671–674]. A common way to measure those forces is to record a force–distance curve. Figure 6.6 shows a force curve as function of the distance between the tip and sample. At the beginning, the tip and sample are not in contact (**a**). The scattering of the curve is caused by mechanical noise. At (**b**) the tip contacts the sample. From here a straight line – according to Hooke's Law – is recorded as presented for the case of hard material. Shortly before contact with the hard surface, contact with the adsorbed water film is detected. As soon as the tip contacts the water film, capillary action pulls the tip towards the surface. This implies that the distance from the jump-to-contact to the hard-surface contact is a

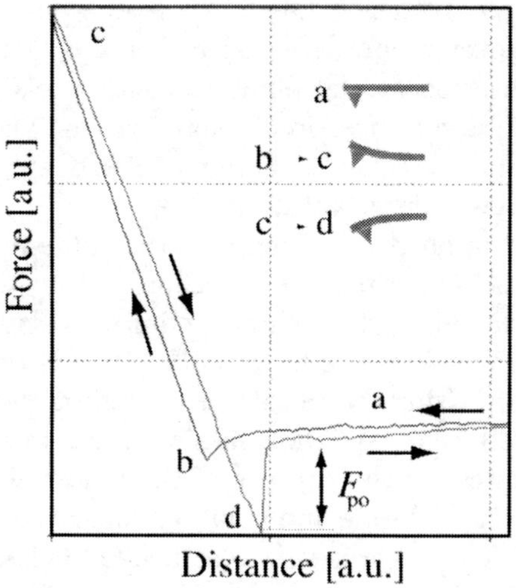

Fig. 6.6. AFM force–distance curve. While approaching the surface the force experienced by the cantilever is monitored. The force–distance curve shows increasing values when the tip is pushed against the surface. In the opposite direction adhesion holds the tip to the surface until the adhesive contact breaks. The negative force at (**b**) is caused by a pulling force due to capillary action.

measure of the thickness of the water film. At (**c**) the motion of the cantilever is reversed. After passing the zero-force line both the tip and sample remain sticking. At a certain point the back-driving force (deflection times cantilever spring constant) equals the force that holds the tip and sample together and the cantilever jumps back. This force is called the pull-off force and is treated as adhesion as described in Sect. 2.4.

Depending on the molecular conditions of the surface (i.e., adsorbates, fluid film) the sticking of the tip and sample is more or less pronounced. This simple and straightforward technique can be applied in air and in vacuum. In air the capillary action is hard to suppress. However, the negative impact on the scanning resolution can be reduced by hydrophobization of AFM tips so as to image at high humidities with decreased sample-tip adhesion [675]. In addition, the application of a lubricant significantly reduces adhesion as shown in experiments with flagellar filaments [676]. The lubrication leads to the hydrophobization of the tip due to hydro-carbon coverage.

The action of van der Waals forces can be reduced by an electric double-layer (see Sect. 2.4.2). Improved atomic resolution was obtained by force compensation due to a surfactant as a watery solution of $C_{16}TACl$ [198]. Under UHV conditions the capillary forces are replaced by van der Waals interactions or the formation of chemical bonds. In a UHV chamber adhesion up to 1000 nN was measured for a Si tip versus Si sample [677].

Despite the simplicity of the method, attention has to be paid to data processing. Because of the tilted cantilever mount, friction makes a contribution to the force–distance curves. Friction at the point of contact leads to the forward motion of the tip, creating an offset in the force curve. The magnitude of the offset can be in the same range as the action of short-range attractive interactions, causing difficulties in interpretation [678].

Force–distance measurements were applied to determine the adhesive force between proteoglycan molecules. A single pair exerts a force of up to 400 pN [70]. Unbinding forces were derived in [679]. The force required to separate the tip and a functionalized bead was found to be 160 pN $\pm$ 20 pN for biotin and 85 pN $\pm$ 15 pN for iminobiotin. Using the cantilever the interactive forces between single DNA strands were measured. The adhesion forces were in the range of 0.83 nN to 1.52 nN, caused by the rupture of the interchain interaction [680]. Forces between a protein molecule and a non-biological surface were measured using a modified AFM. The protein was attached to glass microspheres and the cantilever was then used to dislodge the sphere from hydrophilic and hydrophobic polystyrene substrates [681]. The cantilever was glued to a standard optical microscope where the displacement of the spheres was measured by pushing the sphere with the cantilever leading to a translation of 8 nm. It was shown that proteins adsorb better on hydrophobic than on hydrophilic polystyrene substrates. The forces were in the lower nN range.

Part of the force–distance curve is always the indentation. Here, the tip is pushed into the surface. Whereas hard surfaces show a linear response, soft materials are indented by the tip. Gaub et al. reported on the linear elastic properties of biomolecules by selective indentation [682]. The unfolding of single molecules (giant muscle protein, titin) by pulling shows a sawtooth structure with peaks of about 100 pN due to the breaking of interconnected globular domains.

6.3.2 Nanoscale Friction and Wear

Friction experiments were reported on RNA polymerase sliding [683] determining the frictional drag force by applying the two-dimensional Einstein–Sutherland friction–diffusion relation. The total friction force (translation and rotation) was measured at a value of 4.2 pN. Lieber et al. introduced a lateral force microscope based on chemical force microscopy [591]. It was shown that due to the softness of the biological samples shear forces dominate the chemical forces. When the shear forces are reduced, it was shown that the friction is sensitive to changes in the free energy of the surface [684].

When the forces are chosen high enough (45 µN), the AFM can also serve as a nano-extraction device. This was shown in [685], where after nanosampling and subsequent chemical analysis by mass spectroscopy the binding force of a biotin–avidin pair was determined. It was also shown that microdissection can also lead to wear represented by debris formation due to

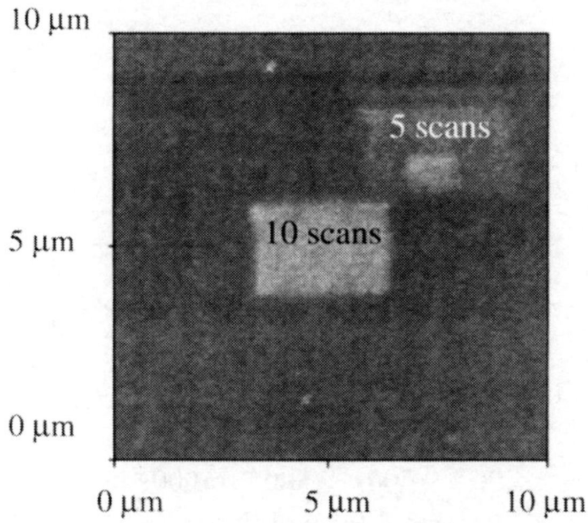

Fig. 6.7. Wear of a thin film. The tip of an AFM was used to scan a Langmuir–Blodget film at high normal force. After the scans the surface was imaged with the same tip at light load.

the cut after first contact [686]. When the forces are low (10 nN – 15 nN) AFM ensures wearless sliding.

Sliding of gelatin as a function of temperature showed a large contribution due to viscoelastic relaxation [687]. The dispersion as well as the mean value of the friction vary with temperature, underlining the statistical nature of friction on the nanometer scale. Higher forces are accompanied by wear, as shown in Fig. 6.7. The wear pattern was obtained with a normal force of 800 nN on a Langmuir–Blodget film of stearic fatty acid. Defect creation was also shown in [688] for VacA oligomers. By means of a line scan a total height of 10 nm was measured, corresponding to the film thickness of the organic film.

6.3.3 Nano-indentation

Nano-indentation is hardness measurement in the range of nanometers and nanonewtons (compare with Sect. 5.2.1). Nano-indentation experiments are performed with the AFM [689]. Similar to the force–distance curves used to measure adhesion [551], the surface is indented by the AFM tip. Very often the silicon or silicon nitride tips are replaced by diamond tips.

Whereas hard surfaces show the expected behavior according to Hooke's law, soft and brittle surfaces reveal a discontinuous indentation caused by the lateral motion of sample atoms [690]. For example, indentation was used to determine the compressibility of a cell wall. By subtracting the cantilever

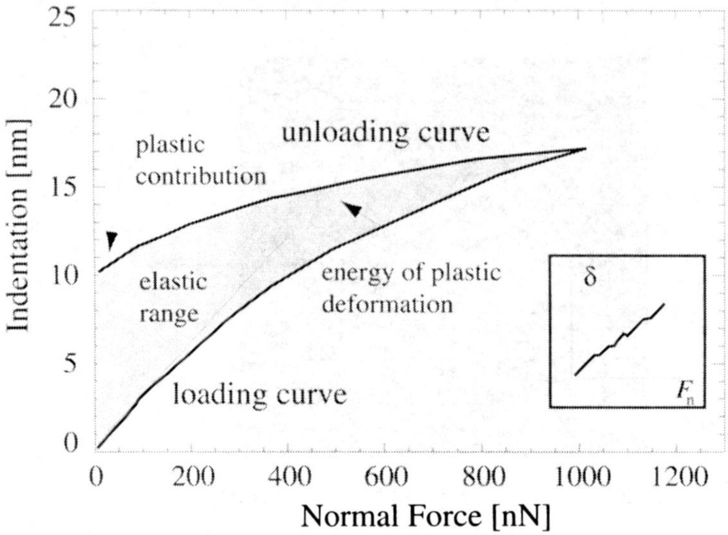

Fig. 6.8. Loading and unloading curves during indentation. The area between the unloading and loading curves equals the mechanical energy of plastic deformation. The inset depicts the indentation δ into a brittle surface. Due to the breaking of the asperities the curve shows steps.

deflection on a bacterium (*Magnetospirillum gryphiswaldense*) from the deflection on the hard substrate, the stiffness of the bacterial cell wall was determined (≈ 0.42 N/m) [616,631,691]. From the recorded force–indentation curves the local hardness, the rigidity and the modulus of elasticity of the sample can be determined. According to the model of Oliver and Pharr [692] the reduced elastic modulus of the sample material is calculated from the slope of the unloading curve, i.e., by dividing the rigidity of the sample material by the projected contact area of the probe (see Fig. 6.8).

7. Microscopy Techniques

To understand the microtribological properties of biological material, information about the surface sculpture and material design are necessary. Additionally, many biological frictional and adhesive systems are supplemented by various secretions delivered in the contact area. Such secretions are products of specialized cells. The biological sciences profited enormously from the development of microscopy techniques. Light microscopy, together with histological and histochemical techniques, aided in understanding the structure and functions of tissues and cells. Scanning (SEM) and Transmission Electron Microscopy (TEM) have contributed to studies at the sub-cellular level of biological organization.

Microscopy is used not only to visualize biological structures, but also to analyze the chemical makeup and physical properties of the bio-materials. Such microscopy is often called analytical electron microscopy. In the last two to three decades, various techniques associated with electron microscopy have been developed in order to prepare and evaluate tissues. These include autoradiography, freeze-fracture, freeze-drying, Critical-Point Drying (CPD), immunocytochemistry, electron diffraction, cryo-electron microscopy, elemental analysis, quantitative microscopy, high voltage microscopy, etc.

This chapter was written to give a short overview of useful techniques which can be applied for studies of biological surfaces, adhesives and friction-active biological materials. The microscopy techniques described are well-known to biologists working on the structure and ultrastructural design of biological materials. However, most tribologists are not familiar with them. Most of the reviewed techniques have been successfully applied by the authors during their own structural and experimental work. More detailed information on microscopy techniques can be found in specialist books on microscopy [693] and histochemistry [694]. This chapter is divided into three parts. In the first part, the different types of microscopy are listed. The second part gives some general insight into the preparation procedures. The third part provides several techniques useful for surface characterization and understanding of material design.

The diversity of modern microscopy techniques is immense. For beginners, it is often very difficult to find an appropriate way to answer all the questions that are asked. In Fig. 7.1, a coarse schematic of possible ways of processing

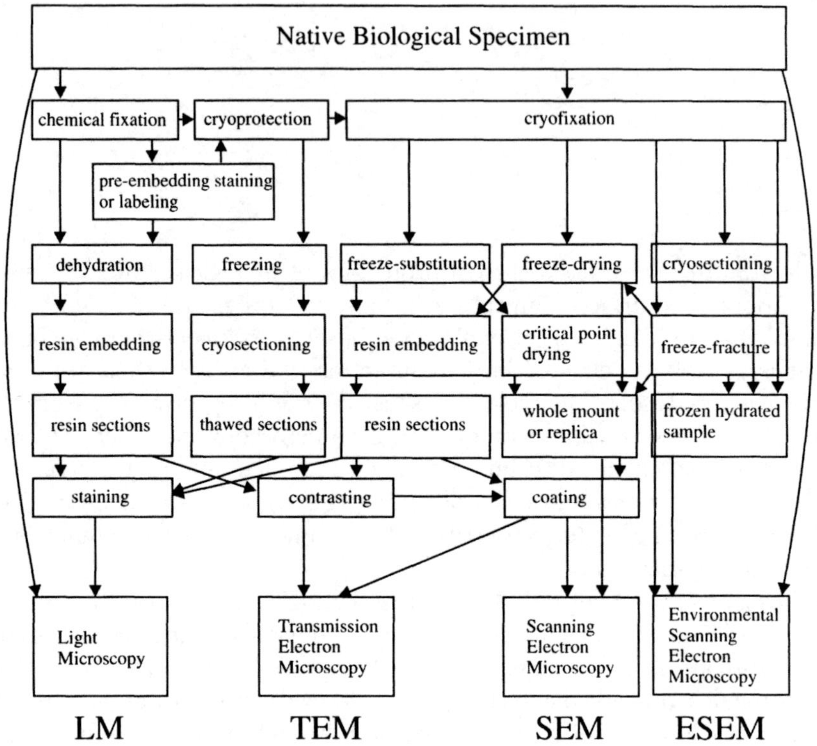

Fig. 7.1. Pathways for specimen preparation for different types of microscopy.

biological samples is provided (the figure was adapted from [323]). In further sections, the authors explain which methods (or sets of methods) can be applied for particular tasks.

7.1 Principles of Microscopy Techniques

7.1.1 Light Microscopy

The general rule of any type of microscopy is that the image formed in a microscope does not exactly correspond to the original object. Therefore, several techniques are usually necessary to obtain different information on the object under consideration. There are two important parameters, resolution and magnification, which are used to quantify the performance of a microscope. The resolution is the smallest degree of separation at which two minute details can still be distinguished from each other. In other words, the resolution characterizes the resolving power of a microscope and is, therefore, the most important criterion to describe the microscope. Magnification is the ratio of image size to object size. It is actually the scale by which the object is related to the image dimensions. Above a scale range, the resolution is proportional to the magnification. However, over a certain magnification, there is no improvement in resolution. This means that any further enlargement provides no new information about the object.

Light microscopy uses visible light and its interactions with the sample for object imaging. Since the lowest wavelength limit of the visible light is about 500 nm, this becomes the most important constraint for image resolution. The next constraint is the aperture of the objective and condenser lenses. The objective aperture is a measure of the size or angle of the cone of light delivered by the illuminating condenser lens to the object plane and of the cone of light emerging from the object that is collected by the objective lens. Light passing through fine details of the object is diffracted into directions different from the original direction of propagation, and the finer the detail involved, the wider the angle over which this light is diffracted. Therefore, the aperture can be defined as a measure of the ability of the microscope to collect diffracted light from fine details of the object. Since the depth of focus is usually quite small in light microscopes, the objects have first to be dissected. The second reason for sectioning a sample is that some biological objects contain dense structures or coverages.

The transparency of most components of biological tissues makes the usefulness of this instrument largely dependent upon the fact that differences in absorption can be created by histological staining. Staining techniques make it possible to distinguish components that would otherwise escape detection. There are several types of light microscopy, such as phase contrast, polarization, interference, and fluorescence microscopy, which are modifications of light microscopy. Each of these modifications has its own advantages for special purposes and also its limitations. Light microscopy is often properly called bright-field light microscopy.

Even for highly transparent, unstained objects, structural information is available in form of phase differences. Phase contrast, interference, and polarized microscopy, are used to convert such phase differences into contrasted intensity differences. These methods are based upon molecular differences of biological media, which influence the index of refraction. The index of diffraction is a measure of the rate of propagation of a light wave through a given medium, and is defined as the ratio of the velocity of light in vacuum and the velocity of light in the medium. Light passing through an object of higher refractive index will be retarded in comparison with light passing through a surrounding medium of lower refractive index. Actually, the total retardation of the light wave is a result of the index of refraction and the thickness of the object.

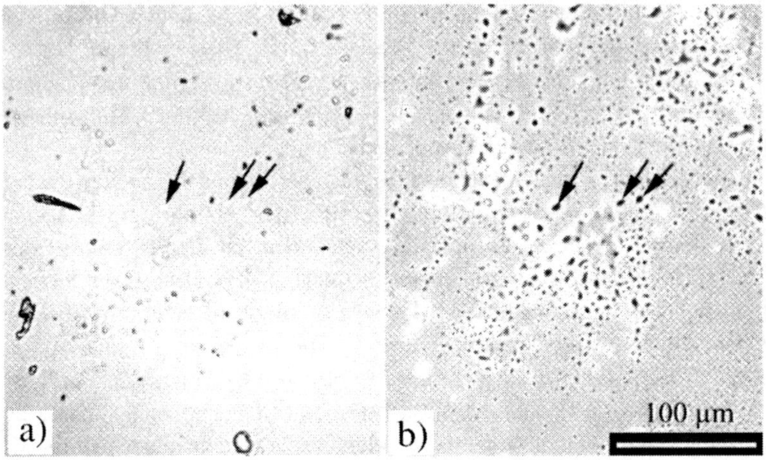

Fig. 7.2. Footprints of the fly, *Calliphora erythrocephala*, on a glass surface as they are viewed in the bright field of the light microscope (**a**) and in the phase-contrast microscope (**b**). The secretory droplets are involved in the fly attachment to smooth surfaces.

7.1.2 Phase-Contrast Microscopy

This modification of light microscopy is capable of converting small differences in refractive index, not detectable by the eye, into intensity differences that can be detected. The basic principle of such a microscope is that light coming from areas corresponding to transparent regions of the specimen with an index of refraction different from that of their surroundings is out of phase with the surrounding light. The bright-field light microscope is not sensitive enough to such phase differences in the image plane of the microscope and shows, therefore, minor differences in intensity between these regions with different optical properties. The phase-contrast microscope is equipped with special apertures and with absorbing and phase-shifting plates. These devices are used to separate diffracted light from the surrounding light and to shift the phase and intensity of the diffracted and surrounding light according to the way they interact in the image plane to produce intensity and contrast differences. In the phase-contrast microscope each objective is designed to work optimally for only a limited range of retardations. The major drawback of the microscope is that halos often occur around the structures because of incomplete separation of the surrounding light from the diffracted light. This technique is very useful for studies of thin layered patterns of fluids, such as insect adhesive secretion (Fig. 7.2).

7.1.3 Interference Microscopy

In the interference microscope, two separate beams of light are sent through the specimen to observe the same refractive details as in the phase microscope. The two projected beams are recombined in the image plane. The first one is focused through the details of the specimen. The other one, called the comparison beam, is focused in an area located above or below the observed details. This device allows the direct quantitative measurements of both retardation and related values of the index of refraction and structure thickness. Such a measurement ability can be used for evaluations of a volume per unit area and for contour mapping of small objects. Nomarski optics or differential-interference microscopy is a modification of interference microscopy, which provides the most excellent results in the imaging of fine transparent details due to the detection of gradients of the refractive index.

7.1.4 Polarization Microscopy

Since most friction-active biological materials, such as cartilage, muscle-connective tissues, vertebrate skin and invertebrate cuticle are natural fibrous composite materials with a preferred orientation of fibers [695], polarization microscopy can serve as a useful tool to study the anisotropy of these materials. When a beam of plane polarized light meets such an anisotropic object, it acts as if it is divided into two beams polarized in planes perpendicular to one another. These beams have different velocities and such a specimen is called birefringent. Objects with a high degree of molecular orientation have a different refractive index depending on the plane of the oscillation of the beam of polarized light. In this case, the refractive index relates to the orientation of the molecules. This technique makes it possible to detect the presence of an ordered molecular arrangement, even if the molecules are not directly visible. Through this, information about structural arrangements at the molecular level can be obtained. For example, in the insect cuticle, where fibers are arranged parallel in one lamella and rotate in the successive lamellae, the preferred orientation of fibers in lamellae can be detected and the period of rotation in the successive lamellae can be measured [519].

7.1.5 Fluorescence Microscopy

In this technique, light of one wavelength is used to illuminate the specimen. The image is the result of emission from molecules that have absorbed the primary exciting light and then re-emitted light of a longer wavelength. A variety of filters that can be placed below the condenser and above the objective provide the desired secondary emission, which has to be detected. In fluorescence microscopy, the specimen looks as if it were incandescent (self-illuminating). The fluorescence microscope requires white-light sources (xenon lamps) and special objectives.

The secondary emission can originate from natural fluorochromes located in the specimen under consideration. For example, resilin, an elastic protein of the insect cuticle, has such an auto-fluorescence at ultraviolet wavelength emissions, convenient for detection by this method [696–698]. In recent decades, fluorescent dyes bound to certain specific components of the sample or those coupled to specific antibodies have been used to localize a wide range of natural substances. This technique demonstrates the highest specificity of detection and the best contrast in comparison with other light microscopy techniques. It is ideal for studies of natural composite materials, in order to show material composition and an arrangement of different composites in the material.

7.1.6 Transmission Electron Microscopy

In this system, the light source is replaced by electrons emitted by a filament. TEM, to some extent, is analogous to light microscopy. The effective wavelength is determined by the voltage by which the electrons are accelerated. Controlled magnetic and electric fields work as a kind of lens. A fluorescent screen is used as a detection device. An image can also be captured on light-sensitive films or plates.

The whole system requires high vacuum, high voltage for its functioning and mechanical stability to prevent image artifacts caused by vibrations. Additionally, TEM requires electron-transparent sections, which have to be one or two orders of magnitude thinner that those used for light microscopy, as well as special treatment and preparation of the samples (1 μm thick sections can be explored). In spite of high complexity and costs, such a system provides structural information which is impossible to obtain by the use of other, even the most modern, methods. This device, which is able to work at magnifications up to 100,000 with increasing resolution, provides another dimension in studies of biological materials [693]. The overall limit of resolution is about 0.5 nm. The main problems of resolution are caused not by limited wavelengths (in the range of 5 pm), but mainly by imperfections of the magnetic and electrical fields used in the electron-beam control. In addition to dissected material, whole thin biological specimens, such as butterfly scales or insect wings, can be studied directly without previous dissection.

7.1.7 Scanning Electron Microscopy

This technique is completely different from light microscopy and transmission electron microscopy. In the previous techniques, the image was created directly in an image plane, whereas in the SEM the image is formed indirectly by sampling information from the specimen surface point by point. The sample, which is located in the column under vacuum conditions, is bombarded with a precisely focused electron beam of about 10 nm diameter. This beam scans the large area of the specimen or the whole specimen in a raster pattern. A cathode-ray tube on the display panel is synchronized with the electron beam in the column and simultaneously scans the screen of the display panel. The result of this is that each point on the cathode-ray tube corresponds to a point on the specimen. A detector placed near the specimen measures the Secondary Electrons (SE) emitted from each point bombarded by the beam.

When the surface of a specimen has irregularities, the intensity of the secondary emitted electrons will vary with the angle at which the electron beam meets each point of the surface. The brightness of the cathode-ray tube is modulated proportionally to the intensity of the SE emission. This will result in an imaging of the specimen point by point on the fluorescent screen of the cathode-ray tube. This image can be recorded by a built-in camera. For the past two years, the analogous information on the intensity of the SE emission has been digitized and used to build an image on a computer screen. This computer-aided imaging offers immense possibilities for further use of the information obtained from the microscope.

Since the beam of the SEM does not penetrate the specimen, even thick samples without dissection can be used. If compared with the TEM procedure, sample preparation for SEM is relatively simple. It usually requires mounting of the dry sample on SEM metal holders and coating with carbon, gold, chrome, palladium or other metals. The second big advantage of this device is its great depth of field, such that it was called a 3D microscope in the early years of the SEM technique. Indeed, by merging images (stereo pair) of the same object at two different tilt angles (usually 6°–8°), one can obtain a three dimensional effect. The resolution of the early SEM was not comparable with that of the TEM. However, in the last 10–15 years, new SEMs have become capable of a resolution down to 0.5 nm. Since the SEM techniques, in general, provide detailed information on surface topography and may also be used in observations of the ultrastructural architecture of biological materials, they are one of the most important tools for studies of friction-active and adhesive biological materials, where surface characterization becomes very important.

Generally, the SEM may be operated in two major modes: high resolution and great depth of field. High resolution demands a small, coherent spot with minimal aberrations and an adequate signal-to-noise ratio. Great depth of field may be achieved by using small apertures, long working distances, and lower magnifications. Since resolution is determined by these two conditions, the spot size may be increased to obtain a better signal-to-noise ratio. These two modes are used to view specimens with two extremes of topography: the first for low-sculptured specimens and the second for highly-sculptured specimens. The SE-detector, which is used for the detection of secondary electrons, reveals information about the topography of an object. Additional information can be obtained from Back Scattered Electrons (BSE), which are elastically scattered electrons with energy levels comparable to the primary electrons of the beam. Because of their high energy, they may be scattered from a greater depth of the specimen. As they interact with the atoms of the specimen, they may generate additional secondary electrons that are some distance away from the original secondary electrons generated by the same beam spot. The backscattered electrons may not be detected by the SE-detector because of their high velocities, and only a special BSE-detector can be used for this purpose. An interesting property of the BSE-detector is that elements with higher atomic numbers appear significantly brighter in the BSE-images than lower atomic numbered elements. Such a property can be used for studies of immuno–gold labelled biological tissues, where the small gold particles (16 nm–18 nm in diameter) can be detected in the BSE-mode of the SEM together with the surface topography. The BSE-technique was successfully applied for studies of porous biological material by the use of uranyl-acetate infiltration at one side of the sample and by detection of uranium ions at the other side [317].

The most important disadvantage of the SEM technique is that the sample has to be fixed, dried out, frozen, and coated. So-called Environmental SEM (ESEM) allows the observation of freshly killed or even live specimens at pressures of about one million times higher than in the conventional SEM. It operates in the standard range of accelerating voltages (0 kV–30 kV) with a resolution of 1 nm–10 nm at a pressure of 6 mbar. The main advantage of the ESEM is that it is differentially pumped so that the gun area has a vacuum in the 10^{-6} mbar range, which is common in most SEMs. The specimen chamber remains at the increased pressure, which is essential for studies of freshly hydrated tissues.

In summary:

- By means of light microscopy the information about inner structures of biological material, its architecture and chemical composition can be obtained. Some applications allow the study of biological substrates in their natural, i.e., hydrated state.
- Image improvement and information about the chemical composition of the biological tissue can be obtained by sophisticated staining techniques.
- TEM is a method to obtain information about the inner structure of a sample with highest resolution.
- SEM images the surface with high resolution and no limitation concerning the depth of the structure. However, the BSE mode of the SEM can provide some information on the material composition.
- Usually, both TEM and SEM require vacuum conditions and dehydrated samples to operate properly. However, ESEM operates with low vacuum and allows imaging of highly hydrated samples at high resolution.

7.2 Preparation Procedures

7.2.1 Fixation

As was shown above, most microscopy techniques work with dead, chemically preserved (fixated) or frozen tissues. Light microscopy requires specimens cut into thin slices, called histological sections, with a thickness between 1 μm and 1000 μm. Such sections are transparent in visible light and can be stained with various dyes in order to increase the contrast of single components. In fact, virtually every step of the preparation procedure affects the quality of the final result. Tissue preparation for microscopy can be divided into the following steps: fixation, washing, dehydration, infiltration with transitional solvents, infiltration with embedding media (wax, paraffin or various resins), embedding, and cutting. The process begins with the living tissue and ends with a tissue that is virtually water-free and preserved in a static state in an artificial medium. This medium infiltrates the tissue, replacing water and makes the tissue firm enough for cutting. There are thousands of preparation road maps (protocols) for both light and electron microscopy. In this section, only the standard methods are given. Some variations useful in studies of friction-active biological materials are given in Sect. 7.3 and Sect. 7.4.

The first goal of specimen fixation is to interrupt the dynamic processes of the living tissue as quickly as possible and, by this means, to stabilize structures with a minimum of change. In other words, an ideal fixative must also stop destructive autolytic processes at the time of fixation. Since there is no ideal fixative, one usually decides on the fixative giving an optimal result with the desired type of tissue. There are two basic types of fixation: the chemical fixation and freezing. As chemical fixatives, 70° ethanol, formaldehyde, and their mixtures with picric acid and acetic acid have been used [699]. Some fixatives tend to build coarsely clumped material in the tissues. Therefore, the best fixatives are those that produce very fine-grained precipitates. Currently, glutaraldehyde is the most frequently used fixative in both light and electron microscopy. Electron microscopy usually requires a two-step protocol with pre- and postfixation. As a primary fixative, an aldehyde (formaldehyde, glutaraldehyde or their mixture) is used. As a secondary fixative, osmium tetroxide solution is commonly used. As a fixative glutaraldehyde has an ability to cross-link proteins by terminal aldehyde groups. Because proteins are present in most biological tissues, the use of glutaraldehyde fixation will cross-link soluble proteins to each other and connect them with skeletal and membranous compounds of cells and tissues. The result of glutaraldehyde fixation is a meshwork held together by glutaraldehyde molecules. The rate of penetration of this fixative into tissues is very slow. Therefore, to obtain rapid immersion fixation, the immersed tissue should be less than 1 mm thick.

Osmium tetroxide reacts primarily with lipid components of tissues. Many cell and tissue components, such as fibrous proteins of the extracellular matrix, can be well stabilized by osmium. In electron microscopy, osmium is an additional component of the stain, because of its high molecular weight. It further functions as a mordant, a substance which is a kind of precursor for later stains or dyes. In the case of TEM, osmium is a mordant for enhancing lead staining. The penetration rate of osmium is even lower than that of glutaraldehyde. However, due to its high volatility, it can be used for “smoking” of initially dry biological samples. As an additional mordant, tannic acid is sometimes used in protocols for natural fibrous composites, such as microfilaments and connective tissues [528]. Presumably, it enhances later staining with heavy metals.

Since plants are often covered by epicuticular waxes, and plant cells are surrounded by a cellulose cell wall, the fixation of plant tissue is more difficult than that of animal tissue. In addition, the low protein content of plant tissue does not allow the building of cross-links between them. Plant vacuoles can dilute fixatives, which causes a decrease of the fixative concentration. Therefore, in "plant protocols", higher concentrations of fixatives and longer fixation times are used compared with those used for most animal tissue. Insect and nematode cuticles and vertebrate bones are animal tissues that tend to have similar requirements.

The quickest possible fixation is to freeze samples in liquid nitrogen or in a propane jet cooled by liquid nitrogen. However, this method has some limitations because of crystal formation within the tissue. The TEM application of this technique is currently improved by the use of cryofixation devices capable of sample fixation at low temperature and high pressure. The results obtained in cryofixation protocols provide more detailed information about lipid-containing tissues and organelles. Most histochemical protocols basically use numerous modifications of the cryofixation technique [694]. This method avoids the extraction of lipid and other constituents in the process of dehydration. It also has the additional advantage of its great speed. The frozen specimens can be immediately dissected by the use of the cryotome or ultracryotome or further chemically fixated. The last modification of the method is called freeze-substitution, where the replacement of ice by alcohol or other solvents is carried out at very low temperatures [700, 701].

7.2.2 Embedding

The washing procedure after every fixation step is extremely important, because it eliminates free unreacted molecules of a fixative that remain within the tissue. Dehydration follows the last washing procedure after fixation. It is a process aiding the replacement of water in biological specimens with a fluid having the ability to be mixed with water and with hydrophobic embedding media as well. The main dehydrating agents are usually ethanol and acetone. The water is replaced gradually in a series of 50%, 70%, 80%, 96% and 100% ethanol–water solutions. Finally, a solvent capable of dissolving the embedding medium (100% ethanol, acetone or propylene oxide) is used. To obtain sections that are sufficiently thin for microscopy examination, it is necessary to infiltrate the tissue after fixation, with a solution of gelatine, paraffin, celloidin, resin or other plastic materials, which are able to solidify so that the tissues and embedded matrix may be sectioned as a whole. The next step is the infiltration of transitional solvents mixed (usually in proportion 1:1) with an embedding medium. Gradually, the resin- or wax–solvent ratio is increased until pure resin or wax is used. The pure resin specimens are transferred into molds, containing the liquid resin to become polymerized. After polymerization, the embedding medium solidifies, which can be dissected by metal, glass, or diamond knives.

As the embedding media, epoxy resins, such as Araldite or Spurr, are used mostly for light and electron microscopy. Wax and celloidin are suitable for light microscopy only [699,702]. Embedding with celloidin, which is used at room temperature, disturbs the specimen less and results in lower deformation of the specimen than both paraffin and wax embedding procedures, which expose tissue to the higher temperatures necessary to keep the wax molten during the infiltration procedure.

7.2.3 Sectioning

The method of sectioning is usually dependent on the embedding medium used in the previous step. Tissues embedded in paraffin or wax may be sectioned rapidly, with a microtome, at a thickness of 10 μm–60 μm. Such sections are used for light microscopy and are called thick sections. Sections with a thickness between 0.5 μm and 10 μm are usually called semi-thin sections. These sections may be stained by any of a number of combination of dyes, which are able to color different tissues or tissue components in an unspecific way, i.e., more or less selective. When a freeze-substitution protocol is applied to the frozen samples, they can afterwards be embedded and cut for light or electron microscopy.

Alternatively, frozen sections may be cut in the frozen condition by the use of a cryomicrotome at a thickness of 10 μm–30 μm or an ultracryomicrotome at a thickness of 60 nm–1000 nm. Ultracryomicrotomy has several advantages, for instance, the short duration of the whole procedure, which can be minimized to several hours. However, the main reason for using cryoultramicrotomy is the possibility of being applied in immunohistochemical protocols and in elemental analyses of substances or ions by x-ray analysis. In many cases, this is the only way to localize some particular substances. Further information about these methods can be obtained from special monographs [366,703]. Resin-embedded sections can be cut with a metal knife at a thickness of 10 μm–60 μm, if the soft mixture of resin is used (Durcupan Fluka) [704]. The hard mixture embeddings are usually sectioned with a glass knife at a thickness of 1 μm–10 μm. Semi-thin sections, after the staining procedure, can be observed under a light microscope.

Ultra-thin sections for TEM (40 nm–100 nm) can be sectioned with a glass or diamond knife using an ultramicrotome. Thicker sections do not allow electrons to pass. Ultramicrotomes are mainly equipped with a thermal advancement mechanism to advance a specimen in precise, repeatable steps toward the knife. The cutting technique is tedious and requires much practice compared to the common microtomes. The ultramicrotomy procedure includes the following steps: (1) trimming the resin block containing an embedded sample; (2) fracturing with glass knives (in the case of a diamond knife this step is omitted); (3) preparing specimen support grids or slot grids by coating them with a formvar film; (4) sectioning of the sample in the ultracryotome; and (5) picking up the sections with a grid. Each of these procedures has, in every laboratory, some particular tricks to improve the quality of sections or to accelerate the sectioning process. Some special types of tissue can, for example, require special slot grids (or single hole grids) allowing the greatest viewing field of the sections.

7.2.4 Histological Staining

Since most living tissues are transparent, details of their inner structure cannot be observed without applying one or more biological stains. Most staining protocols have developed empirically. More or less unspecific stains, like haematoxylin, eosin or toluidine blue just increase the contrast of the whole section. To demonstrate other constituents of fixed cells, tissues, or extracellular materials, special staining methods are used. For example, fatty components can be demonstrated if fat solvents are avoided in fixation and dehydration. Fats are colored with sudans (black or red) or nile blue. The interplay between both the fixation type and the staining type can often influence the overall result of staining. The mechanism of stain binding is not always clear. However, in many examples anionic and cationic stains form salt linkages with ionizable radicals of biological substances. These constituents are amphoteric because they can ionize either as bases or as acids depending on the pH value and several other environmental conditions. The type of protein behavior (acid or base) depends on its overall charge, i.e., the net sum of all positive and negative charges, at the pH value, at which the staining is carried out. Every protein has its isoelectric point, namely a pH value at which the number of positive charges equals the number of negative charges. At its isoelectric point a particular protein is stained poorly, but above and below this pH value a reasonable staining can be obtained. Above the point, the protein will bind basic dyes, such as methylene blue or basic fuchsin, whereas below the point, it will bind acidic dyes, such as eosin, orange G or light green [699]. Generally, proteins differ to a wide extent in their amino–acid composition and, therefore, in their isoelectric points. Depending on the value of the isoelectric point, diverse tissue components tend to be preferably stained by basic stains (basophilic components) or by acidic stains (acidophilic components). However, only a limited number of staining procedures involves electrostatic (salt) linkages between dye and tissue.

7.2.5 Histochemistry

Histochemistry is a special application of chemical and/or physical methods to histological preparations that allows identification of chemical substances in tissues or cells. The challenge of histochemistry is to localize a chemical process and provide information, which cannot be obtained from common biochemical procedures, by the use of tissue homogenates. Chemical methods permit identification of lipid classes, staining of carbohydrates or the localization of numerous proteins. Physical methods are restricted to the emission of ionizing particles from unstable isotopes incorporated in tissue components (autoradiography), absorption of x-rays of selected wavelengths by various elements (historadiography), emission of x-rays of specific wavelengths by elements bombarded by an electron beam (electron microprobe analysis).

As previously mentioned by [702], an ideal histochemical procedure (1) should not change the position of substances under consideration during the preparation procedure; (2) should be specific for the substance studied; and (3) should be able to intensively stain the reaction products, making it possible to visualize them under a light microscope.

Among physical methods, autoradiography (or radio–autography) is most widely used. It also gives good localization of specific substances. The first step is to inject a substance with a radioactive isotope label into a living tissue, which then binds to some targeted areas of tissues. The tissue sample, containing radioactive material, is sectioned and coated with a photographic emulsion for a period of days or weeks. After the development process, the emulsion can be observed in a light microscope. Labelled areas appear dark, because the black spots of the developed emulsion correspond to the location of radioactive substances. A highly sensitive method for localization of specific proteins of polysaccharides in tissues is the fluorescent antibody technique. This method uses the fact that the body recognizes foreign protein substances, called antigens, and produces highly specific molecules (antibodies). After binding a fluorescent dye to such an antibody, an ideal stain with a unique specificity can be created.

7.2.6 Contrasting Technique for TEM

In the light microscope, image contrast is achieved by various stains which color some particular details of a biological structure more than others. Since in an electron microscope no visible light is used, no color exists in the image. In TEM, the image contrast is generated by the differences in the kinetic energies of electrons transmitted through the section. The image contains, therefore, areas ranging from very bright to quite dark. The darker areas are associated with specimen areas having a higher density, whereas brighter areas come from regions of lower density. Dense areas of the specimen can absorb electrons to a higher extent so that such electrons are subtracted from the resulting image. Normally, unstained biological sections have a contrast, that is not sufficient to permit recognition of structural details.

To increase the contrast of ultra-thin sections, several contrasting substances are used. All these substances are salts of heavy metals. As mentioned above, the postfixation procedure, which usually involves osmium tetroxide, can increase image contrast and can set a kind of precursor for further staining. This substance reacts selectively with particular constituents of the biological tissue. The two most often used contrasting substances are lead-citrate and uranyl-acetate. Both types of stains are non-specific stains. In this section, only information on the positive contrasting method is given, when the density of a specimen is increased in contrast to the relatively electron-transparent background. There are two alternative methods for applying contrasting substances to sections: pre-embedding and post-embedding. In the pre-embedding method, tissues are immersed in a buffered 0.5%–1% solution of uranyl-acetate. In this case, a staining reaction takes place prior to the embedding procedure. In the post-embedding method, trimmed specimen blocks or sections on copper grids are contrasted. Usually, the post-embedding staining is a two-step procedure. The first step is uranyl-acetate staining and the second one is lead-citrate staining. Lead-citrate provides a high contrast of sections, but is easily precipitated while in contact with CO_2. For this reason numerous techniques exist for carrying out the staining procedure in a low CO_2 atmosphere or for preparing CO_2-free water. Between the two steps of staining and after the final step, the specimens must be carefully rinsed in a jet of bi-distilled water and, finally, air-dried prior to examination.

7.2.7 Critical-Point Drying

As mentioned above, conventional SEMs work mostly with dry samples. Since most biological substances are relatively soft, they can shrink or collapse during air-drying. Such drying artifacts can influence the results of studies of biological surfaces. The most frequently used procedure to prevent surface deformation of soft samples is critical-point drying.

The Critical-Point Drying (CPD) procedure is a way by which water can be removed from biological tissues. It depends on the fact that above a certain critical pressure and temperature both liquid and gas phases occupy the same volume. At this point there is no longer any surface tension and the liquid phase in the tissue can pass to the gas phase without causing distortion. Since the critical-point for water is at 324°C and 22 MPa, it becomes necessary to replace the water from the tissues first by another solvent with a more appropriate critical-point. The most commonly used critical-point drying liquid is CO_2, which is normally used after ethanol or acetone dehydration.

In the first phase of the CPD procedure the dehydrant (ethanol or acetone) is repeatedly replaced with a CPD liquid (transitional fluid). This procedure is usually carried out in a pressure chamber corresponding to the pressure needed to keep the CPD liquid in a liquid state. In the second phase, the chamber with a sample flowing in a CPD liquid is completely sealed and the temperature is increased. The heat increase results in an increasing pressure inside the chamber. The transitional fluid reaches a critical-point after which the specimen dries out without surface deformation. After achieving the critical-point the heat has to be maintained at the critical temperature to prevent condensation of the vapor back into the liquid. The dried specimen can be mounted on the SEM stub or first fractured or cut with a razor blade to expose the inner structure.

The CPD procedure can be successfully applied to soft-bodied animals, plants, and their parts. Very good results have been obtained for cell cultures and semi-thin resin sections with removed resin, and for macerated biological composites, such as vertebrate connective tissues and insect cuticle (see below). To minimize damage to specimens during the CPD procedure, different special holders are used. For pieces of tissues, the most useful are porous teflon tubes, which allow good exchange of fluids and gases.

Some types of tissues, however, may exhibit significant shrinkage, ranging from 10% to 60%. Impurities in the transitional fluids may cause extraction of foreign materials on the surface. The first problem can be partially solved by a freeze-substitution procedure prior to the CPD. The second problem can be solved by checking a control specimen dried under air conditions.

7.2.8 Freeze-Drying

The second method for drying samples without forming deformation artifacts is to freeze tissues rapidly and to dehydrate them at low temperature in high vacuum. This procedure is called freeze-drying and has a lower shrinkage rate compared to the CPD procedure described above. The frozen specimen is transferred to a chamber capable of maintaining temperatures below $-80°C$ and a vacuum at a pressure of 10^{-4} mbar–10^{-3} mbar. Ice will sublimate to vapor and will be continuously removed by the vacuum system. As a rule, fresh tissues are not used for freeze-drying, but those previously dehydrated in ethanol are used instead. Usually, extremely rapid freezing rates are required to avoid crystal formation that can lead to tissue damage. At high freezing rates water undergoes a change into a solid amorphous state.

The specimen is maintained under vacuum/cold conditions as long as it is necessary for drying. Normally, it takes from several hours up to several days. After drying the specimen is gradually warmed up to room temperature, mounted on a SEM holder and sputter-coated with gold, palladium, carbon, etc.

7.2.9 Cryofracturing and Freeze-Fracturing

For many structural studies it is important to obtain some insight into the inner three-dimensional design. There are several methods to do this. The first one is to dry tissue and fracture it. This method can only successfully be applied for relatively hard, stable structures such as tanned arthropod cuticle or dried wooden samples. According to an additional method, called cryofracturing, the tissues are frozen and fractured at low temperatures. Then the sample has to be dehydrated, dried according to the most appropriate drying procedure described above, mounted on a SEM stub, coated, and observed in SEM. The latter method demonstrates good results also with soft tissues, cells and even with sub-cellular structures.

Freeze-fracturing is another method for examining the fractured surface of specimens in TEM. In fact, this is a replicating technique providing 3D information on the ultrastructural architecture of tissues and cells. An initial procedure is applied to fixate small pieces of tissue in 0.5%–3% glutaraldehyde and, after a washing procedure, infiltrate them with 20%–30% glycerol serving as a kind of cryoprotectant. To break the frozen sample into two halves, a special freeze–fracture apparatus is needed. The apparatus consists of a chamber, mounting table with the sample, movable knife (razor blade) and electrodes for Platinum–Carbon Coating (PCC). The knife fractures the ice surrounding the sample in a random manner. When the fracture reaches living tissue, the tissue splits into two complementary surfaces (faces) with their back sides embedded into the ice. Usually, the fracture is situated close to the most stable structures. In the case of cells it is the cell membrane. In the case of integument it can be the border area between the extracellular material and the layer of epidermal cells.

The next step is to produce a replica of each of the two complementary surfaces. This is a platinum–carbon cast for observation in the TEM. The PCC source is usually located at an angle of about 45° above the sample. The lower the angle, the higher is the relief contrast that will be reached. After thawing, the specimen is transferred to a sodium hypochlorite solution to dissolve all organic material from the replica. The replica is washed several times to remove residues of solvent and dissolved tissues, picked up on a mesh grid, dried, and examined in the TEM.

Replica regions, which appear dark, face the source of the platinum–carbon coating. Less-coated areas are more transparent for electrons and, therefore, appear electron-lucent in the TEM micrographs. The relief obtained in the PCC method is often very difficult to interpret because it is hard to say which structures are convex and which are concave. In the following section another application of the PCC procedure for the characterization of material properties will be described.

7.2.10 Freeze-Substitution

Freeze-substitution is a technique to dissolve frozen water with an organic solvent at low temperature ($\approx$ -80°C) [323]. The speed of dehydration depends on the temperature and the water-absorbing capacity of the solvent at that temperature. An ideal solvent should dissolve ice without dissolving lipids from tissues and keep the cell structures in their initial condition. Often additional fixatives are added to the solvent in order to stabilize biological structures. In other words, the sample will be fixated in a frozen condition.

However, the fixatives work very slowly at that low temperature and the fixation time usually has to be prolonged. An additional possibility is to carry out the fixation process in several steps at increasing temperature (-80°C, -60°C, -40°C, -20°C). After freeze-substitution, the samples can be either embedded in resin for sectioning (light microscopy or TEM) or dried (CPD or freeze-drying) and coated for SEM .

In summary:

- To study inner structures, biological specimens need to be fixated in order to be imaged properly.
- Water has to be removed or replaced under particular conditions to prevent material deformation.
- Embedding techniques are used to stabilize the specimen for sectioning.
- Different kinds of sectioning procedures are used to obtain nanometer- or micrometer-sized thin layers of biological specimens for microscopy.
- By means of staining, chemical agents bind to selected types of tissue. This procedure increases the contrast of imaging and shows the distribution of particular structures in the biological tissue.
- Histochemical staining is a tool for identifying chemical constituents (proteins, lipids) in the sample material.
- Heavy-metal substances are used in TEM to increase the contrast of the image by changing the kinetic energy of the electrons due to selective absorption.
- CPD and freeze-drying are the most useful methods for drying the sample with minimal deformation artifacts.
- Samples can be fractured in the frozen condition. A replica of the structure is used for further investigation.

7.3 Microscopy Methods Used in Surface Characterization

Both friction and adhesion are results of the interplay between different forces (capillary adhesion, viscosity, intermolecular interactions, chemical bonds, electrostatic forces) in the contact area between two surfaces or one surface and a medium. The relative contribution of these forces to the resulting force depends on the profiles of both contacting surfaces, the mechanical properties of the materials and the presence and nature of the liquid film between them.

Detailed information about the profile sculpture and material design of co-opted surfaces may help in understanding the frictional behavior of these systems. In recent years, our knowledge about the contours of biological surfaces has advanced considerably. Various experimental methods have been developed and applied to visualize and measure the size and shape of the irregularities of the animal integument and plant surfaces. Additionally, many treatments have been developed to study the ultrastructural architecture of natural composite materials. This section describes several microscopy methods previously used by the authors to characterize both microsculpture and ultrastructure of insect attachment systems.

7.3.1 Surface Contour in Biological Surfaces

The SEM is the most powerful tool for the qualitative and quantitative examination of surface microsculpture. The irregularities, occurring on biological frictional surfaces, usually range from 10 nm to hundreds of micrometers (Fig. 7.3). This range of sizes can be studied at magnifications of 100–70,000. The most important disadvantage of the SEM method is that structures appear in images not really three-dimensional. In other words, a tilting angle may strongly influence the linear parameters selected for measurements. Therefore, in order to provide repeatable measurements of surface irregularities, the sample should be tilted so that the desired structures become oriented approximately in one plane. Often density of irregularities has to be measured. In this case, one has to be sure that the surface plane is oriented parallel to the imaging plane. Otherwise, structures will appear denser than in reality.

Measurements can be made by the use of modern computer software providing image analysis. The authors obtained the best results with AnalySIS (Soft-Imaging Software GmbH, Münster, Germany), and Sigma Scan (SPSS). The AnalySIS software package contains a so-called Stereo module, which is thought to obtain z-values from two stereo pairs. However, its capacity is restricted to very simple structures.

Surface outgrowths can also be measured from TEM images. However, this method is not fast enough for routine work and has some other difficulties. Ultra-thin sections, prepared for TEM, can only occasionally find the longitudinal axes of surface outgrowths. Therefore, the measured values will always be significantly lower than in reality. To avoid this, a special measurement protocol has to be selected where only several maximum values per section are selected for further statistics. On the other hand, TEM enables stereoscopic pairs of micrographs to be obtained by using a tilting object holder. TEM stereoscopy can provide some insight into the three-dimensionality of sectioned structures.

One of the most important tasks of microscopy is to provide information about structures in their natural state. Since most microscopy procedures require fixation, dehydration, drying, and sectioning, diverse artifacts are unavoidable. Even relatively stable surface structures can become changed during such treatments. Another challenge is to examine the natural surface at different developmental stages of the same organism. The dental wax cast technique is a simple SEM method, which makes repeated observations of the same developing surface possible. This method uses a two-component dental wax (President light body) to obtain negative casts from a living surface at

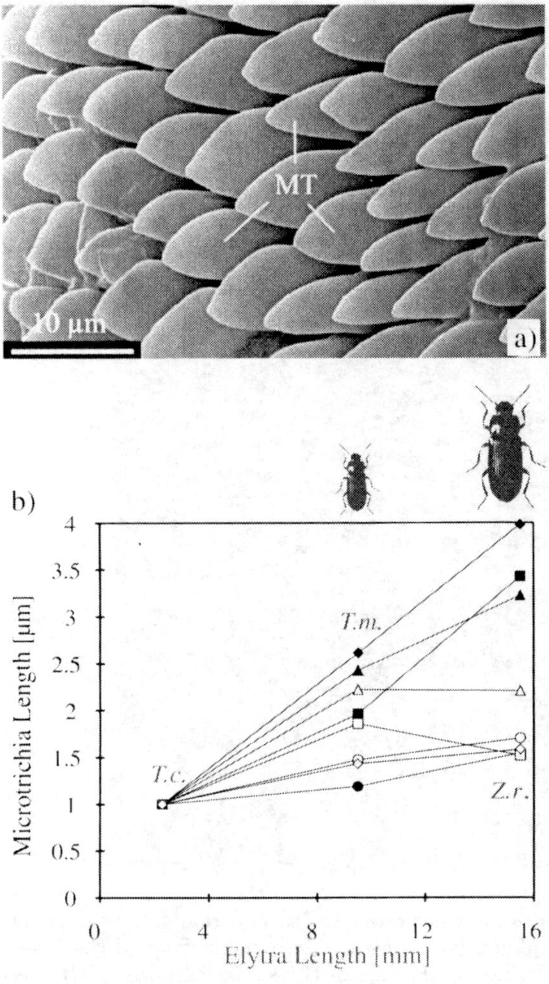

Fig. 7.3. Scale effects on the surface microsculpture in beetles from the family Tenebrionidae. (**a**) *Zophobas rugipes*, microtrichia composing the dorso-medial thoracic field. (**b**) Dependence of the length of microtrichia of eight fields of microtrichia on the elytra size. Data on microtrichia length are recalculated relative to those in the smallest species studied. *T.c.*, *Tribolium castaneum*; *T.m.*, *Tenebrio molitor*; *Z.r.*, *Zophobas rugipes*.

room temperature. After hardening, the wax casts are filled out by Spurr's low-viscosity resin [705], polymerized, coated with gold–palladium and examined in an SEM. This method demonstrated good results with relatively long cuticle outgrowths (10 µm–30 µm) as well as with small structures (0.02 µm) (Fig. 7.4, for details see [462]).

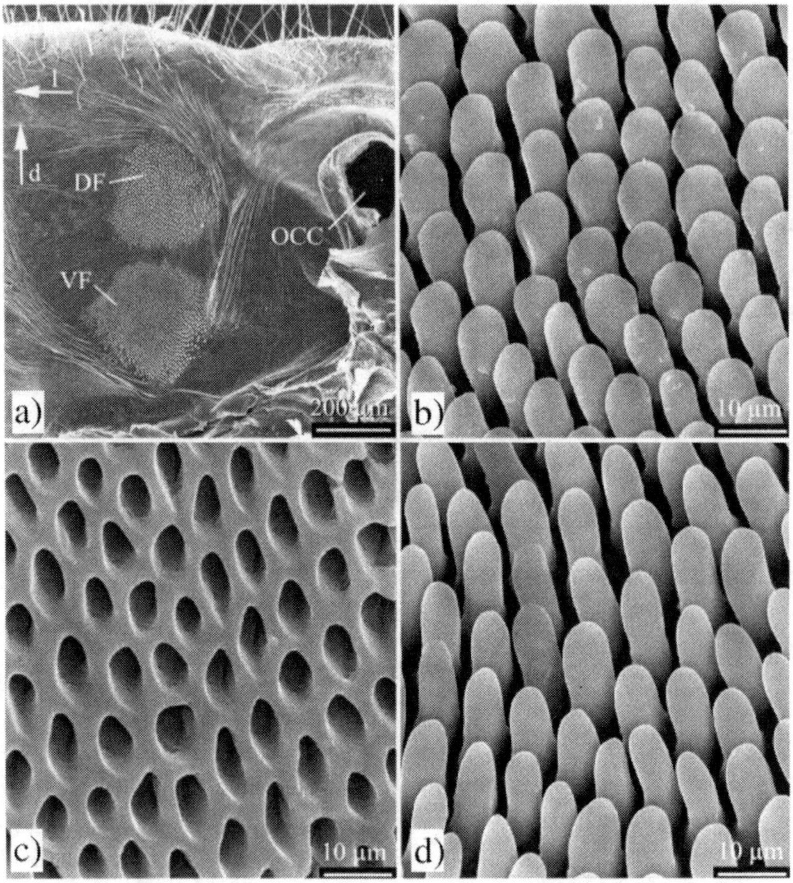

Fig. 7.4. Dental wax cast method for surface replicas. (**a**) Left rear side of the head of a damselfly, *Pyrrhosoma nymphula.* (**b**) Ventral microtrichia field of the head-arresting system, air-dried. (**c**) "Negative" replica of the surface shown in (**b**). (**d**) "Positive" replica of the surface shown in (b) by the use of the "negative" shown in (**c**). d, dorsal direction; DF, dorsal field; l, lateral direction; OCC, occiput; VF, ventral field.

7.4 Special Techniques for Studies on Material Design

7.4.1 Ultrastructural Architecture of Micro-outgrowths

Work on the material structure of the frictional surfaces requires a rough view of the material structure without further details. The most appropriate method for this is a quick stain of semi-thin sections with methylene blue [706] or toluidine blue [699]. Animals are usually fixed and embedded in a low-viscosity resin as described below for the standard TEM procedure, cut at 2 µm. Serial semi-thin sections may also be used for 3D reconstructions of the particular structure or surface as a whole.

The remarkable diversity of microsculptures on biological surfaces has been described in past decades. In taxonomic literature, the following terms are used to name diverse cuticular protuberances: cornuti, signa, spines, chetae, bristles, spicules, darts, denticles, hairs tubercles, scobinate patches, graniculi, virga, pyle, pubescence, tomentum, dust, pollen, etc. [707]. These structures usually have different structure and origin, even in the same group of living organisms. A general classification was proposed for Arthropods into which a variety of modifications fits quite well. Their principle is based on the origin of such processes [708] (Fig. 7.5). There are four main types of processes: (1) the multicellular processes in which the epidermal cells are not differentiated, simply underlying the process; (2) the multicellular processes in which the epidermal cells are differentiated into several specialized cells; (3) the unicellular processes; (4) the sub-cellular processes, where there is more than one projection per cell [355]. Smooth cuticle as an additional type of irregularity may be taken into account.

To understand diverse functional aspects of these structures, more information is often required about the ultrastructural architecture of the material. In this section, the authors provide several protocols, which were applied in our previous works on the structure of the insect cuticle [472, 709, 710]. To show the structure of the composite material within sub-cellular processes, several protocols, running almost parallel, can be applied.

For the standard TEM procedure, parts of the insect body are fixed for 12 h at 4°C in 2.5% glutaraldehyde (in 0.01 M phosphate buffer at pH 7.3), and postfixed for 1 h in 1% osmium tetroxide in phosphate buffer at 2°C. After washing, the preparations are stained for 1 h at 4°C in 0.1% aqueous uranyl-acetate solution, washed, dehydrated, and embedded in a low-viscosity resin [705]. Ultra-thin sections are picked up on copper grid slots coated with formvar film and observed in TEM. Sections are stained with uranyl-acetate

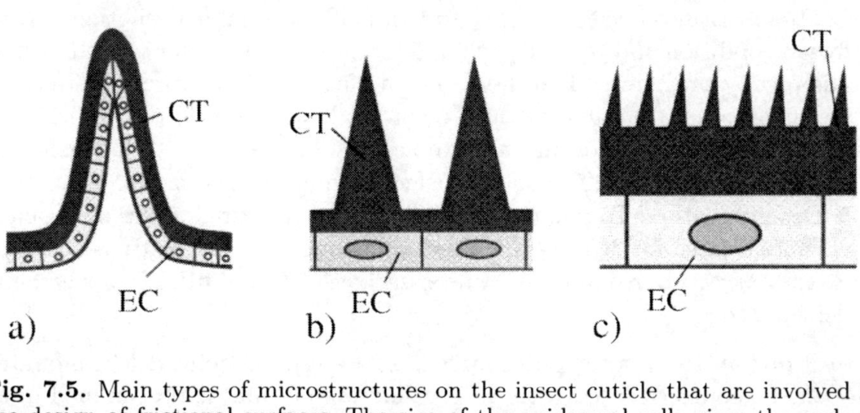

Fig. 7.5. Main types of microstructures on the insect cuticle that are involved in the design of frictional surfaces. The size of the epidermal cells gives the scale of the structure dimensions. (**a**) Multicellular process. (**b**) Unicellular processes. (**c**) Sub-cellular processes. CT, cuticle; EC, epidermal cells.

and lead-citrate. To study the mechanical differences of the components of this composite material, the shadowing method (see Sect. 7.4.4) was used.

For the SEM procedures, parts of the insect body can be fixated in various fixatives, such as 2.5% glutaraldehyde (in 0.01 M phosphate buffer at pH 7.3) or alcoholic Bouin [699] for 24 h. After fixation, the desired pieces are cut out and preserved in 70% ethanol. To obtain information on structural components of the cuticle, five different procedures can be applied: (1) fracturing of dried or frozen cuticle; (2) deproteinization of the cuticle; (3) semi-thin sectioning of the cuticle, embedded in Spurr's resin, with further resin removal; (4) freeze substitution; (5) dechitinization.

In detail:

- 1. Fracturing of cuticle. This simple method showed good results on stable tanned cuticles. The dehydrated elytra and pieces of cuticle cut out of the thorax are air-dried and fractured using a razor blade. Cuticle can also be frozen in liquid nitrogen or propane, fractured, and then air-dried.
- 2. Deproteinization. To remove the protein cover of microfibrils the fractured cuticle pieces can be immersed in 20% NaOH at 100°C for 5 h–6 h, washed and dehydrated (Fig. 7.6).
- 3. Sectioning of resin-embedded structures. Dehydrated cuticle previously fixed in glutaraldehyde and osmium tetroxide as described for TEM procedure was embedded in a low-viscosity resin and sectioned using a diamond knife (1μm–2 μm). Sections were picked up on pioloform-covered glass plates, treated with Maxwell's solution [711] for 2 min–5 min to remove the resin, washed and dehydrated (Fig. 7.7).
- 4. Freeze-substitution. The freeze-substitution technique is a standard method in cell science [700,701]. The authors applied this method to understand the behavior of biological material under load. Usually, frozen biological material can recover to a certain extent after being thawed. The OsO_4-substitution was necessary to fixate all material deformations in the frozen condition and to keep such a deformation in the course of the whole microscopy procedure. The shock-frozen pieces of tissue from different experiments were fractured using a razor blade and transferred to 0.5% osmium tetroxide solution in absolute acetone at -80°C for 48 h, washed in absolute acetone at -20°C and transferred to absolute ethanol.
- 5. Dechitinization. To remove chitin fibrils, the fractured pieces of the cuticle were treated for 24 h–72 h with an aqueous solution of chitinase (Merck, *Streptomyces sp.*, lyophilized 8 units/mg) at 37°C and pH 5.5, washed and dehydrated.

All preparations, after procedures 2–5, are critical-point dried, mounted on holders, sputter-coated with gold–palladium (10 nm) and examined in an SEM at 20 kV.

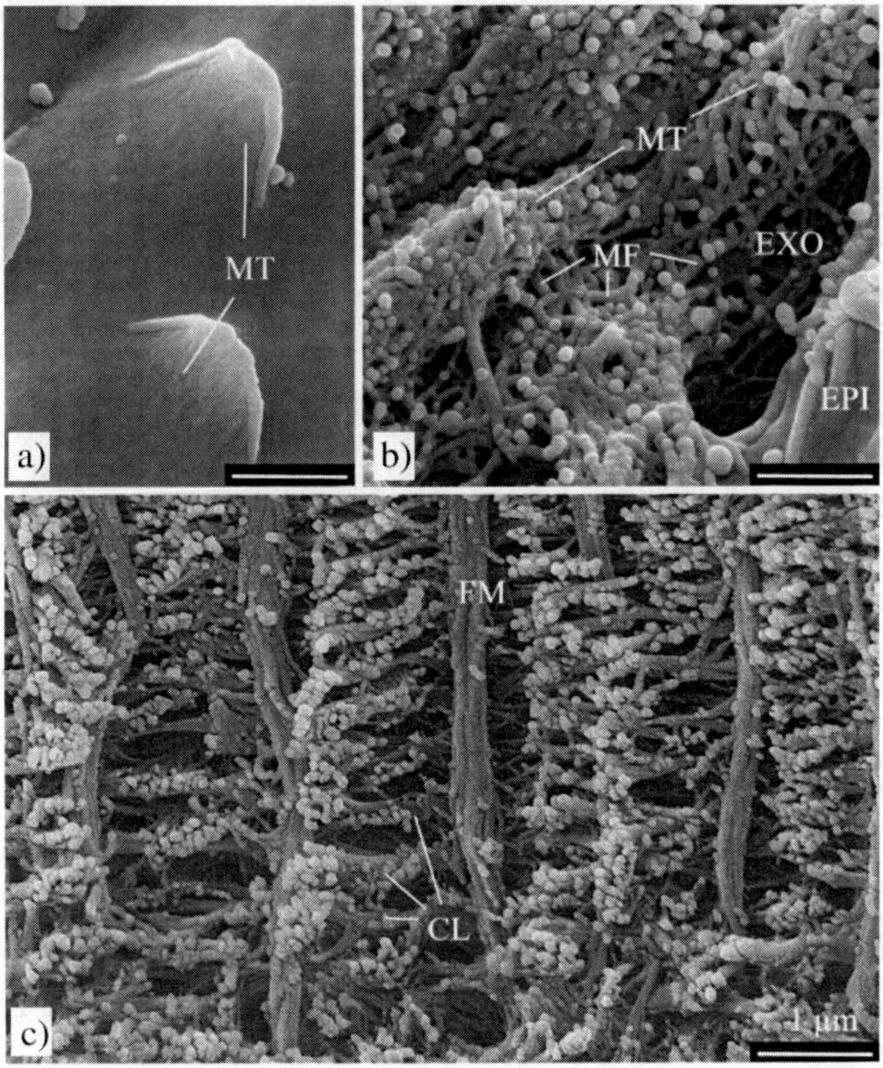

Fig. 7.6. NaOH maceration technique applied to the unguitractor plate of the damselfly *Lestes barbarus* (**a**), (**b**) and elytra cuticle of the beetle *Tenebrio molitor* (**c**). (**a**) Surface of an intact cuticle; (**b**), (**c**) NaOH treated cuticle. CL, cuticle layers; EPI, epicuticle; EXO, exocuticle; FM, pore canal filaments; MF, microfilaments; MT, microtrichia.

7.4.2 Pore Canals

Frictional or anti-frictional biological materials often have protruding pore canals transporting adhesive or lubricating substances to the cuticle surface. The density and shape of the canals can be of importance for understanding

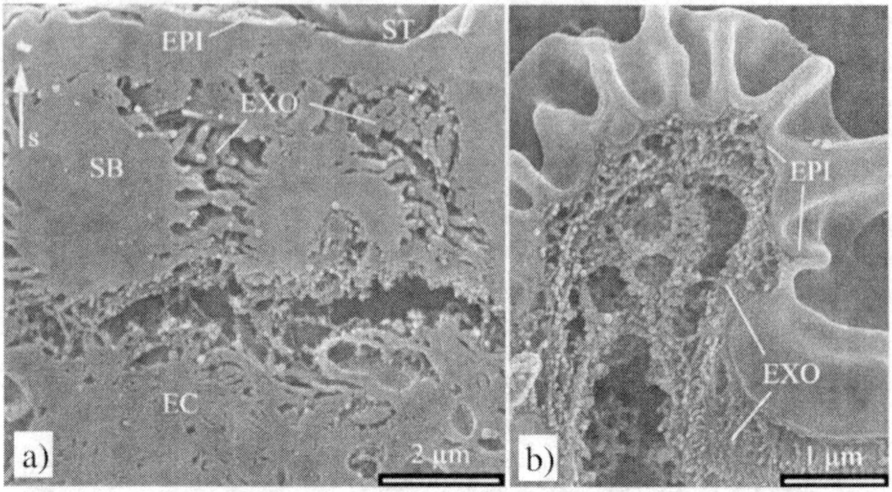

Fig. 7.7. Maxwell's methods of the resin removal from semi-thin sections. (**a**) Exocuticle of the adhesive setae bases in the beetle, *Rhagonycha fulva.* (**b**) Ultrastructure of the neck membrane in the damselfly. EC, epidermal cells; EPI, epicuticle; EXO, exocuticle; s, direction to the cuticle surface; SB, basis of the seta; ST, seta.

the frictional properties of a particular material. The following procedures, described in the previous section, can be used for SEM study of porous materials: (1) fracturing of dried cuticle (Fig. 7.8b); (2) deproteinization of the cuticle (Fig. 7.8a); (3) dechitinization of the cuticle; (4) semi-thin sectioning of the cuticle embedded in Spurr's resin.

An additional method, giving 3D information, is the preparation of casts of the porous canals. Insect pore canals are extremely small compared to the avian egg pores or most of the plant pores. For this reason, low-viscosity casting material was used. Dehydrated pieces of cuticle were embedded in a low-viscosity resin, fractured, and treated with NaOCl for 2–5 days to remove the organic material [712, 713]. After treatment the preparations were washed and dehydrated, sputter-coated with gold–palladium, and examined in an SEM. Ultra-thin sections, coated at an angle of 15° with carbon–platinum, gave very good results in studies of porous substrates.

7.4.3 Fluids Occurring in the Contact Area

The adhesion of the attachment hairs (setae) of beetles to clean glass has been demonstrated previously [503]. Coverslips were cleaned in absolute ethanol and absolute acetone. Tarsi cut off freshly killed animals were carefully placed on the coverslip using fine forceps. Freshly attached pulvilli can also be studied in a fluorescent microscope in ultraviolet light (DAPI filter). Air-dried preparations give good results in SEM. Such an approach may aid in the

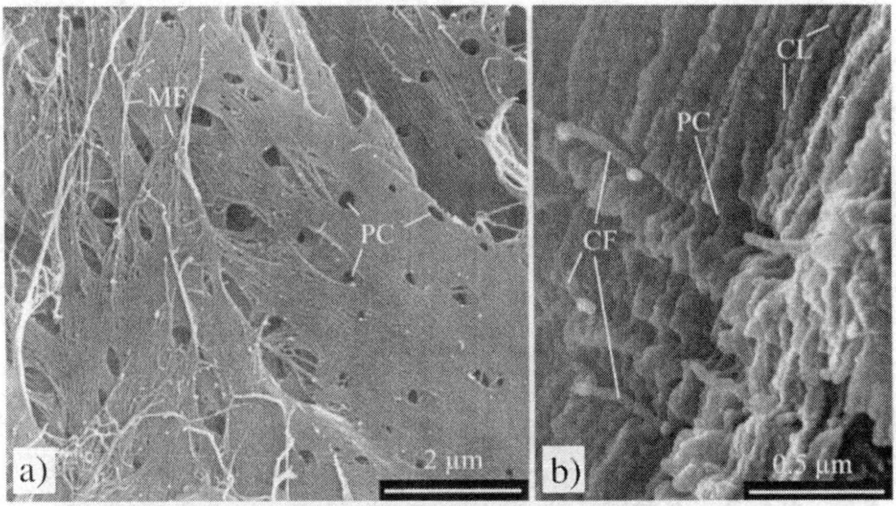

Fig. 7.8. Two methods applied in studies on the pore canals in the neck cuticle of the damselfly *Lestes barbarus*. (**a**) NaOH maceration technique showing microfibril arrangement around canals. (**b**) Fractured cuticle. CF, cuticle microfibrils; CL, cuticular layers; MF, microfibril; PC, pore canals.

calculation of a contact area and gives results relatively close to the real contact area. A set of SEM experiments comprising washing in organic solvents and preparing Spurr casts together with imaging fresh control samples can provide evidence about the presence of epidermal secretions on the cuticle surface (Fig. 7.9, from [710]). Since secretory substances are often essential in the attachment of flies, beetles, crustaceans, molluscs and other animals on smooth substrates, the study of these substances became more and more important. In the case of insect pulvilli, the footprints can be obtained by a careful and short pressing of the leg against the glass. The size and area of single droplets can be quantified from phase contrast photographs. It seems that the basic substance of these prints evaporates in the course of several hours. What remains can be examined by a variety of histochemical methods after fixation in glutaraldehyde vapors. For analytic methods, such as gas-chromatography, only freshly emerged individuals (not having walked) are required to prevent artifacts.

7.4.4 Material Properties of Fibrous Composites

TEM and SEM help to describe structural details down to the finest dimensions of an organism. Structural elements, such as macromolecules, can be visualized using several advanced techniques [311]. Other goals of TEM and SEM are to provide information on the material properties of biological tissues in the context of a living system. Many biological tissues are crystalline

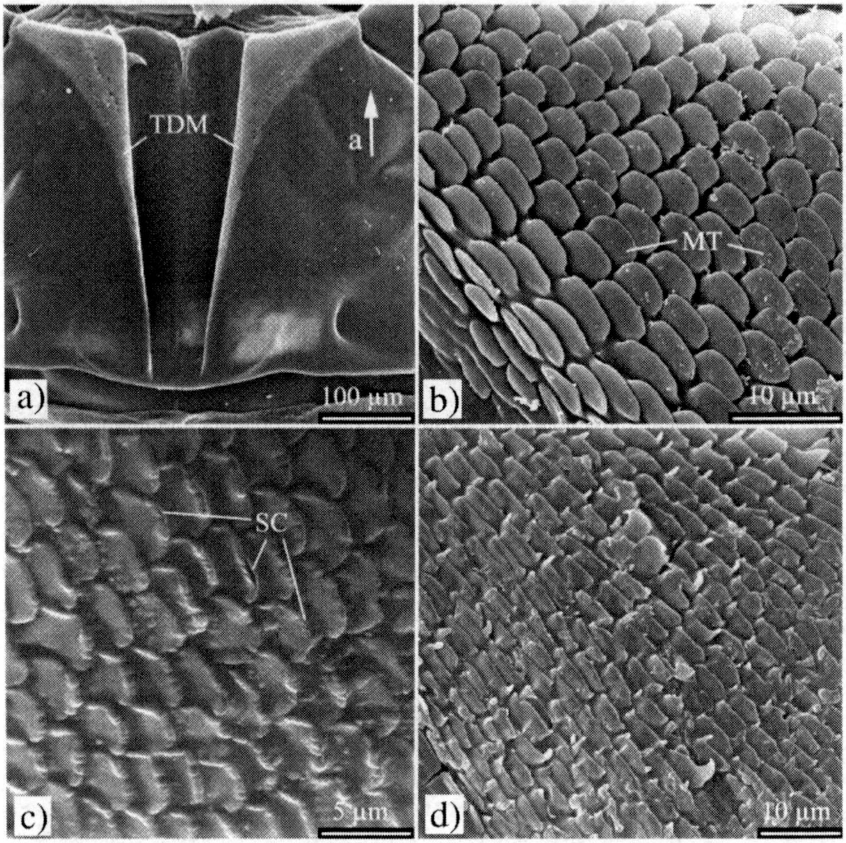

Fig. 7.9. Set of experiments providing evidence about the presence of an epidermal secretion on the cuticle surface. (**a**) Dorso-medial thoracic field of microtrichia (TDM) in the tenebrionid beetle, *Tribolium castaneum*. (**b**) Cuticle washed in organic solvents (ethanol, acetone). (**c**) Intact cuticle (uncoated preparation) viewed covered with a secretion. (**d**) Dental wax cast of the living cuticle showing impurities on the surface. a, anterior direction; MT, microtrichia; SC, secretion.

composites, such as mollusc shells [314,322,714], vertebrate bones [713,715, 716] or fibrous composites, such as the plant cell wall [717], and the animal integument and its derivatives [472,519]. The latter are usually penetrated by canal systems of different origins [718]. Such material contains two or more components with different mechanical properties, so that the resulting material has new mechanical properties [719]. Components with similar electron-optical densities are sometimes barely distinguishable in the TEM because of a very low contrast, even in heavily stained specimens. Using standard TEM and SEM preparation procedures, it is often difficult to determine the difference in mechanical properties between single components of the composite. There are several microscopy protocols that can be applied.

Using freeze-fracturing and freeze-etching techniques and SEM, it is possible to obtain fractures corresponding to the marginal areas between two components with different mechanical properties [720]. SEM studies of Spurr resin semi-thin sections with removed resin may provide information about the 3D–structure of biological composite materials [472,711].

The method of platinum–carbon coating of ultra-thin sections for TEM and SEM can be applied to the study of biological composites, in order to obtain information about mechanical properties of components. The method uses a unidirectional shadowing of ultra-thin sections and aids in understanding of the mechanical properties of material components. The changes in profile of ultra-thin resin-embedded sections during different stages of preparation reflect the material properties of the various components: harder regions deform less than softer ones. Such changes in the section profile can be visualized with the help of carbon–platinum coating and is used as evidence for specific material properties in particular regions of natural composite materials. This method increases the relief contrast, improves the 3D view of structures, and, in combination with standard TEM and SEM procedures, gives clear demonstrations of the material design.

Pieces of the cuticle are embedded, sectioned and contrasted according to the standard TEM-preparation procedures described above. To observe changes in the profile of the upper and rear surface of the ultra-thin section, the sections are picked up on copper grids with either the upper or rear surface facing downwards, carbon–platinum sputter-coated at an angle of about 15° and examined in both TEM and SEM. The distribution of chitin crystallites in the insect cuticle, and the ultrastructure of the pore canals, specialized for the transport of epidermal secretions to the cuticle surface, have been studied previously [721]. The surface of the resin block did not reflect any particular regions of the embedded material. However, after the cutting and mounting procedures, the section profile had strongly changed. The ultra-thin section profile was three dimensional and symmetrical in the upper and rear surfaces. Such changes of the plane surface into a set of hills and valleys can remain invisible if the electron densities of regions with different mechanical properties are similar. The PCC procedure supported the visualization of these regions in SEM and TEM (Fig. 7.10).

In TEM images, structures with dense shadows on the side of the coating direction and with lucent shadows on the opposite side appeared as hills. Structures with an opposite pattern of shadows appeared as depressions. The harder material remains less deformed after the ultra-thin sectioning and mounting procedure if compared with softer material (Fig. 7.11). TEM observations of the resin-embedded sections often do not reveal the presence of regions of different electron densities in a composite material. This can happen when the more electron-dense material is less dense than the other material. Both the sectioning and the mounting procedures can cause changes in the section profile such that the harder areas become thinner, if compared

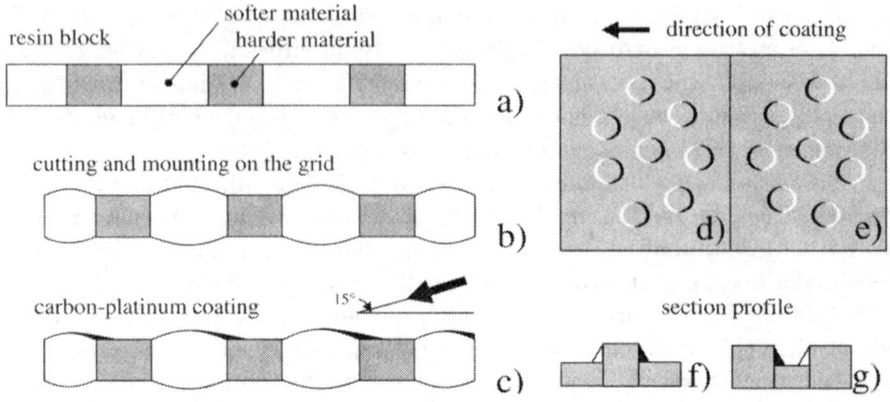

Fig. 7.10. Schematic of platinum–carbon coating method of ultra-thin sections. (**a**) The profile of the resin embedded section of a composite material consists of softer (white areas) and harder (grey areas) regions. (**b**) Changes in the section profile after cutting and mounting procedures; note the hill and valley profile of the section. (**c**) Carbon–platinum coating of the section results in the deposition of electron-dense particles on one side of the hills (black areas). The other side of the hills contains electron-transparent shadows. (**d**), (**e**) Schematic of a possible TEM view, from above, onto the surface of a resin-embedded section of a composite material containing regions of differing electron densities. Even if the electron density of the whole field appears uniform because of the different depths of the softer (**d**), (**f**) and harder (**e**), (**g**) materials, the PCC method can resolve such differences, because of the different distribution of the carbon–platinum coating material (black) and shadows (white).

with softer areas. This behavior of the section can result in similar electron densities of the different areas of the specimen. The use of a coating procedure will enhance the contrast of the specimen. This contrast will reflect the different areas not according to their electron densities, but to their profiles. An increase in the relief contrast improves the 3D view of structures and can give a very clear demonstration of the material design when combined with the stereo pair method. An additional advantage of the method is that the coated sections can be used both in TEM and SEM. The method does not need any special preparation procedure. Ultra-thin sections already used in TEM studies can be coated and used again in TEM or SEM studies.

The last two sections can be summarized as follows.

- Since in biological microtribology the mechanical properties of volume material must not be neglected, detailed insight into inner structures is necessary.
- The correct determination of the contact area depends on the proper understanding of the shape and function of the outermost parts of the biological systems.

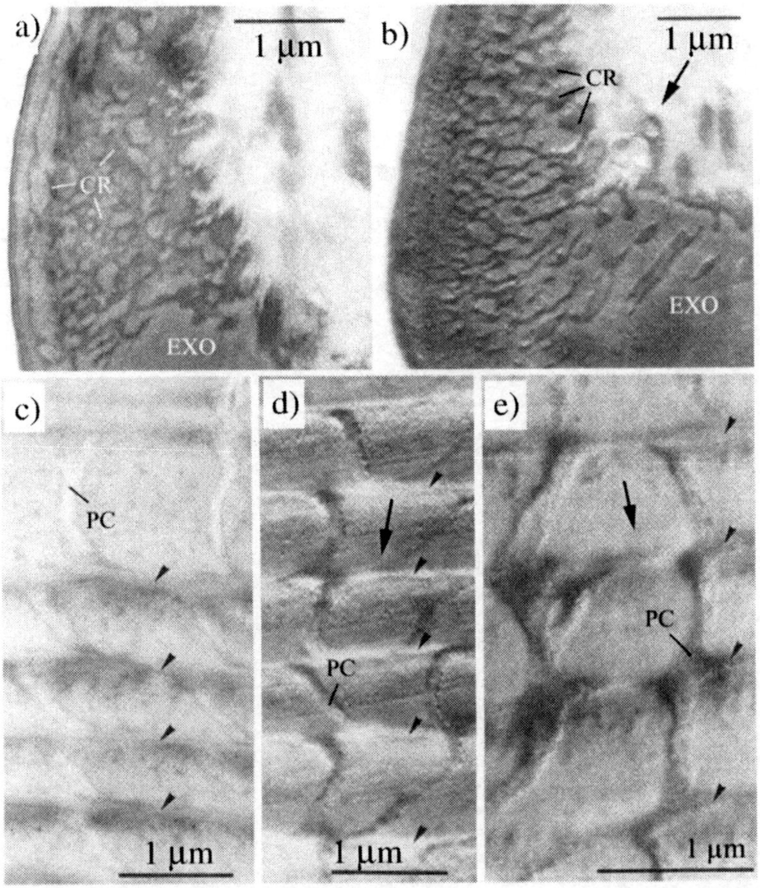

Fig. 7.11. Carbon–platinum coating technique for ultra-thin sections. Comparison of TEM images of uncoated (**a**) and coated (**b**) sections of the specialized hard cuticle of the fly prestomal teeth. Corresponding images show similar areas of cuticle. (**c**)–(**e**) Pore canals of the tarsus cuticle in the leg of the grasshopper, *Omocestus viridulus*, as is viewed in uncoated (**c**) and CP-coated ultra-thin sections (**d**), (**e**). Black arrows indicate directions of coating; black arrowheads show location of lamellae with the same orientation of microfibrils. CR, chitin crystallites; EXO, exocuticle; PC, pore canals.

- The knowledge of the inner structure aids in understanding the mechanical behavior under load.
- Adhesion is controlled by contact area and/or secretion of adhesives. Microscopy reveals the location where the adhesive is administered and provides an estimate of the amount of liquid that is supplied to the surface.
- Many bio-materials can be regarded as complex composites. Microscopy can be used to investigate the orientation and structure of fibers.

Part IV

Case Studies

8. Samples, Sample Preparation and Tester Setup

Biological motion systems are highly adapted micromechanical units. Due to the long evolution period these systems exhibit an optimum combination of mechanical (stability, elasticity) and tribological (friction, adhesion, wear) properties. Therefore, biological systems are excellent candidates as model systems for the design of new materials and for micromechanical applications.

Macroscale applications of materials with particular frictional properties have been extensively studied and the results of these investigations have been incorporated in existing industrial products, such as, for instance, Velcro fasteners or tire materials. There is a growing need for new types of materials with controlled friction and adhesion. This is especially true for the development of new types of adhesives.

Presently, the majority of tribological research on bio-systems is restricted to the lubrication mechanisms of human and animal joints. To the best knowledge of the authors, this study is the first attempt to correlate the microtribological properties of a biological system with its material structure.

Nature offers an overwhelming variety of different surfaces, ranging from rough materials like bark to very smooth materials, such as, for instance, the leaves of many plant species. The attachment system of a cricket – as our example of a biological microsystem (see Fig. 8.1) – is very well adapted to all

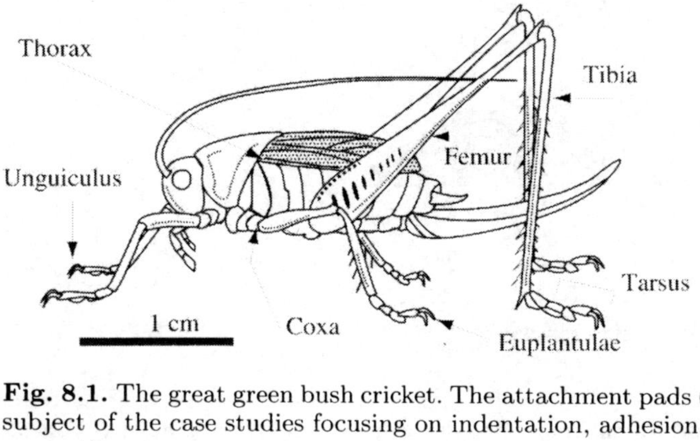

Fig. 8.1. The great green bush cricket. The attachment pads (euplantulae) were the subject of the case studies focusing on indentation, adhesion, friction and material properties.

of these surfaces. Attachment systems, where attachment–detachment performances are often very fast, are perfectly adapted to achieve an optimized contact to the substrate in the shortest time.

This part of the book presents case studies [722–725] on the indentation, adhesion and friction of the attachment pads of living and dead insects. Profile changes of the surface and the orientation of cuticle microfibrils were tested by means of scanning electron microscopy after freeze-substitution.

8.1 A Biological Microsystem

The attachment system of the great green bush cricket *Tettigonia viridissima* is designed to hold the insect on a rough surface using small claws at the end of each leg (see Figures 8.2 and 8.3a). To adhere to a smooth surface, however, the insect uses attachment pads. These pads have a cross-sectional shape of a hemisphere. The cuticle of the pads is a natural friction-active material with a specific inner and outer structure. Each end section of the leg (tarsus) contains 3–4 pads (euplantulae). The biggest are 1.7 mm–2.5 mm wide.

In order to obtain detailed structural information, the pads needed to be imaged in their natural state. To accomplish this, the pad of a living insect was lightly pressed against a smooth surface and frozen in this condition with liquid nitrogen. As the reference, the authors used pads not contacting the substrate. The shock-frozen pads were then fractured using a razor blade and transferred to 0.5% osmium tetroxide solution in absolute acetone at -80°C for 48 h, washed in absolute acetone at -20°C, transferred to absolute ethanol and critical-point dried. An additional technique was used to obtain information about the orientation of inner structures. Semi-thin sections (0.5 μm–2.0 μm) of pads embedded in Spurr resin were sectioned using

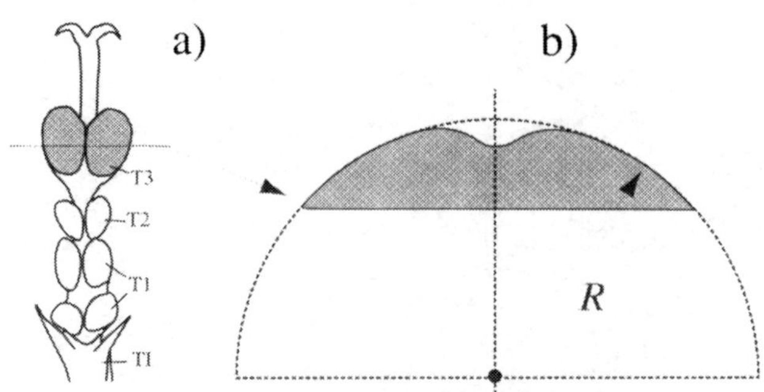

Fig. 8.2. Schematic of the attachment pad of the *Tettigonia viridissima*. (**a**) Ventral view of the tarsus of the third pair of legs. TI, tibia, T1–T3, leg tarsomeres. The pad of the most distal tarsomere (grey colored area) was used in the experiments. (**b**) Cross-section of the pad. R, radius of pad curvature.

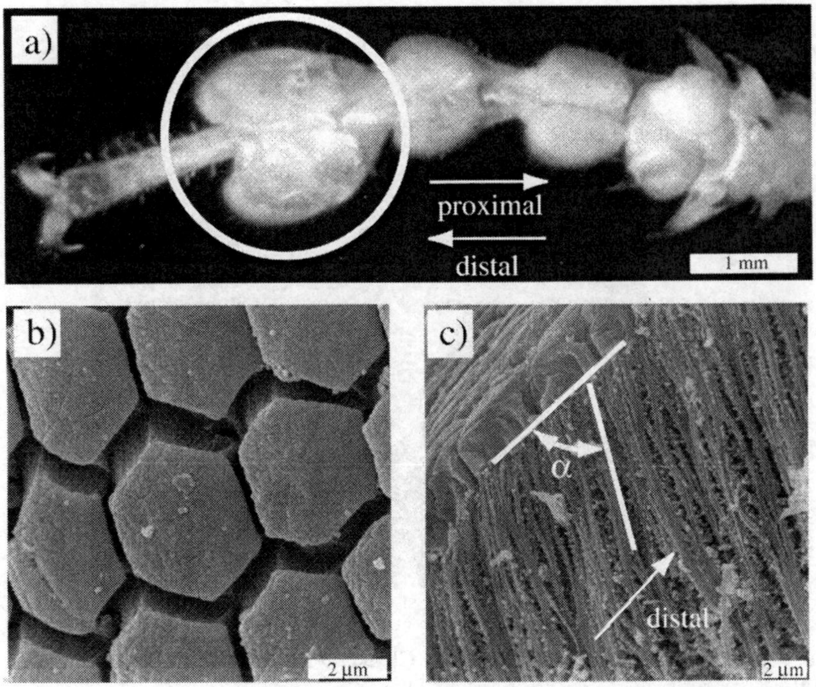

Fig. 8.3. (**a**) Last segment of the leg. The circle denotes the attachment pad studied here. The terms proximal and distal describe the leg motion directed towards or away from the insect body. (**b**) Top view of the surface microstructure of an attachment pad. (**c**) Cross-section of an attachment pad. In the upper left corner the surface microstructure can be seen.

a diamond knife. The sections were picked up on pioloform-covered coverslips, treated with Maxwell's solution for 2 min–5 min in order to remove the resin, washed in absolute ethanol, and critical-point dried. For details see Sect. 7.2. All preparations were mounted on holders, sputter-coated with gold–palladium (10 nm) and examined in a Hitachi S-800 scanning electron microscope at 20 kV.

The outer structure of the attachment system is composed of hexagonal outgrowths of highly flexible material. If no pressure is exerted, the surface is patterned by regular hexagons with a diameter of about 5 μm (Fig. 8.3b). This structure is conserved when the pad is pressed against the smooth surface. Each hexagon is separated from its neighbor by a trench with a width of about 1 μm resulting in a ratio of area covered by hexagons to the area of hexagon plus trenches of 7/8.

The inner construction is characterized by a dendrite-like structure branching in the proximity of the surface, Fig. 8.3c. The main rods of the dendrites contact the surface at an angle α of about 70°. The remaining volume is filled up with a liquid (haemolymph). The rods (diameter, 1.12 μm) split into even

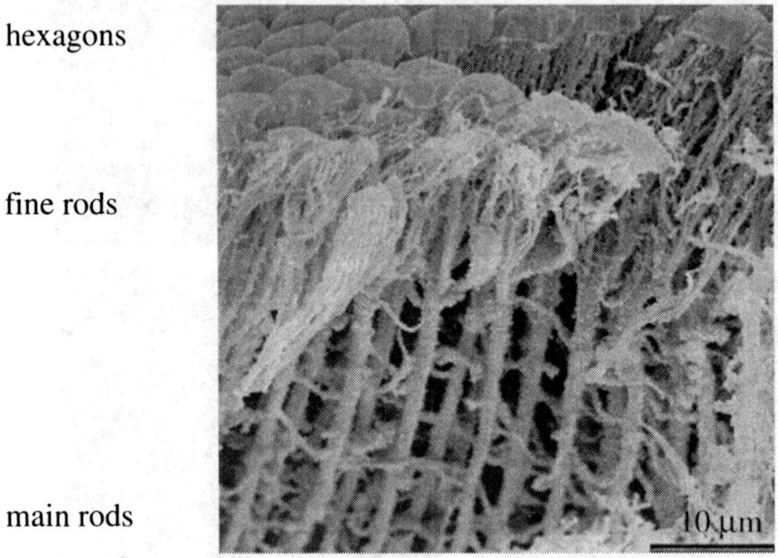

Fig. 8.4. The inner structure of the pad. Clearly visible is that the main rods branch into finer rods close to the inner surface.

more flexible branches (diameter, 0.08 µm) close to the inner surface (see Fig. 8.4). In sections and fractures, the uppermost layer resembles a thin film with a thickness of 180 nm. The thickness and the non-fibrous structure correspond to the epicuticle, the outermost layer of insect skin.

8.2 Sample Aging

One of the most important issues for testing biological systems is sample aging. When tests are performed with parts cut off from the insect body, the samples undergo rapid structural changes caused by the loss of water. To underline this effect, the mass of the detached parts was measured as a function of time. Pieces of the leg – with and without pads – were severed. To prevent leakage of haemolymph, the cut ends were sealed with a droplet of wax. Judging from the morphology, it was expected that parts containing the attachment pads would evaporate a higher amount of water than other parts of the leg. Therefore, the evaporation was measured for both types of material. The initial mass was taken as 100%. Figure 8.5 shows the results for the parts with and without pads. It can be seen that after 1 hour about 25% of the weight is lost. This has a drastic impact upon the microstructure of the pad, and the hexagonal structure changes significantly.

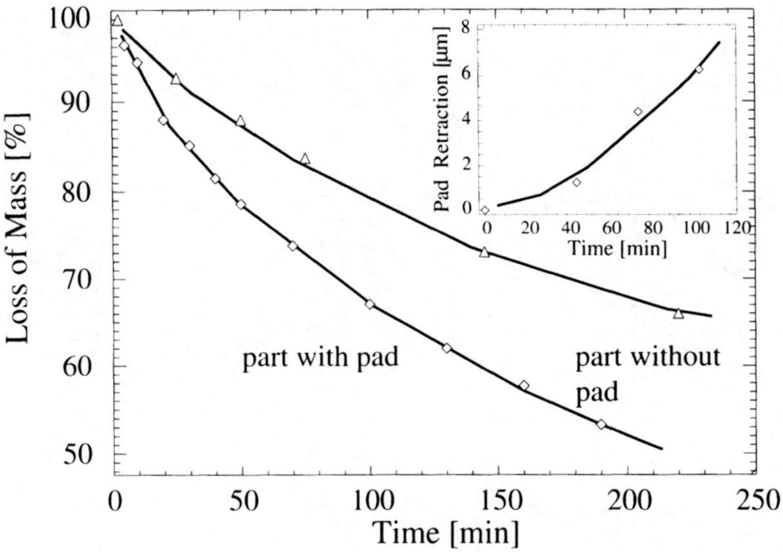

Fig. 8.5. Loss of mass versus time for a part of the leg that contains the attachment pad and one without. The inset depicts the retraction of the pad due to loss of water.

The impact of drying was experienced in friction tests performed with cut legs. When the pad was approached by a flat counter sample, the point of engagement was taken as a direct measure of the structural change. In other words, when at the beginning of the experiment engagement occurred at z_0 and later on engagement was possible at z_1, then the difference of both values was taken equal to the retraction of the pad as shown in the inset of Fig. 8.5. A retraction by 6 µm results in a change in normal force of 600 µN (with k_z=100 N/m). This change in normal force with time leads to different results and makes comparison and analysis difficult. As a consequence, all following tests were carried out with living insects. However, it has to be pointed out that freshly cut legs showed the same tribological behavior as the legs of living insects. For instance, the pads of freshly severed tarsi showed the same pull-off force as the living pads. This means that the experiments properly reflect the response of the material. The superposition of the material properties and the active response of the insect was also probed and will be shown later.

9. Case Study I: Indentation and Adhesion

This study focusses on the attachment performance of the attachment pads. By means of force–distance curves the forces are evaluated quantitatively. Also, since contact between the flexible pad and the rigid surface is always accompanied by an indentation of the pad, indentation is discussed prior to the effects of adhesion.

9.1 Measurement Procedure

During contact formation of a pad and a flat sample the pad undergoes an indentation. When the pad rests in contact, viscoelastic relaxation was measured and is discussed in Sect. 11.2. To pull the pad and sample apart, force is necessary to break the contact, since adhesion represents an energetic minimum. These phenomena (indentation and pull-off) were probed by the microtribometer introduced in Sect. 5.1.5.

Measurements of indentation and adhesion of the pad were performed with legs of living insects. A piece of silicon wafer was attached to the spring to serve as the upper sample, with its polished side facing the pad. The insect was immobilized and attached to the sample holder shown in Fig. 9.1a. One tarsus of a foreleg was firmly fixed with wax and was the lower sample (Fig. 9.1b). The measurements were made on the most distal pads of the tarsus.

To measure indentation the sample holder (carrying the insect) was moved toward the flat sample. After the two samples remained in contact for a predefined amount of time (resting time, Δt) under a certain normal force, the disengagement process was initiated. This process ended with the breaking of the adhesive contact and manifests as the pull-off force (Fig. 9.1c).

In a reference experiment to investigate the deformation of wax under the applied force, the pad was replaced by the Si ball with approximately the same shape and diameter as the pad. The experiment showed that the hardness of the wax is much greater than that of the pad. Hence, the layer of wax was considered rigid and undeformable, not affecting the results obtained from the pads in this particular force range.

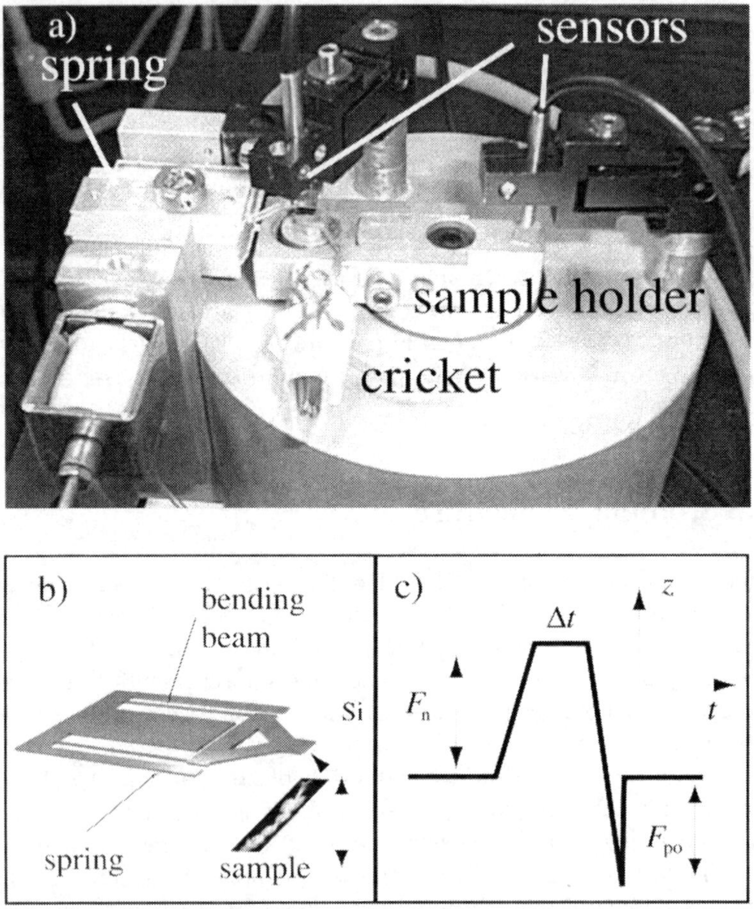

Fig. 9.1. Indentation and adhesion measurements. (**a**) Photograph of microtribometer. The pad of a living insect is attached to the sample holder, and the upper sample, a silicon flat, is attached to the spring. Driven by a motor, the sample holder moves the insect to the Si sample. The deflection of the glass spring is monitored by a fiber-optic sensor. Another sensor monitors the motion of the sample holder. (**b**) Detailed schematic of both samples. (**c**) Schematic of a typical curve recorded versus time. F_n, normal force; F_{po}, pull-off (adhesion) force; Δt, resting time.

9.2 Indentation

From our experience, resulting from sample handling as well as from observations of the insect in its natural habitat, it was deduced that the pad material must be very flexible in its response to mechanical interaction. This assumption was strongly supported by the following micro-indentation experiments. The pad was pressed against a matrix of circular bumps with a diameter of 6 μm and a height of 5 μm (see Fig. 9.2a, b). In this state the pad was

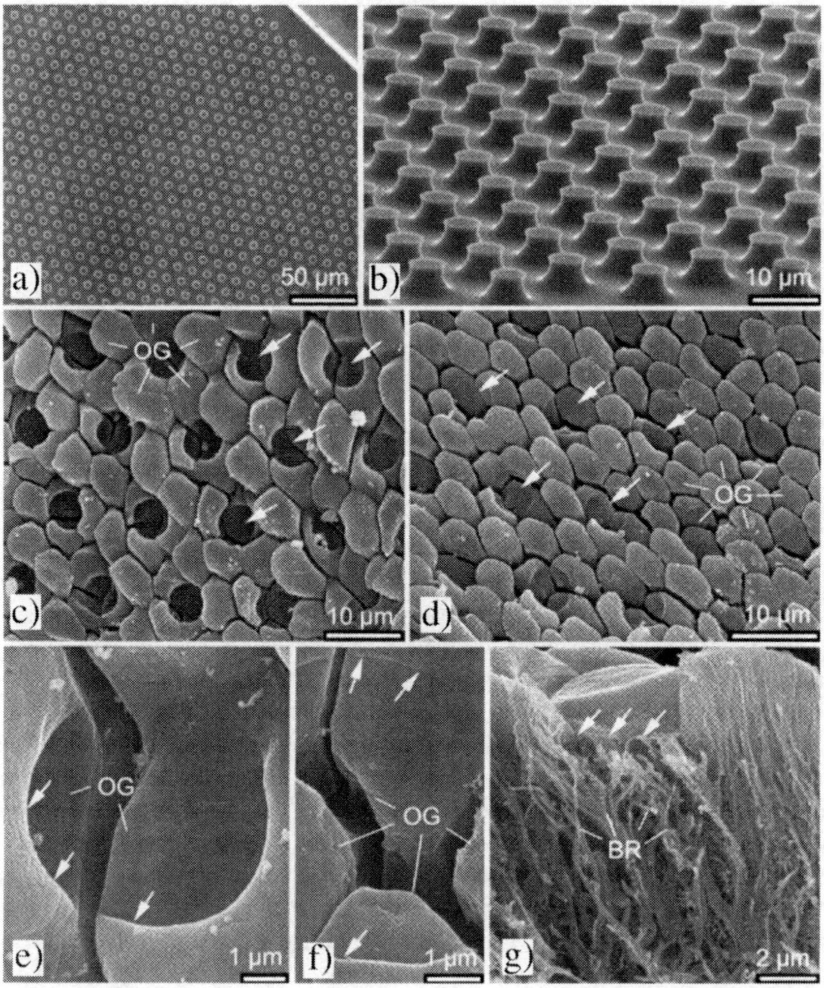

Fig. 9.2. Micro-indentation experiments. (**a**) Structured Si surface, top view. (**b**) Same, perspective view. (**c**), (**e**) Pad surface deformation at high normal force. (**d**), (**f**) Pad surface deformation at moderate normal force. (**g**) Fracture of the pad in the area of indentation. The arrows indicate the deformation. BR, branched rods; OG, outgrowth.

frozen, sputtered with gold and imaged by SEM. Figure 9.2c-g clearly shows the indents.

The test was repeated several times without freezing. After multiple indentations the pad was again frozen and imaged. The SEM images showed a completely unaffected hexagonal pad surface, proving that the pad material is extremely flexible and capable of replicating a given surface profile. Since roughness is a multi-scale property, the pads are able to replicate several

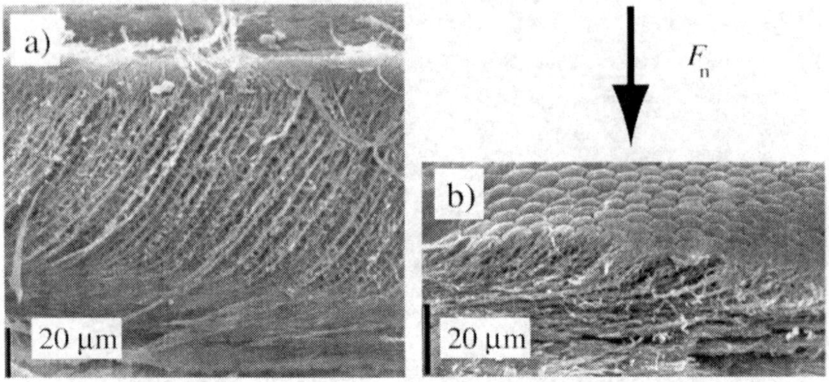

Fig. 9.3. (**a**) Cross-section of a pad without contact with a surface. (**b**) Pad during compression.

length scales. Large structures, as shown for the Si bumps, are replicated by means of the hexagonally patterned skin supported by the main rods. Subtle changes in the profile (sub-micron range) are replicated by the surface of single hexagons with their underlying network of fine bush-like structures. This means that the geometric contact area in indentation is equal to the real contact area, since the trenches disappear as shown in Fig. 9.3b.

To measure the indentation of the whole system, the pad was imaged in its uncompressed shape. Then, the pad was pressed against the smooth silicon surface and imaged again (compare Fig. 9.3a, b). The level of compression is comparable with the real-life situation when the weight of the insect rests entirely on the pad. Under compression the hexagonal structure develops perfectly, mediating the maximized holding force.

The indentation of the pad was then determined quantitatively by measuring the deflection of the spring in the z-direction (see inset of Fig. 9.4). The pad approaches the silicon sample with a velocity of about 10 μm/s. The sensor measures a certain distance before contact. After contact formation this distance decreases. The deflection of the spring is equal to the displacement of the sample holder minus the indentation of the pad. As a reference, the experiment was repeated with the silicon ball instead of the pad. Here, the indentation can be neglected due to hardness and the deflection of the spring is equal to the displacement of the sample holder. The difference between this curve – a straight line according to Hooke's Law – and the curve of the soft material equals the indentation.

From these data the real contact area was calculated. As a first approach it was assumed that no adhesion exists between the pad and the Si surface. Thus, the area of contact A_c was calculated by using Hertzian contact mechanics. As discussed in Sect. 2.3.4, this model can be used to describe the contact formation between an elastic sphere and a planar substrate. It is ap-

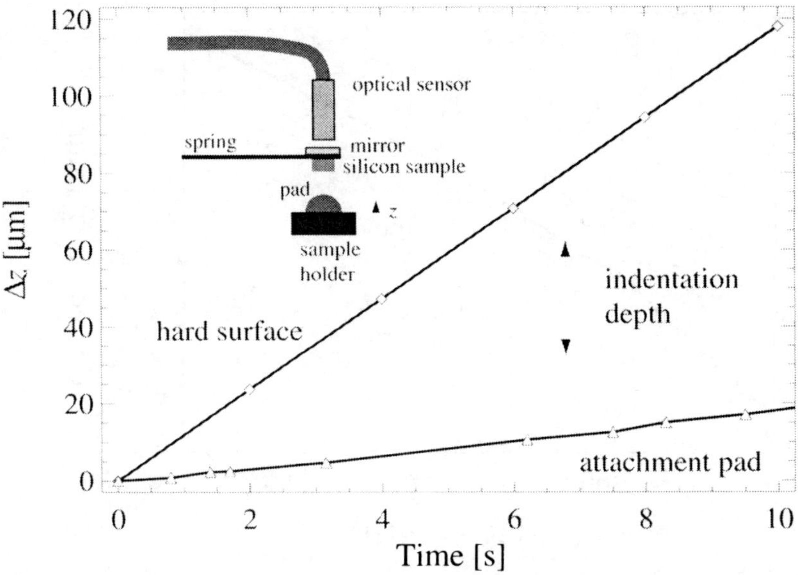

Fig. 9.4. Sensor signal obtained from the spring deflection due to contact between the soft pad and the hard Si surface. Inset: experimental setup. The distance between the sensor head and the mirror is measured optically. A deflection can only be measured when the samples are in contact.

plicable if the indentation of the pad (assumed to be an elastic sphere) is small in comparison to the radius of the pad. The indentation of the pad δ was well modelled by (9.1), which is derived by combining (2.20) and (2.22):

$$\delta^3 = \frac{1}{R}\left(\frac{F_\mathrm{n}}{K}\right)^2 . \tag{9.1}$$

Typical data for indentation versus applied force are shown in Fig. 9.5. The indentation saturated at an applied force of 800 μN. For all pads used in the experiments, the indentation ranged from 40 μm to 80 μm. For normal forces lower than 500 μN Fig. 9.5 shows a good fit after assigning to the effective elasticity (K) a value of 40 kPa.

For higher normal forces the measured curve deviates from the curve obtained by (9.1) due to viscoelastic contributions. In this range Hertzian contact mechanics fails. A medium indentation of 60 μm is a large value in the framework of contact mechanics. However, compared with the value of the pad radius (2.86 mm) the indentation can be considered small so that the approximation used in the Hertz model still holds. The structural investigations proved that the assumption of treating the pad as an elastic sphere is also reasonable. The pad recovers completely after indentation. No plastic contribution was detected.

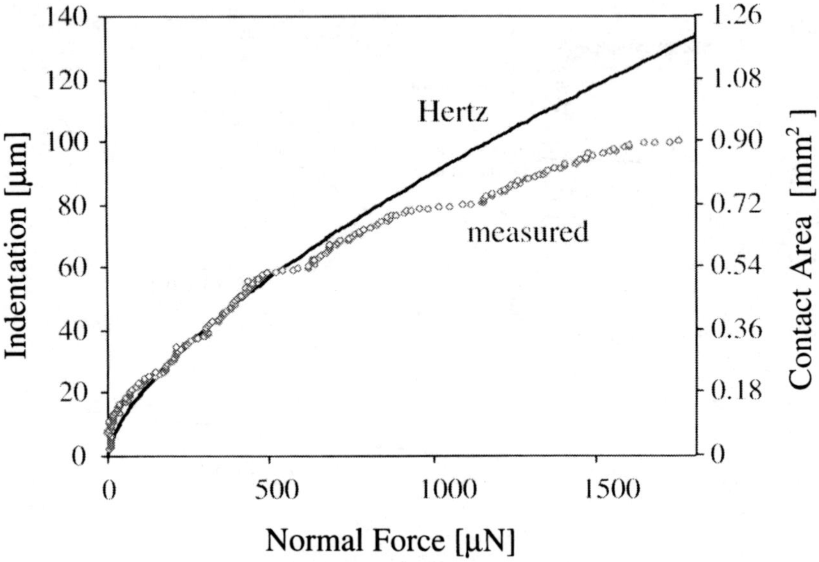

Fig. 9.5. Indentation and contact area calculated by means of Hertzian contact mechanics.

The indentation experiments can be summarized as follows.

- The pad material has a soft constitution. The effective elasticity is comparable to that of soft rubber. This property always allows the insect to flatten out the pad to achieve maximum contact.
- The pad recovers completely from indentation.
- Since the inner architecture of the pad consists of flexible tree-like structures, it is assumed that this construction is capable of adjusting the pad perfectly to the microroughness at different length scales.
- Although the pads relax when subjected to compression, the contact area seems to remain constant, which indicates that the pads are designed to maintain the optimum holding force.

9.3 Adhesion

9.3.1 The Attachment Process

The great green bush crickets have an average weight between 0.5 g and 1 g. Since each of the six legs has one major pad, the normal force ranges between 817 µN and 1634 µN. The measured pull-off force remained low as long as a critical normal force of about 800 µN was not exceeded. For higher normal forces F_{po} remained constant, as shown in Fig. 9.6. In this state the insect

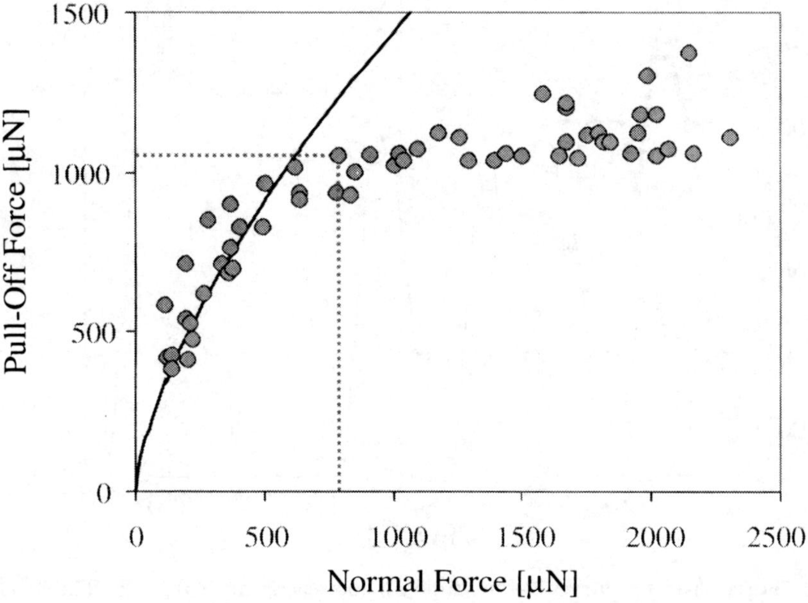

Fig. 9.6. Pull-off force measured on the pad of a living cricket. For small normal forces the pull-off force increased with increasing applied force and saturated for an applied force higher than 800 μN (dotted line). The saturated pull-off force was about 1100 μN. Based on the assumption that the pull-off force is proportional to the contact area caused by the applied force, the data were fitted in accordance with the Hertz theory.

adheres safely to the substrate and gravity is balanced for the upside-down position. All tests showed no dependence on Δt.

Material deformation is one of several factors responsible for the area of real contact between solids. Data on material deformation of the pad aided in understanding the factors influencing the adhesive properties of the pad. Figure 9.5 shows that the indentation has a rapid initial increase and a saturation for higher normal forces. With increased contact area it was harder to indent the pad. At applied forces higher than 800 μN maximum contact between pad and silicon surface was achieved accompanied by a saturation of the adhesion force of about 1100 μN. The corresponding indentation at this load was 75 μm, which is comparable with the thickness of the pad ($\approx$ 100 μm). For higher normal forces the pad behaved almost like a hard sample. For this condition, the indentation did not increase further. Before maximum contact was obtained, both the contact area and adhesion force increased with increasing applied normal force.

An insect situated upside-down under a smooth surface, needs to balance gravity with a pull-off force between 800 μN and 1600 μN according to the individual weight. In our experiments the pull-off force of the pad ranged

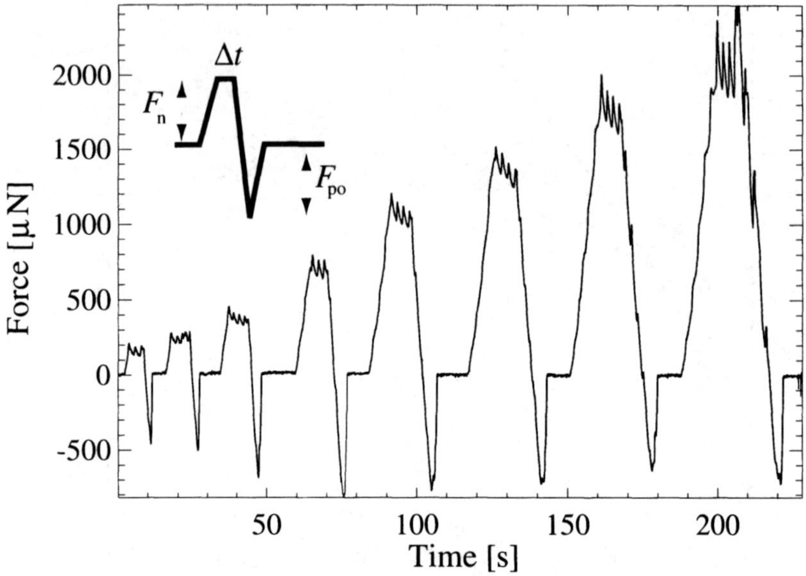

Fig. 9.7. Force distance curves versus time for increasing normal force. The inset defines normal and pull-off force as well as resting time.

from 700 μN to 1100 μN. This force is barely sufficient to hold the insect in this position. To achieve higher adhesion our observations suggest that the insect can also employ the other attachment pads on each leg. These additional pads are used only when it is necessary, e.g., for the case of wind acting on the insect.

Analyzing the force curves in Fig. 9.7, it was found that the separation of the pad from the silicon surface was recorded as a single jump-off. The retracting part of the force curve was basically linear. This shows that the contact area did not change during retraction, and the separation was basically a one-step action. Figure 9.7 shows that the pull-off force remained constant for normal forces corresponding to the weight of the insect and up.

9.3.2 The Detachment Process

Two mechanisms are conceivable for the process of detachment. The first mechanism could be a rolling process (compare with Sect. 3.5.10). The pad is peeled off the surface like an adhesive stripe that is pulled off the wall at one end. The second mechanism has to do with the ability of the insect to react to stress. Our measurements showed that the response of the insect – manifested in an up-and-down motion of the pad during the resting time (see Fig. 9.7) – increases with increasing normal force. The motion is periodic and is caused by the heartbeat of the insect. It is interesting to note that the cricket possesses a central heart and six further hearts in each leg. These

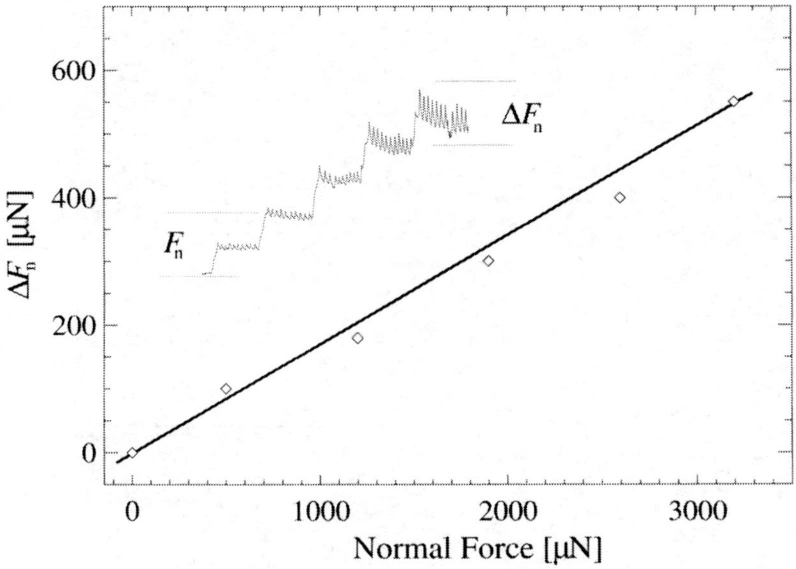

Fig. 9.8. Response of the pad to increasing normal force. ΔF_n is the amplitude of the response. F_n is increased step-wise.

micropumps make it possible to pump haemolymph to each leg separately. Figure 9.8 shows how the pad responds to increasing normal force. To achieve a maximum pull-off force of 1100 μN the pad must be indented by a distance of 75 μm. In this range the variation of the indentation can be up to 10 μm. Thus, it can be argued that a possible detachment mechanism could also be caused by the inflation of the pad, resulting in a decreased contact area and, therefore, a lowered pull-off force. As a result, the pad can detach easily.

This also implies that our pull-off experiments probed the passive behavior of the insect. The detachment, however, is an active process. An adhesion force exists only when the pad is flat. As soon as the pad regains its spherical shape, the adhesive contribution goes to zero. This also explains why Hertzian contact mechanics fits the indentation curves so well.

9.3.3 Adhesive Properties of the Secretion

The attachment system of *Tettigonia viridissima* is supplemented by a secretion produced by epidermal cells and transported onto the pad surface through the pore canals of the pad cuticle (see Sect. 4.5.1). This section shows that the secretion is a necessary contribution to adhesion.

The adhesive properties of the secretion were studied on footprints deposited on clean silicon and glass surfaces (see Fig. 9.9a). The pad of a freshly cut leg was pressed on a silicon surface and then removed. This procedure was repeated several times with the same pad to obtain a sufficient amount

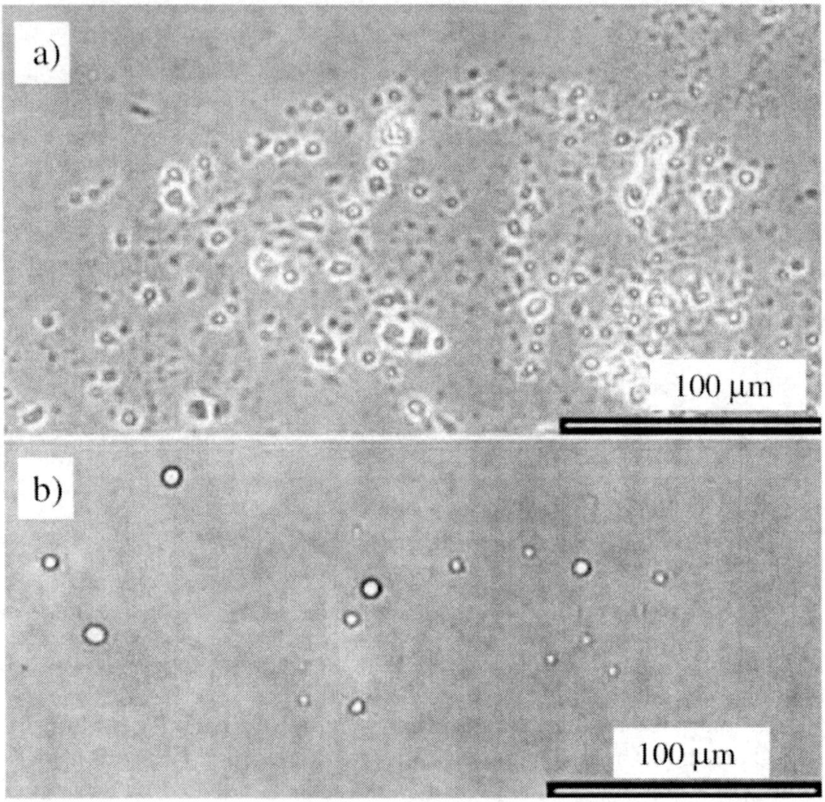

Fig. 9.9. Footprints of the *T. viridissima*. (**a**) The inhomogeneous distribution of secretion on a glass surface (cover slip). (**b**) The droplets that occurred in the footprint preparations in distilled water.

of secretion. The silicon flat was glued to the spring to act as the upper sample. The silicon ball was attached to the sample holder. Before probing the secretion, ball and flat sample were sequentially cleaned with acetone and ethanol in an ultrasonic bath. No adhesion force between the ball and the clean silicon surface was measured, indicating that the water coverage was too low to initiate capillary bridging.

The adhesion measurements of the secretion were carried out at eight locations on the silicon flat. Depending on the selected location, the pull-off force increased or decreased monotonically or remained constant as a function of the normal force (see Fig. 9.10). The increase or decrease in pull-off force was normally between 100 µN–400 µN. In a few cases, only scattered data were obtained.

The reason for the noisy data was the inhomogeneous distribution of the secretion. Figure 9.9a shows that the secretion does not cover the substrate homogeneously. This means that the ball contacts a location with a thick

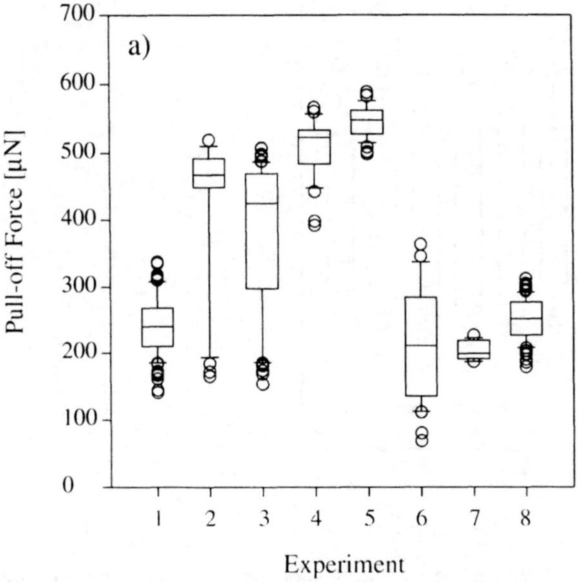

Fig. 9.10. Average pull-off force at eight locations measured with the silicon ball. Due to the inhomogeneous distribution of secretion, the pull-off force was location-dependent.

coverage in one experiment and in the next experiment a location with thin or no coverage. Upon contact the ball deforms the secretion, resulting in a changing contact area and, thus, in different pull-off forces. Figure 9.11 shows the distribution of the pull-off forces as a function of the normal force.

Studies investigating the chemical composition of insect footprints and cuticle lipids indicate that the liquid that covers the pad is mainly composed of hydro-carbons with hydrophobic surface properties. That is why the secretion forms droplets when exposed to water, as shown in Fig. 9.9b. Therefore, capillary action due to water can be neglected. However the hydro-carbons do wet the pad surface (and the hydrophilic Si surface) and form a capillary neck.

In addition to capillarity, an attractive force can also be caused by dispersion forces when the distance between both samples is lower than about 10 nm. In both cases, the contact area plays a crucial role for the strength of adhesion.

In summary:

- The attachment is achieved by pad deformation with a maximum pull-off force of 1100 μN.

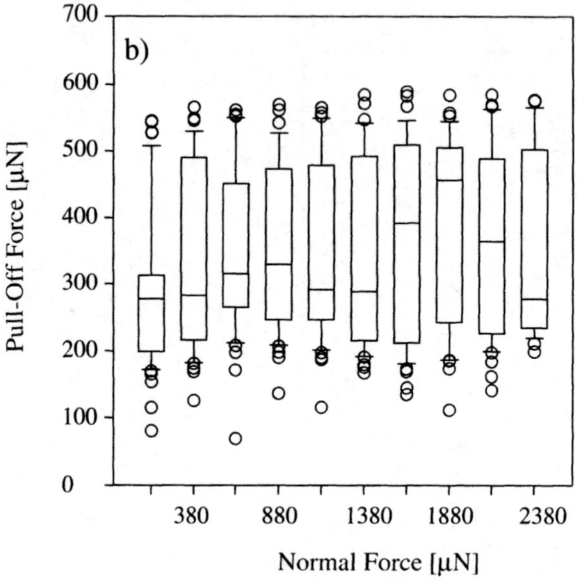

Fig. 9.11. Pull-off force data obtained at eight locations pooled together and plotted versus normal force.

- The adhesive secretion is essential for the attachment of the pad due to capillary bridging.
- Water is kept out of the attachment process because of the hydrophobicity of the pad secretion.
- The detachment is most probably achieved via the inflation of the pad and the resulting loss of contact area.

10. Case Study II: Friction

10.1 Test Setup

For friction experiments, the interaction was measured between the attachment pad and the flat silicon surface. The leg containing the pad was attached to a sample holder that was oscillated in the tangential direction over a length of 30 µm by a piezo-transducer (x-piezo) at a constant velocity (30 µm/s). The silicon sample – a 5 mm by 5 mm piece cleaved from a silicon wafer – was attached to the double-leaf spring as shown in Fig. 10.1. For these experiments the interferometer-based tester was used because of its higher force resolution (see Sect. 5.1.4).

Friction causes the spring to deflect and is measured by the single-beam laser interferometer. The measured length is then multiplied by the spring constant to obtain the tangential force. The laser interferometer has a resolution of 1 nm. Therefore, with a spring constant of 50 N/m a tangential force range from 50 nN to 1.5 mN was covered. The normal force, adjustable in a range between 500 nN and 1 mN, was set by a second piezo-transducer (z-piezo) acting in the normal direction. This deflection was calibrated by means of a fiber-optic length sensor placed above the spring (see inset of Fig. 9.4). During the tests the normal force was set by expanding the z-piezo. The spring constant in the normal direction was 100 N/m. By means of mechanical micropositioners both samples were first approached manually. The final engagement was achieved by expanding the z-piezo. When the first tangential deflection of the spring was detected, contact was established with an average error of 50 nN (due to mechanical noise). The normal force can be increased by increasing the control voltage of the piezo-transducer. The complete tester is shown in Fig. 10.1.

To immobilize the insect, its body was attached to the holder connected to the x-piezo using adhesive tape. The body was positioned upside-down (the back was attached) and oriented in a horizontal direction along the x-axis. The leg tarsus was firmly connected to the sample holder using wax. The silicon surface was positioned parallel to the sample holder in a way that limited the contact of the leg to the attachment pads, as encircled in the inset of Fig. 10.1. In this condition the whole insect was moved together with its leg section by periodic expansion and retraction of the x-piezo. The force acting on the leg was set according to the natural level of mechanical

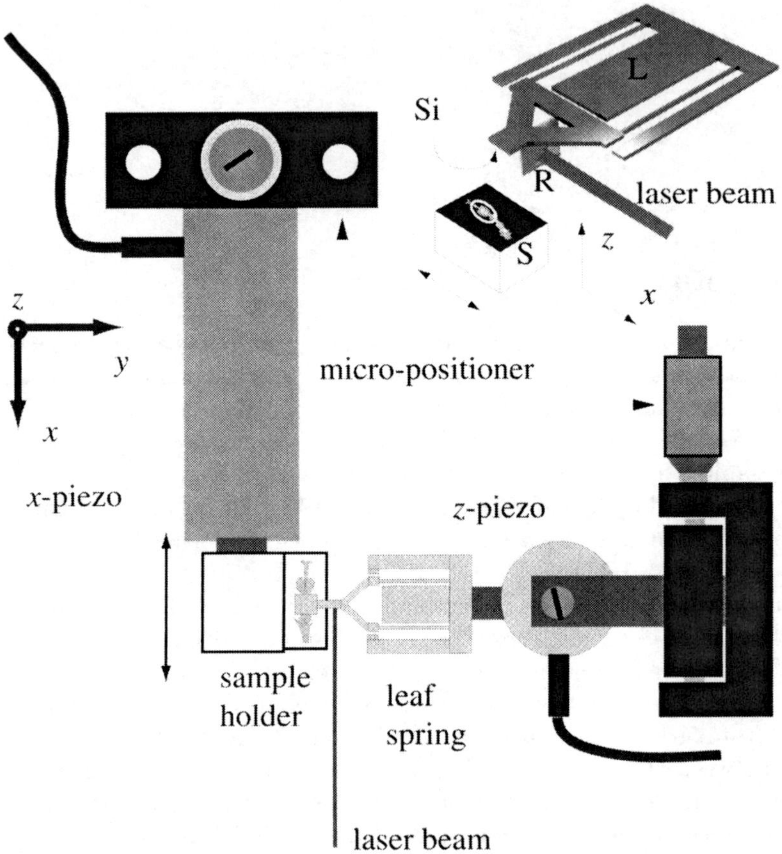

Fig. 10.1. Schematic of the tester (top view). S, sample, i.e., the last segment of the leg. The body of the insect is not shown. L, leaf spring. The Si sample is located at the lower side of the triangular section of the spring. R, reflector for the laser beam.

interaction. It is, therefore, assumed that the pads were not damaged during the tests.

The complete tester was housed in an environmental chamber where both temperature and relative humidity were kept constant. All experiments were carried out in air at the prevailing temperature of 25°C and a relative humidity of 58%.

10.2 Results

By oscillating the sample along the x-axis (distal-proximal) at low normal force, the authors were able to measure the frictional properties of the pad

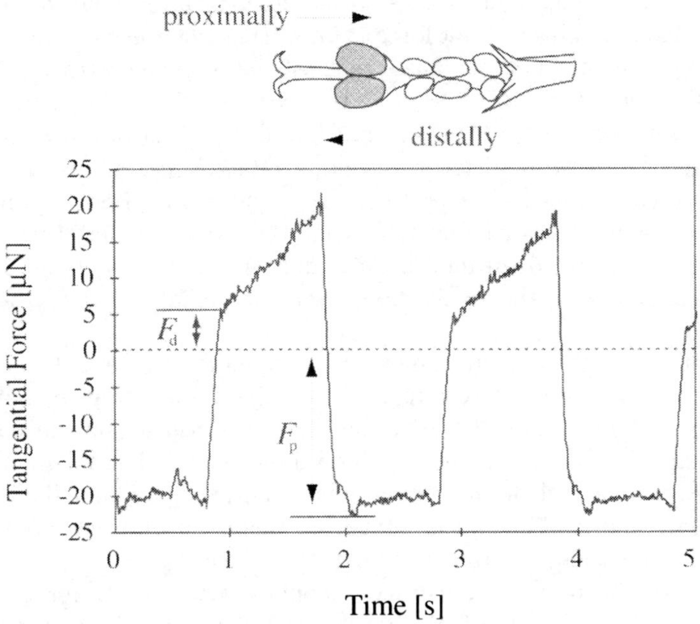

Fig. 10.2. Frictional behavior of the pad. Friction measured in different directions. F_p, friction force measured during pad movement toward the center of the insect body (proximal direction). F_d, friction force in opposite (distal) direction.

surface in both directions (F_d and F_p). The selection of low normal force (about 80 μN–100 μN) was necessary to measure the initial processes of contact formation of pad and counter surface without getting into viscoelastic deformation of the pad. As static friction, the force between zero and the onset of sliding was used (see Fig. 10.2). The experiments showed that static friction during proximal movement was larger and stable compared with friction during distal movement. For example, the static friction force in the distal direction was about 5 μN and in the proximal direction 23 μN. During distal movement, dynamic friction slowly increases.

10.2.1 Load Dependence and Friction Hysteresis

The friction tests were initiated by oscillating the pad while letting the Si sample approach. If dry samples, for instance in the hydrophobic SILICON model system, are brought into contact, then the friction force increases gradually. When the dry samples are replaced by hydrophilic samples, a distinct jump to contact can be observed (see Fig. 2.41). A similar behavior was measured for the pad contacting the Si surface. This indicates that liquid bridges between pad and surface induce capillary forces.

The results can be seen in Fig. 10.3. Besides the jump into contact the negative normal force (pulling force) also underlines the concept of capillarity. The capillary necks suddenly break at a normal force (that is equal to the pull-off force) of -450 µN. This pull-off force is lower than the force obtained in the adhesion experiment, since the flat sample oscillates during retraction. Rearranging (2.42), the interfacial energy for the hydro-carbon-covered surface can be calculated as $\gamma = F_{\mathrm{po}}/4\pi R$=30 mN/m. The pad has a radius of about 2.86 mm and for the pull-off force a value of 1100 µN was assumed.

Due to the action of capillarity the friction force versus load curves exhibit a clear hysteresis. Figure 10.3 demonstrates that the hysteresis has larger values for proximal than for distal motion. Proximal means that the pad is pulled towards the center of the insect body, whereas distal refers to the opposite direction.

For low normal forces the indentation experiments indicated a small contact area leading to low friction forces in both the distal and the proximal directions. For loads higher than 600 µN, the pads establish an optimized contact area accompanied by a maximum pull-off force. The pad and sample form a permanent contact and sliding ceases. As a result, the pad and silicon sample move synchronously. This means that from this point on the friction cannot be measured for the given range of tangential force.

The formation of the permanent contact happens very rapidly for the case of the contact with the smooth Si sample. To slow down the contact formation in order to visualize the process, the smooth surface was replaced by the patterned Si surface, as introduced above. The pattern consists of equally spaced circular bumps with a surface finish like that of regular wafer

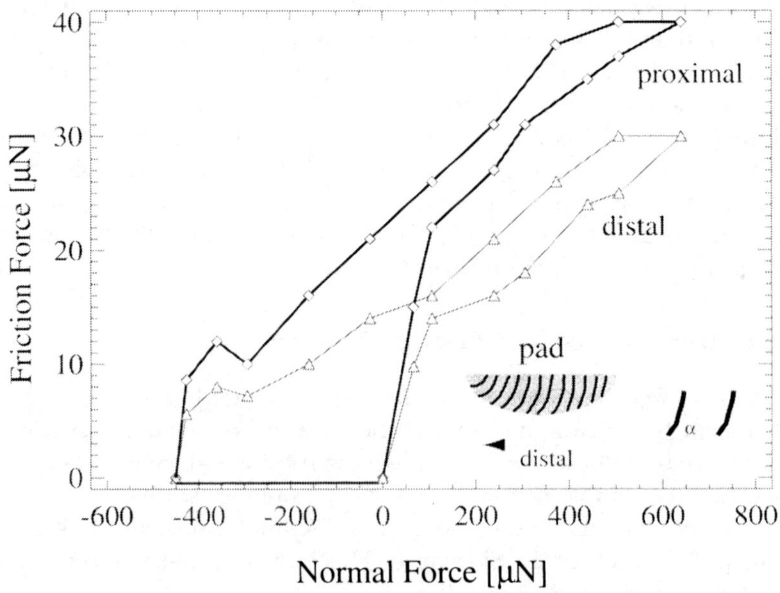

Fig. 10.3. Friction force versus normal force. Inset: pad with dendrite system.

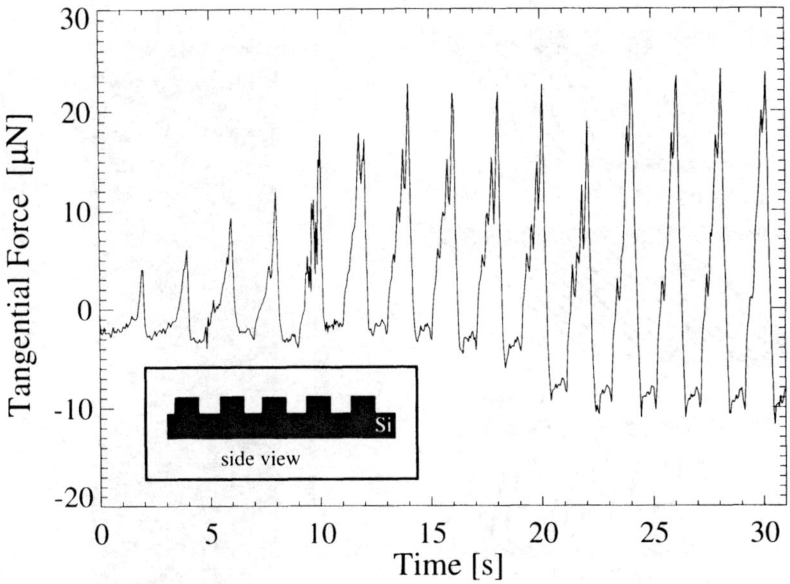

Fig. 10.4. Increasing friction force due to the time-dependent formation of the contact between the pad and a structured Si sample. The inset shows the patterned Si surface. This piece corresponds to a length of about 70 μm. The diameter of the Si bumps is approximately the same as the diameter of one hexagonal section of the pad skin.

silicon. After exceeding 600 μN in normal force the propagation of the contact front was observed. This is shown in Fig. 10.4. The normal force was set slightly above 600 μN and was kept constant. As a result, the contact formed gradually and was followed by increasing friction. For higher normal forces the time to establish the contact decreased significantly.

10.2.2 Anisotropy of Friction and Inner Structure

The anisotropy of friction can be explained by considering the information obtained from the micro-indentation experiments. Since the rods have an initial predetermined slope of 45–70°, the normal force – due to the weight of the insect – will decrease the slope to 5–10° (shown in Fig. 9.3a, b). Leg movements directed proximally also decrease the slope. As the rods deform, the shape of the hexagonal outgrowth also changes. It is assumed that the rods together with its branches and the flexible cuticle ensure optimal contact to the substrate by adjusting the pad to micro- and mesoscale asperities.

If the pad moves distally, some critical force is needed to overcome the rod stiffness to bend the rods towards the other direction (Fig. 10.5c). Thus, the friction force increases slowly, since energy is dissipated continuously due to the bending of the rods. This hypothesis also explains why anisotropy increases with increasing normal force. However, at a certain normal force, the

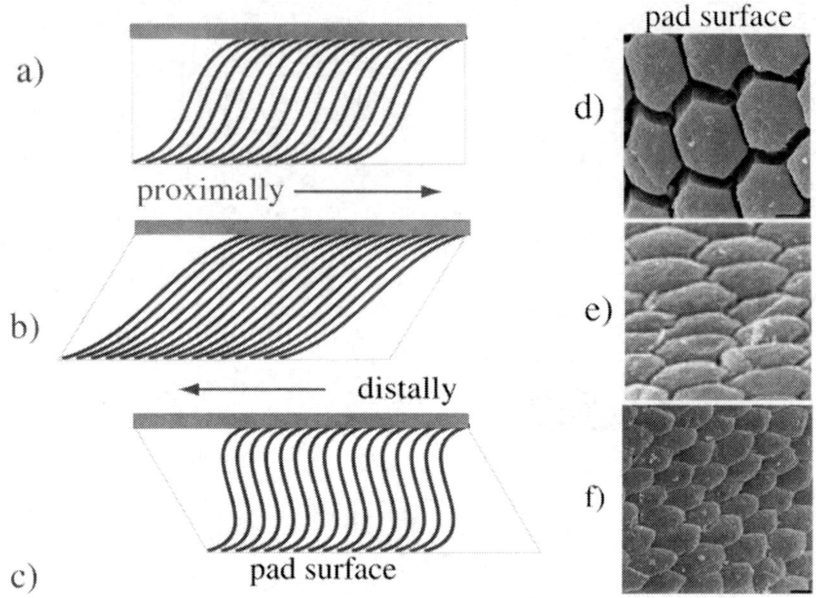

Fig. 10.5. Inner structure of pad under frictional stressing in the distal and proximal directions. (**a**) Normal shape of dendrite structure. (**b**) Deformation during friction in proximal direction. (**c**) Deformation during friction in distal direction. (**d**) Pad surface without frictional stressing. (**e**) Pad surface for friction in proximal direction. (**f**) Pad surface for friction in distal direction.

anisotropy no longer increases. It is assumed that at this normal force the reorientation of the rods towards the opposite direction will be limited. Presumably, the anisotropy in friction force is of importance during fast movements, such as jumping and landing, when friction forces act in different directions.

The pads have a function comparable to that of a clutch. Friction under low normal forces leads to the bending of the flexible structures inside the pad to achieve an optimized contact area. The bending mechanism is supported by the action of the normal force and the proximal leg motion, which is always present when the pad contacts a surface. After exceeding the critical load of about 600 μN permanent contact is established and sliding stops. The attraction that is necessary to balance gravity results from a combination of adhesion provided by capillarity due to a liquid secretion and the formation of a self-mating mechanical contact between the pad and any irregular surface.

The findings are summarized as follows:

- Friction in the initial process of contact formation is necessary to achieve an optimized contact with the surface.
- Friction is anisotropic. Higher friction was measured for pad motion towards the center of the insect body.

- Friction is caused by capillarity, pulling the pad and counter surface together.
- At a critical normal force the pad achieves permanent contact with the substrate. From this point onwards friction ceases, since sliding stops.

11. Case Study III: Material Properties

11.1 Elastic Properties

Since the tribological properties of a biological material are determined not only by the surface but also to a great extent by its inner structure, the mechanical response of the whole pad was investigated to determine its elastic and viscoelastic properties.

Elastic deformation of the pad occurred during the approaching period (see Fig. 11.1). This deformation was well fitted by the Hertz model for normal forces less than 600 μN–700 μN (Fig. 9.5), although the shape of the pad was not exactly spherical and the material properties were not isotropic.

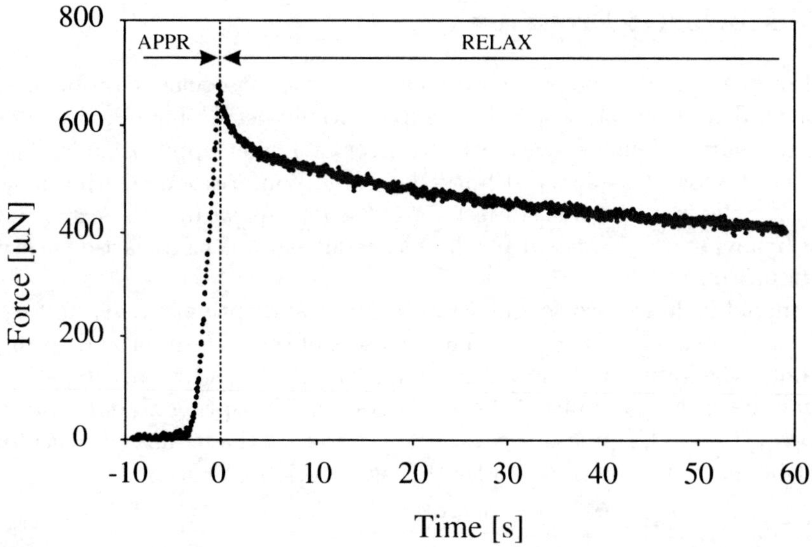

Fig. 11.1. A typical force curve recorded for the evaluation of the mechanical properties of the pad. The force curve includes two distinct parts: the approaching process (APPR) and the relaxing period (RELAX). In the approaching process, the pad was rapidly approached to the hard silicon surface to obtain an interacting force (applied force), Generally, it took 4 s–6 s to reach an applied force of 600 μN–700 μN. After that the pad was kept in contact with the silicon surface, and the interacting force slowly relaxed.

Taking the relation (2.20) introduced in Sect. 2.3.4, the normal force relates to the indentation as:

$$F_n = \frac{4}{3}\frac{E}{1-\nu^2}a\delta \,. \tag{11.1}$$

An applied force between 600 μN and 700 μN resulted in an indentation of 40 μm–80 μm. When such big forces were applied, the resulting deformation could not be described by Hertz theory anymore. This indentation corresponds to the maximum contact between the pad and the silicon surface.

To calculate the pad elasticity, Hertz theory was applied only to the initial part of the indentation curve. Indentation follows (9.1) when the pad shape was assumed to be a part of a sphere (Fig. 8.2). Taking 2.86 mm for the average radius of the pad curvature (1.5 mm for the pad radius and 100 μm for the pad thickness), an effective elastic modulus of 21.9 kPa–64.1 kPa was found by fitting the data of the indentation versus normal force. When a Poisson ratio of the pad material of 0.5 is assumed (generally accepted for cell material), an elastic modulus ranging from 15.3 kPa to 45.0 kPa was obtained $[E = 0.75 \cdot K(1-\nu^2)]$. For higher normal forces, the deformations deviated from those predicted by the Hertz theory.

11.2 Viscoelastic Properties

In addition to elasticity, the pad material also revealed viscoelastic properties. When the indentation was stopped, an exponentially decreasing relaxation of Δz was measured. This decrease was stronger at higher applied loads. The pad material showed a higher stiffness for rapid compression. During slow compression the material was subject to relaxation. According to Sect. 2.3.3 the pad behavior was modelled by the basic elements of viscoelastic theory as shown in Fig. 11.2.

The model includes two sets of serial dashpot and spring (η_1, E_1 and η_2, E_2) and an additional spring (E_0). The two sets of serial dashpot and spring reflect both the rapid and slow force relaxation. All the springs together determine the elastic modulus of the whole system. The spring E_0 determines the stress of the pad for time $t \to \infty$. It is assumed that the strain ε is basically constant in the resting period. Under this assumption one obtains

$$\sigma(t) = \varepsilon(E_0 + E_1 \mathrm{e}^{-t/t_1'} + E_2 \mathrm{e}^{-t/t_2'}) \,, \tag{11.2}$$

where two time constants are given as t_1' and t_2'. The stress curve (Fig. 11.3) was obtained from the interacting force in the resting period divided by the contact area of the pad and the silicon surface. By fitting this curve by the Kelvin model, the time constants, the viscosities and the elastic moduli were obtained.

Force curves with an applied normal force of 600 μN–700 μN were used to calculate the viscosity. The contact area A_c, caused by the applied force, was derived from the Hertz theory:

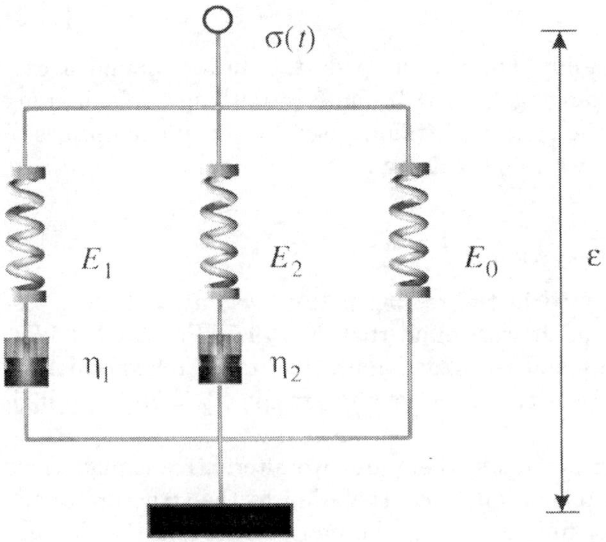

Fig. 11.2. The Kelvin viscoelastic model with two sets of dashpot and spring responsible for both rapid and slow relaxation.

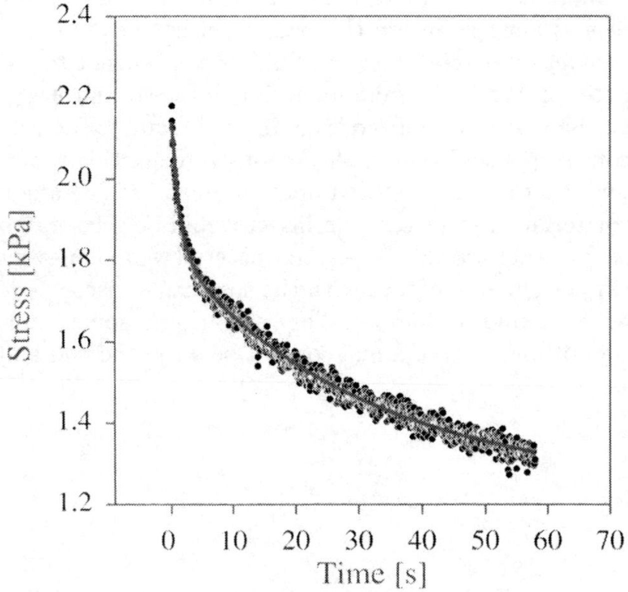

Fig. 11.3. The stress versus time curve. These data, derived from the resting period of a force curve, were used for the calculation of the viscoelastic properties of the pad. There are two kinds of relaxation: a rapid one (at the beginning) and a slow one. The curve was fitted (line) by the Kelvin viscoelastic model.

$$A_c = \pi a^2 = \pi R\delta \, , \tag{11.3}$$

where a is the contact radius. For various pads the authors found a corresponding contact area ranging from 0.31 mm^2 to 0.91 mm^2. Combining (2.10), (2.22) and (11.1) the constant strain caused by the applied forces of 600 μN–700 μN and lower was obtained from

$$\varepsilon = \frac{\sigma}{E} = \frac{F_n/A_c}{E} = \frac{4}{3}\frac{1}{1-\nu^2}\frac{1}{\pi}\left(\frac{\delta}{R}\right)^{1/2} \, . \tag{11.4}$$

The stress curve, obtained in the resting period, was fitted by the viscoelasticity model of Kelvin. It was found that E_0=19.7 kPa, E_1=5.1 kPa, E_2=9.6 kPa, η_1=6.8 kPa·s and η_2 =294 kPa·s. The rapid relaxation time ($t_1' = 1.322$ s) has a low-viscosity η_1, the slow relaxation ($t_2' = 30.7$ s) a high viscosity η_2.

For the two relaxation processes, there are two alternative explanations as follows. (1) The rapid relaxation t_1' is attributed to the structure of the endocuticle underlying the fibrous part of the pad cuticle, and the slow relaxation t_2' is attributed to the surface structure. (2) The rapid relaxation is the property of the fiber structure of the pad only, including the dendrites of the surface structure and the endocuticle. The slow relaxation is due to the interaction of these fibers with the fluid inside of the fibrous cuticle and due to the haemolymph and air sacs inside the pad. In either of these two explanations, the slow relaxation is related to the fluid or air, but not to the solid phase of the pad material. While the functional significance of the elastic properties of the pad can be easily recognized, it is more difficult to find an explanation for the viscous response. When pads demonstrate elastic properties only, one would expect that energy is stored upon landing, as is common in other cuticle-related materials. In this case, an insect would be accelerated again just after landing. In other words, an elastic material would prevent the insect from getting in permanent contact with the surface. A viscoelastic material, however, serves as a kind of damper. The relaxation response aids in surface replication and optimization of a real contact between the pad and the underlying substrate.

12. Outlook

This book was conceived as a first approach to discover the micromechanical properties of biological systems. The outcome of further studies could be used to foster new design philosophies for MEMS.

For instance, when object manipulation and micropart handling are performed on the microscale, adhesion becomes increasingly important with decreasing object size. When mechanical grippers are used, the holding force can be set to an appropriate value. The release of the micropart, however, is commonly accomplished by the action of gravity after opening the gripper jaws. As the dimensions shrink, adhesion decreases by a power of two, while gravity decreases by a power of three, causing problems for the disconnecting process as well as for the manipulation accuracy.

It was shown that adhesion is the sum of several contributions all arising from atomic or molecular interactions. The major contribution to adhesion in the microrange arises from capillary forces due to adsorbed water. The water forms capillary bridges holding both parts together. In addition, capillary condensation contributes strongly when microslits are formed. This is always the case for micrygrippers. One way to control capillary action is to generate hydrophobic, i.e., non-polar surfaces. These surfaces do not wet; thus no water can be adsorbed. Another way is to reduce the real area of contact. This can be achieved by microstructuring the gripper jaws. Since adhesion is a function of the contact area, the attracting forces can be reduced significantly. A third way is to learn from bio-systems where adhesion and surface topography are controlled in order to achieve optimized attachment properties.

What important conclusions can be drawn from optimized biological systems to improve MEMS? First of all, the attachment force (gripping force) is optimized for the weight of the mobile system. The use of viscoelastic materials ensures the maximal gripping force at the lowest contact area due to micro- and mesoscale contact formation. Since the pads have a hydrophobic hydro-carbon coverage, the action of adsorbed water leading to capillarity is suppressed. This is very important, since the action of water cannot be controlled by the insect. In addition, many plant surfaces are covered by a water-repelling wax layer.

The resulting holding force arises from a combination of intimate contact and the action of capillarity due to the controlled application of hydro-carbons. However, van der Waals forces (dispersion force) as a cause of adhesion cannot be ruled out, since the flexible parts of the attachment organs could perhaps be capable of achieving intimate contact with a spacing lower than 10 nm. In this case, both surfaces are in reach of electronic interaction and mutual polarization is possible.

In contrast to mechanical microsystems having a fixed geometry, biological microsystems can actively change geometry and topography according to the needs of the prevailing environmental conditions.

One of the main differences from technical systems is the permanent self-maintenance of the tribological contact in bio-systems. The state of lubrication is always monitored and, if necessary, the lubricant is replenished. Furthermore, the tribological contact is kept free of debris (pad-licking behavior of the crickets).

These findings may also have an impact on macroscale systems. MEMS could be part of macromechanical systems to monitor the physical conditions (temperature, lubricant thickness, lubricant chemistry) and to manage the maintenance of the system.

Our investigations have shown that the tribology of biological microsystems has to consider contributions that are mostly negligible on the macroscale. For instance due to the flexibility of the materials not only are the topmost atomic layers important for friction and adhesion, but so also are the underlying structures. This means that the study has to combine surface science to discover the effect of the surface, materials science to investigate the contribution of the bulk material and micromechanics to determine the action of inner structures and the role of elasticity and viscoelasticity. In addition, a strong input of chemistry and bio-chemistry knowledge is necessary to analyze the secretion.

This high degree of interdisciplinarity is an inevitable requirement but also a rewarding challenge.

Part V

Appendix

A. Contact Models

This appendix lists the friction force formulae derived by combining the Bowden Tabor model ($F_\mathrm{f} = \tau A_\mathrm{c}$) with the appropriate contact model. In addition, it lists the model parameters, which determine the applicability of the contact models.

Hertz Model

$$F_{\mathrm{f,Hertz}} = \pi\tau\left(\frac{RF_\mathrm{n}}{K}\right)^{\frac{2}{3}} \tag{A.1}$$

$$F_{\mathrm{a,Hertz}} = 0 \tag{A.2}$$

DMT Model

$$F_{\mathrm{f,DMT}} = \pi\tau\left(\frac{R}{K}\right)^{\frac{2}{3}}\left(F_\mathrm{n} + 4\pi\gamma R\right)^{\frac{2}{3}} \tag{A.3}$$

$$F_{\mathrm{po,DMT}} = -4\pi\gamma R \tag{A.4}$$

JKR Model

$$F_{\mathrm{f,JKR}} = \pi\tau\left(\frac{R}{K}\right)^{\frac{2}{3}}\left(F_\mathrm{n} + 6\pi\gamma R + \sqrt{12\pi\gamma R F_\mathrm{n} + (6\pi\gamma R)^2}\right)^{\frac{2}{3}} \tag{A.5}$$

$$F_{\mathrm{po,JKR}} = -3\pi\gamma R \tag{A.6}$$

Maugis Model (analytic approach according to [726])

$$F_{\mathrm{f,M}} = \pi\tau a_{0,\psi}^2\left(\frac{\psi + \sqrt{1 - \frac{F_\mathrm{n}}{F_{\mathrm{po,M},\psi}}}}{1+\psi}\right)^{\frac{4}{3}} \tag{A.7}$$

$$F_{\mathrm{po,M},\psi} = -3\ldots-4\pi\gamma R \tag{A.8}$$

$\psi \Rightarrow 0,\ \lambda_{\mathrm{m}} \Rightarrow 0 \longrightarrow$ DMT Model
$\psi \Rightarrow 1,\ \lambda_{\mathrm{m}} > 5 \longrightarrow$ JKR Model

Parameters

Derjaguin parameter

$$\beta = \frac{64}{3\pi}\left(\frac{\gamma^2 R}{\pi K^2 s^3}\right)^{\frac{1}{3}} \tag{A.9}$$

Tabor parameter

$$Z = \left(\frac{16R\gamma^2}{9K^2 s^3}\right)^{\frac{1}{3}} \tag{A.10}$$

Maugis parameter

$$\lambda_{\mathrm{m}} = 2\sigma_0 \left(\frac{R}{\pi\gamma K^2}\right)^{\frac{1}{3}} \tag{A.11}$$

Relations between the parameters

$$\lambda_{\mathrm{m}} = 0.4Z \tag{A.12}$$

$$\lambda_{\mathrm{m}} = 1.1570\beta \tag{A.13}$$

Validity ranges of models

$\beta, Z, \lambda \Rightarrow 0\ (\lambda_{\mathrm{m}} < 0.1) \longrightarrow$ DMT Model
$\beta, Z, \lambda \Rightarrow \infty\ (\lambda_{\mathrm{m}} > 5) \longrightarrow$ JKR Model

B. Capillary Theory

In Sect. 2.4.3 a simple approximation ($F_c = 4\pi\gamma R$) was introduced to calculate the capillary force between the ball and the flat sample. Throughout the book cases were shown where this equation did not hold. This appendix, therefore, presents a detailed derivation of this equation and shows where simplifications are introduced.

The starting point of the derivation is Fig. B.1 showing the meniscus between the ball with radius R and the flat surface. The meniscus is approximated by a segment of a circle with radius r_1. This is the first simplification: assuming that the ball and flat have identical wetting properties and thus, equal contact-angles Θ. The capillary neck has the height b. According to (2.40) the capillary force can be expressed by:

$$F_c = \Delta p \pi x^2 \ , \tag{B.1}$$

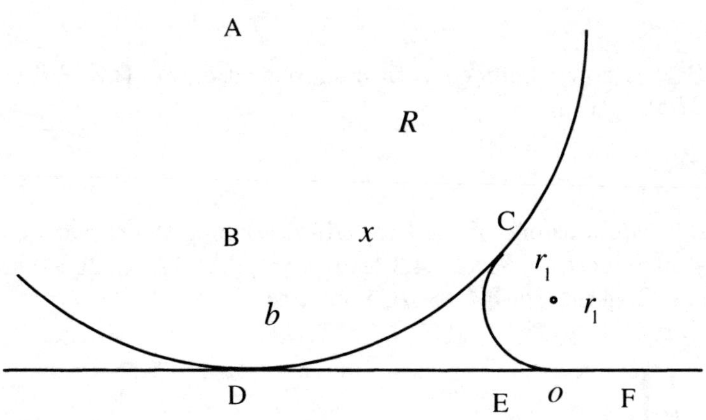

Fig. B.1. Capillary theory

where πx^2 is the area of the ball that is wetted. Considering the geometric relations in Fig. B.1 one obtains for the triangle ABC:

$$x^2 = 2Rb - b^2 \tag{B.2}$$

and for triangle ADF:

$$(R + 2r_1)^2 = (x + o)^2 + R^2 . \tag{B.3}$$

Triangle CEF gives

$$b^2 + o^2 = 4r_1^2 . \tag{B.4}$$

This, of course, is a further simplification, holding only for small values of b; this means for thin liquid film thickness. In addition, the following relation is incorporated:

$$\frac{o}{b} = \frac{o + x}{R} . \tag{B.5}$$

Combining (B.5) and (B.3) one obtains:

$$4Rr_1 + 4r_1^2 = \frac{o^2}{b^2} R^2 . \tag{B.6}$$

Now (B.4) is inserted into (B.6) giving after division by R^2 and multiplication by b^2:

$$\frac{b^2 r_1}{R} + \frac{r_1^2 b^2}{R^2} = r_1^2 - \frac{b^2}{4} . \tag{B.7}$$

The terms on the left-hand side can be neglected, since R is much larger than b and r_1, and one obtains:

$$b \approx 2r_1 . \tag{B.8}$$

Equation (B.1) is now combined with a simplification of (B.2) ($x^2 = 2Rb$, since $b^2 \ll 2Rb$):

$$F_c = 2\pi R\gamma \frac{b}{r_1} . \tag{B.9}$$

Equation (B.9) contains one of the two radii describing the curvature of the meniscus as introduced in Sect. 2.4.3 [$\Delta p = \gamma(1/r_1 + 1/r_2)$]. It is assumed that $r_2 \gg r_1$. Together with (B.8) one obtains:

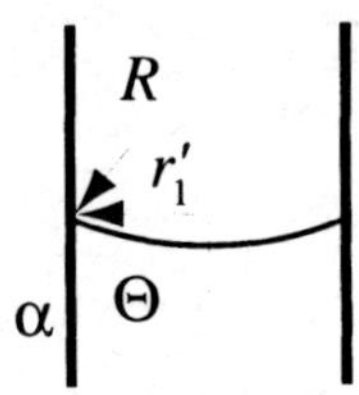

Fig. B.2. Angular relation.

$$F_c = 4\pi R\gamma \, . \tag{B.10}$$

This equation holds only if the radius r_1 applies to a sphere. For non-spherical form one obtains (B.11), using the angular relation $\cos\Theta = r_1'/R$ as shown in Fig. B.2:

$$F_c = 4\pi R\gamma \cos\Theta \, . \tag{B.11}$$

C. Glossary

abdomen	posterior part of insect body
acanthae	cuticle outgrowths originating from single epidermis cell lacking sockets
achene	seed of plants
adipose	fatty
aeshnid	dragonfly species from the family Aeshnidae
ambulacral	part of the ambulacrum, radial zone or groove on the surface of an echinoderm, containing podia or tube feet
Annelida	animal phylum including worms with segmented body
anole	lizard bearing attachment pads
anoline section	group of lizards belonging to family Iguanidae
antennae	paired movable appendages of the head in insects and crustaceans
antennules	anterior pair of the antennae of crustaceans
Anthozoa	a class of marine cnidaria comprising the corals, sea anemones, and related forms
apex	terminal region of the structure
aphid	insect from family Aphididae, plant lice
apical	located closely to the tip of a structure
apodems	an ingrowth of the integument to which muscles attach
arboreal	living on trees
arolium	attachment pad between claws of an insect
arthrodial membrane	membranous cuticle joining body segments in Arthropoda
Arthropoda	lineage of animals with jointed outer skeleton and jointed segmental appendages (spiders, crustaceans)
ascidians	marine animals belonging to the class Ascidiacea
Asteroidea	the class of Echinodermata comprising the starfish
Asteronyx loveni	echinoderm species, belonging to Ophiuroidea
Balanus balanoides	barnacle species, crustacean
barnacle	marine crustaceans belonging to the group Cirripedia

basal disk	part of the body opposite to the oral side in some Cnidaria; responsible for the attachment to substrates
Bivalvia	group of Mollusca with a two-valved shell
Blattaria	insect order including cockroaches
brachii (brachial)	relating to the arm or a process like an arm
Branchiobdellidae	branchiobdellids, animal group belonging to the phylum Annelida
branchiostegal	bony processes supporting the membranes enclosing the gill chamber in most fish
Brassica	latin for cabbage
byssus	bundle of tough threads of sclerotized protein, secreted by the foot in many molluscs used in anchoring
the	animal to the substrate
calyx	cup-shaped biological structures
canine	animal belonging to Canidae, for example, dogs and wolves
Carabidae	family of beetles, ground beetles
cephalenteron	anterior thickened part of the larva in ascidians
cercopid cicadas	leafhopper species of insects (family Cercopidae)
Cercopidae	family of leafhoppers
Cercopis vulnerata	leafhopper species (family Cercopidae)
chetae	cuticle outgrowths originating from two epidermis cells bearing sockets
Chiroptera	latin for bats
cilium	a minute hair-like process
ciliary	relating to cilia
Ciliata	a large group of unicellular organisms possessing cilia as locomotory devices
Cirripedia	group of marine crustaceans
cladoceran	crustaceans belonging to the group Cladocera
clavate	gradually thickening near the distal end
Cnidaria	phylum of invertebrate animals
Coleoptera	beetles or insect order including beetles
Colocasia	a genus of Asian aroid plants
corium	(or dermis, cutis) inner layer of vertebrate skin made of white fibrous connective tissues
coxa	the most basal segment of the leg
Crematogaster	ant species
cribellate	spider silk produced by a specialized spinning organ called cribellum
Crustacea	primary marine group of Arthropoda with two pairs of antennae

cuboidal epithelium	part of multilayered vertebrate integument
cui	vertebrate animal belonging to the family Caviidae (Rodentia)
cuticle	the non-cellular outer layer of the integument
cuvierian tubules	a glandular tubules that can be extruded from the cloaca
Cycropteridae	family of the hill stream fish
decapod	crustaceans belonging to the group Decapoda
denticle	little outgrowth on the surface
Dermaptera	insect order including earwigs
deuterophlebiid	fly belonging to the family Deuterophlebiidae
Diplodactilidae	lizard family
Diptera	insect order including butterflies and moths
distal	located far from the point of attachment or origin
dorsal	belonging to or situated near or on the back of an animal
duo–gland	adhesive system based on two-component adhesion, one of its parts usually consists of two types of cells or two distinct glands
ecdysis	the act of melting the outer cuticle layers in insects and crustaceans
Echeneidae	family of the hill stream fish
Echinodermata	phylum of invertebrate animals with bilateral symmetry (as larvae) and radial symmetry (as adults)
echinoids	sea urchins and related forms
ecribellate	spider silk produced by spinning organs different from cribellum (see cribellate)
elytron, elytra	a leathery or rigid forewing especially in beetles
empodial pulvillus	median attachment pad on the leg of several groups of flies
epicuticle	outermost layer of the cuticle
epidermis	cell layer (or layers) underneath the integument
Eudiscopus	genus of bats (Chiroptera)
euphausid	crustaceans belonging to the group Euphausiacea
euplantulae	flexible pad-like structures at the ventral side of tarsomeres
eversible pretarsal bladder	eversible, balloon-shaped adhesive structure at the distal end of the tarsus in insects from the order Thysanoptera
extensor	a muscle that serves to extend a limb or other part of the body (antagonist to flexor)

femoral	belonging to or related to the femur
femur	insect leg segment located between the trochanter and tibia
Flagellata	uni-cellular organisms with an elongated filiform structure (flagellum) used for locomotion
flexor	a muscle that serves to bend a limb or other part of the body (antagonist to extensor)
flukes	parasitic worms
follicular	belonging to or provided with follicles, small simple or slightly branched glands
fossula spongiosa	spongy sole, hairy flexible attachment organ on the forelegs of bugs from the family Reduviidae
fulgoroid	leafhoppers from the suprafamily Fulgoroidea
galea	part of insect mouth-parts, the outer maxillary lobe
galeal	derivative of the galea, part of the galea
gastropods	snails and slugs, molluscs belonging to the group Gastropoda
gastrotrichs	Gastrotricha, phylum of worm-like invertebrate animals with terminal adhesive tubes
Geotrupes	genus of the dung beetles
glandular	containing glands
Glischropus	genus of bats (Chiroptera)
glycans	polysaccharids
glycocalyx	outermost part of the integument in Echinodermata
Gobiesocidae	family of the hill stream fish
Gobiidae	family of the hill stream fish
gomphid	dragonfly species from the family Gomphidae
haemolymph	the fluid of the insect body, insect blood
hemidesmosomes	little outgrowths on the surface of adhesive pads of tree frogs
Hemiptera	insect group including bugs, leafhoppers, leaf lice
herbivorous	feeding on living plants; phytophagous
Heteroptera	insect order including bugs
Histriobdellidae	histriobdellids, animal group belonging to the phylum Annelida
Holothuria forskali	echinoderm species belonging to Holothuroidea
holothuroid	belonging to Holothuroidea
Holothuroidea	group of echinoderm animals
Homalopteridae	family of the hill stream fish
Homoptera	group of insects including leafhoppers and plant lice
hyaline	transparent

Hydra	genus of invertebrate animals belonging to Cnidaria
Hymenoptera	insect order including bees, wasps, ants
Hyperolius	genus of tree frogs belonging to the family Rhacophoridae
Hypodermatinae	sub-family of parasitic flies belonging to the family Oestridae
Iguanidae	lizard family
integument	insect skin, body wall including cuticle and layer of epidermis
invertebrates	animals lacking a spinal column
Lepidoptera	insect order including flies
lestid	dragonfly species from the family Lestidae
levator	muscle that serve to raise a body part
Macaranga	South–Asian plant species with specific interrelationships to some ant species
Machilidae	family of very archaic insect group belonging to Archaeognatha
mandible	insect jaws, one pair of the three pairs of mouth-parts
mandibular	part of the mandible
maxilla	part of insect mouth-parts
maxillary	belonging to maxilla
mecopteran	belonging to the insect order Mecoptera, scorpionflies
mentum	part of insect mouth-parts
microtrichia	cuticle outgrowths originating in large number from one epidermis cell and lacking sockets
microvilli	tiny outgrowths on the cell surface
microvillous	covered by microvilli
Mollusca	molluscs, phylum of invertebrate animals
Monogenea	group of parasitic worms
monogenean	belonging to Monogenea
mucopoly-saccharide	any of a class of polysaccharides that bind water to form thick gelatinous material
mucous	covered with mucus or similar viscous matter; secreting or containing mucus
mucus	viscid secretion rich in mucins
Mytilus edulis	species of mussel, a bivalve mollusc
Myzocallis schreib.	aphid species
Myzopoda	genus of bats (Chiroptera)
nektonic	free swimming aquatic animals
Nelumbo	latin name for the plant genus, to which belong lotus

Nematoda	phylum of invertebrate animals
non-apocritan hymenopterans	Symphyta, group of insects from the order Hymenoptera lacking sting, plant wasps
odonate	an insect species belonging to the order Odonata: dragonflies and damselflies
ommatidium	an individual unit of the compound eye in insects and crustaceans
oocyte	an egg before maturation
Orthoptera	insect order including grasshoppers, crickets, etc.
orthopteran	an insect species belonging to the order Orthoptera: grasshoppers, crickets
osteoarthrosis	degenerative changes in the bone and cartilage, wear of joint surfaces
ovipositor	egg-laying apparatus, forming part of the external genitalia of a female insect
papilla	small protuberances
papillate	covered by papillae, little outgrowths
pectoral	relating to or situated on the breast
pedal	part of or belonging to the foot
perimicropylar	located around the micropyle of an egg. Micropyle is the egg opening participating in the egg fertilization.
peristomium	the region around the mouth in invertebrates
perspex	trademark of an acrylic plastic consisting essentially of polymerized methyl methacrylate
phagolysosomes	organelles from the A-cell type of the cartilage
pharynx	a differentiated part of the alimentary canal between the mouth and esophagus in animals
phragma	transverse, plate-like invaginations of the skeleton increasing the area available for muscle attachment
phylum	a direct line of descent within a group of organisms, presumably from a single point of origin
phylogeny	evolution of phyla, groups of living organisms
Pipistrellus nanus	species of bats (Chiroptera)
pleopods	an abdominal appendage of a crustacean
podia	tube feet in Echinodermata
pollination	process of plant fertilization, transferring pollen from a stamen to an ovule
pollinators	insects, agents of pollination
Polychaeta	a group of worms from phylum Annelida
postcervical sclerite	sclerites located in the posterior region of the neck
postero-lateral	located posterior on one side of the body

pretarsus	most distal segment of the insect leg
proboscis	an extended, tubular mouth structure, as in many Diptera, Hymenoptera, Lepidoptera, Hemiptera
procuticle	deeper layer of the cuticle (the cuticle minus epicuticle)
proteoglycans	substances containing protein part and a polysaccharid chain
prothorax	anterior portion of the intermediate part of the insect body, bearing the forelegs
Protozoa	unicellular organisms
protozoan	belonging to Protozoa
protractor	a muscle that prolongs or extends a part
proximal	close to the point of attachment or origin
pterothorax	part of the insect thorax bearing the wings
pulvilli	smooth or hairy paired lateral membranous lobes ventral to the claws in an insect
Quercus	(*Q. robur, Q. ilex*) oak species
quinones	either of two isomeric cyclic crystalline compounds $C_6H_4O_2$ that are derivatives of dihydro-benzene
reduviid	bug belonging to the family Reduviidae
Scincidae	lizard family
sclerites	a hardened (sclerotized) plate of the cuticle
sebum secretion	oily lubricating material secreted by the sebaceous glands located in the substance of the corium and opened into the hair follicle of the skin
secretory granules	portions of secretion
Septifer bifurcatus	species of mussel, a bivalve mollusc
sensilla	simple sensory receptor consisting of one cell or a few cells, especially a hair-like epithelial cell projecting through the cuticle of arthropods
setae	cuticle outgrowths originating from two epidermis cells and bearing sockets; similar to chetae
setose	covered by setae
Sida crystallina	crustacean species belonging to Cladocera
Sisoridae	family of the hill stream fish
skink	lizard bearing attachment pads
stratified columnar epithelium	layer of vertebrate epidermis consisting of a single row of cells
stratum compactum	innermost layers of the vertebrate epidermis
stratum germinativum	one of inner layers of vertebrate epidermis consisting of a single row of cells

syrphids	flies belonging to the family Syrphidae
tarsal	part of the tarsus
tarsomere	sub-unit of the tarsus in the legs of insects
tarsus	terminal region of the leg in insects
taxon	the name of an animal group in a formal system of nomenclature
teleostean fish	division of fish
tenacity	adhesive strength per unit area
tendon	part of the muscle by which it is connected to the skeleton
Tenebrio molitor	beetle species from the family Tenebrionidae
thorax	intermediate part of the insect body bearing the legs and wings
Thyroptera	genus of bats (Chiroptera)
Thysanoptera	insect order including thrips
tibia	insect leg segment located between the femur and tarsus
tipulid	flies belonging to the family Tipulidae
trabeculae	supporting processes of the structure
tracheae	the internal, cuticular tubes of insects
Tribolium cast.	beetle species from the family Tenebrionidae
trichoid complex	trichogen and tormogen cells secreting chetae (setae) in insects
trochanter	insect leg segment located between the coxa and femur
Tuberculoides ann.	aphid species
turbellarians	group of flatworms
Tylonycteris	genus of bats (Chiroptera)
unguitractor	terminal plate of the tendon of claw flexor muscle
ventral	belonging to or situated near or on the side opposite to the back of an animal
vertebrates	animals possessing a spinal column
vespertilionoid	bats belonging to the suprafamily Vespertilionoidea
viscid	having an adhesive quality
volar	belonging to or part of vola, palm of the hand or sole of the foot
Xironodrilus b.	animal species belonging to Branchiobdellidae

D. List of Symbols

a	radius of the contact circle
a_0	contact radius at zero load
$\hat{a}_0$	dimensionless contact radius at zero load
a'	lattice constant
c	concentration
d	distance
$d_{\mathrm{H_2O}}$	thickness of water film
e	electron charge
f	frequency
f_{D}	dipole relaxation frequency
f_{l}	frequency of light
f_{ss}	stick/slip frequency
f_0	zero-friction frequency
g	gravity constant
h	Planck constant
$\hbar$	$= \frac{h}{2\pi}$
h_{c}	height of cantilever
i	running variable
j	imaginary unit
k_t	spring constant for torsion
k_x	spring constant for tangential direction
k_z	spring constant for vertical direction
k_{B}	Boltzmann constant
l	length
l_{c}	length of cantilever
l_0	length of unstrained body
m	mass
m_{e}	electron mass
n	order of diffraction
p	partial pressure
p_{s}	saturation pressure
Δp	Laplace pressure
r	radius
r_{i}	radii

r.h.	relative humidity
s	separation between two solids
t	time
t'	time constant
Δt	time difference
v	sliding velocity
v_{crit}	critical velocity
v_{rel}	relative velocity
w_{c}	width of cantilever
$x_{\max}$	maximum distance in x-direction
x_{s}	distance to first slip
Δx	difference in distance in x-direction
$z_{\max}$	maximum distance in z-direction
$z_{\min}$	minimum distance in z-direction
z_{pv}	peak-to-valley distance
z_0	initial distance
$\bar{z}$	mean value of all $z(x)$
Δz	difference in distance in z-direction

A	Hamaker constant
A_{c}	real contact area
A_{g}	geometric contact area
A_{l}	liquid-immersed contact area
D	diameter
E	Young's modulus
$E(t)$	time-dependent Young's modulus
G	shear modulus
F	force
F_{a}	adhesion force
F_{c}	capillary force
F_{edl}	electronic double-layer force
F_{f}	friction force
F_{d}	friction force in distal direction
F_{fd}	dynamic friction force
F_{fs}	static friction force
F_{n}	normal force
F_{p}	friction force in proximal direction
F_{po}	pull-off force
F_{t}	tangential force
F_{vdW}	van der Waals force
F_{fv}	viscous force
$\hat{F}_{\mathrm{n}}$	dimensionless normal force
ΔF_{f}	stick/slip amplitude

$F_n^{'}$	resulting normal force
ΔF_n	difference in normal force
H	hardness
HU	universal microhardness
HV	Vickers microhardness
K	effective elastic modulus
L	distance between aperture and mirror
L_p	projected length
L_r	relative length
L_s	sampling length
N	number of samples
R	radius of ball
R_a	average roughness
Re	Reynolds number
RMS	root-mean-square roughness
T	absolute temperature
T_p	period time
W	energy
W_a	work of adhesion
W_b	binding energy
W_{el}	electronic energy
W_{kin}	kinetic energy
W_{ph}	phononic energy
W_{pot}	potential energy
Z	Tabor parameter

α	angle
β	Derjaguin parameter
γ	interfacial energy
δ	indentation
ε	strain
ϵ_e	dielectric constant of liquid
ϵ_0	dielectric constant of vacuum
η	viscosity
κ	damping constant
λ	wavelength
λ_D	Debye length
λ_m	Maugis parameter
μ	coefficient of friction
μ_s	static coefficient of friction
μ_d	dynamic coefficient of friction
ν	Poisson ratio
ξ_i	surface charge densities

π	pi
ρ	density
σ	stress
$\sigma(t)$	time-dependent stress
$\hat{\sigma}$	stress overshoot
σ_0	initial stress
τ	shear stress
χ_i	ionic valence
ψ	fit parameter
ω	frequency

$\Delta\Gamma$	Gibbs potential
$\Psi(0)$	potential
Φ	analyzer work function
Θ	contact-angle

References

1. U. Dürig, G.K. Binnig: Tribol. Lett. (to be published) (2000)
2. H. Guckel, K. Skrobis, T.R. Christenson, J. Klein, S. Han, B. Choi, E. Lovell, T. Chapman: J. Micromech. Microeng. **1**, 135–138 (1991)
3. M. Madou: *Fundamentals of Microfabrication*, 1st edn. (CRC Press, Boca Raton, 1998)
4. I. Fujimasa: *Micromachines: A New Era in Mechanical Engineering* (Oxford University Press Inc., New York, 1996)
5. Y. Zhou, B. Nelson: "Adhesion force modeling and measurement for micromanipulation", in *SPIE Conference on Microrobotics and Micromanipulation*, ed. by A. Sulzmann, B. Nelson (SPIE, Boston, 1998), pp. 220–222
6. D. Riemer: personal communication (2000)
7. R. Maboudian: Surf. Sci. Rep. **30**, 207–269 (1998)
8. K.F. Böhringer, K. Goldberg, M. Cohn, R. Howe, A. Pisano: "Parallel microassembly with electrostatic force fields", in *International Conference on Robotics and Automation (ICRA)* (IEEE, Leuven, Belgium, 1998)
9. Y. Rollot, S. Régnier, J.C. Guinot: Int. J. Adhes. Adhes. **19**, 35–48 (1999)
10. M.P. de Boer, T.A. Michalske: J. Appl. Phys. **86**(2), 1–11 (1999)
11. K. Komvopoulos: "Surface texturing and chemical treatment methods for reducing high adhesion forces at micromachine interfaces", in *SPIE Conference on Materials and Device Characterization in Micromachining* (1998)
12. K.E. Drexler: *Nanosystems: Molecular Machinery, Manufacturing, and Computation* (John Wiley & Sons, Inc., New York, 1992)
13. T. Nakano: Jpn. J. Tribology **37**(11), 1343–1350 (1992)
14. Z. Jiang, C.J. Lu, D.B. Bogy, T. Miyamoto: J. Tribol. **117**, 328–333 (1995)
15. M. Scherge, J.A. Schaefer, O. Mollenhauer, F. Spiller: Mater. Res. Soc. Symp. Proc. **569**, 139 (1999)
16. F. Spiller, T. Gubi, O. Mollenhauer, M. Scherge: Konstruktion **51**, 11 (1999)
17. M. Scherge, X. Li, J.A. Schaefer: Tribol. Lett. **6**(3-4), 215–220 (1999)
18. M. Scherge, X. Li, J.A. Schaefer: Tribologie + Schmierungstechnik **3**, 24–28 (1999)
19. M. Scherge, J.A. Schaefer: Polym. Prepr. **39**(2) (1998)
20. M. Scherge, X. Li, J.A. Schaefer: Mater. Res. Soc. Symp. Proc. **522**, 481–486 (1998)
21. M. Scherge, J.A. Schaefer, O. Mollenhauer: "The Role of Water on the Micro-Tribology of MEMS", in *MICRO SYSTEM Technologies 98*, ed. by H. Reichl, E. Obermeier (VDI Verlag, Berlin, 1998), pp. 630–632
22. M. Scherge, J.A. Schaefer: "Surface modification and mechanical properties of bulk silicon", in *Tribology Issues and Opportunities in MEMS*, ed. by B. Bhushan (Kluwer Academic Publishers, Netherlands, 1998), pp. 463–470
23. M. Scherge, J.A. Schaefer: Tribol. Lett. **4**(1), 37–42 (1998)
24. M. Scherge, H. Büchner, G. Jäger, J.A. Schaefer: J. Opt. **29**, 1–5 (1998)

25. J.N. Israelachvili: *Intermolecular and Surface Forces*, 2nd edn. (Acacemic Press, San Diego, 1992)
26. M. Morita, T. Ohmi, E. Hasegawa, M. Kawakami, K. Suma: Appl. Phys. Lett. **55**(6), 562–564 (1989)
27. J.J. Gilman: Science **261**, 1436–1440 (1993)
28. J.A. Schaefer, J. Anderson, G.J. Lapeyre: J. Electron Spectrosc. **38**, 21–28 (1986)
29. J.A. Schaefer, J. Anderson, G.J. Laupeyre: J. Electron Spectrosc. **44**, 373–382 (1987)
30. M.R. Houston, R.T. Howe, R. Maboudian: J. Appl. Phys. **81**(8), 3474–3483 (1997)
31. J.A. Friedrich, G.W. Neudeck, S.T. Liu: Appl. Phys. Lett. **53**(25), 2543–2545 (1988)
32. T. Kosugi, H. Ishii, Y. Arita: J. Vac. Sci. Technol. A **15**(1), 127–132 (1997)
33. R.S. Sayles: Tribol. Int. **29**(8), 639–650 (1996)
34. J. Sugimura: Jpn. J. Tribology **39**(3), 313–321 (1994)
35. J.A. Greenwood: "Problems with surface roughness", in *Fundamentals of Friction: Macroscopic and Microscopic Processes*, ed. by I.L. Singer, H.M. Pollock (Kluwer Academic Publishers, Netherlands, 1992), pp. 57–76
36. K. Johnson: *Contact Mechanics* (Cambridge Univ. Press, Cambridge, 1994)
37. B. Bhushan, G. Ganti: Wear **180**, 17–34 (1995)
38. B. Mandelbrot: *The Fractal Geometry of Nature* (Freeman, New York, 1977)
39. Y.C. Fung: *Biomechanics: Mechanical Properties of Living Tissues*, 2nd edn. (Springer-Verlag, Berlin, 1993)
40. Y.M. Haddad: *Viscoelasticity of Engineering Materials* (Chapman & Hall, London, 1995)
41. I.G. Goryacheva: *Contact Mechanics in Tribology* (Kluwer Academic Publishers, Netherlands, 1998)
42. H. Hertz: Reine und Angewandte Mathematik **92**(2), 156–171 (1862)
43. A. Fogden, L.R. White: J. Colloid Interface Sci. **138**(2), 414–430 (1990)
44. K.L. Johnson: "A continuum mechanics model of adhesion and friction in a single asperity contact", in *Micro/Nanotribology and its Applications*, ed. by B. Bhushan (Kluwer Academic Publishers, Netherlands, 1997), pp. 151–168
45. R.G. Horn, N.J. Israelachvili, F. Pribac: J. Colloid Interface Sci. **115**(2), 480–492 (1987)
46. B. Derjaguin, V. Muller, Y. Toporov: J. Colloid Interface Sci. **53**(2), 314–326 (1975)
47. V. Muller, V. Yushchenko, B. Derjaguin: J. Colloid Interface Sci. **92**(1), 92–101 (1983)
48. V. Muller, V. Yushchenko: J. Colloid Interface Sci. **77**(1), 91–101 (1980)
49. D. Tabor: J. Colloid Interface Sci. **73**(1), 294 (1980)
50. D. Tabor: J. Colloid Interface Sci. **67**(2), 380 (1978)
51. D. Tabor: J. Colloid Interface Sci. **58**(1), 2–13 (1977)
52. B. Derjaguin, V. Muller, Y. Toporov: J. Colloid Interface Sci. **73**(1), 293 (1980)
53. B. Derjaguin, V. Muller, Y. Toporov: J. Colloid Interface Sci. **67**(2), 378–379 (1978)
54. A.E. Lee: J. Colloid Interface Sci. **64**(3), 577–579 (1978)
55. U.D. Schwarz, H. Bluhm, H. Hölscher, W. Allers, R. Wiesendanger: "Friction in the low-load regime: Studies on the pressure and direction dependence of fractional forces by means of friction force microscopy", in *Physics of Sliding Friction*, ed. by B.N. Persson, E. Tosatti (Kluwer Academic Publishers, Netherlands, 1996), pp. 369–402

56. M. Enachescu, R. van den Oetelaar, R.W. Carpick, D. Ogletree, C. Flipse, M. Salmeron: Tribol. Lett. **7**, 73–78 (1999)
57. D. Maugis: J. Colloid Interface Sci. **150**(1), 243–269 (1992)
58. D. Maugis, B. Gauthier-Manuel: J. Adhes. Sci. Technol. **8**(11), 1311–1322 (1994)
59. D. Maugis: Langmuir **11**(2), 679–682 (1995)
60. I.N. Sneddon: Int. J. Eng. Sci. **3**, 47–57 (1995)
61. H.M. Pollock: "Surface forces and adhesion", in *Fundamentals of Friction: Macroscopic and Microscopic Processes*, ed. by I.L. Singer, H.M. Pollock (Kluwer Academic Publishers, Netherlands, 1992), pp. 77–94
62. L.O. Heim, J. Blum, M. Preuss, H.J. Butt: Phys. Rev. Lett. **83**(6), 3328–3331 (1999)
63. W.N. Unertl: J. Vac. Sci. Technol. **1**, 1779–1786 (1999)
64. A. El Ghzaoui: J. Appl. Phys. **85**(2), 1231–1233 (1999)
65. J.N. Israelachvili: "Adhesion, friction and lubrication of molecularly smooth surfaces", in *Fundamentals of Friction: Macroscopic and Microscopic Processes*, ed. by I.L. Singer, H.M. Pollock (Kluwer Academic Publishers, Netherlands, 1992), pp. 351–385
66. P. Mächtle, C.A. Helm: Thin Solid Films **330**, 1–6 (1998)
67. R.A. Quon, R.F. Knarr, T.K. Vanderlick: J. Phys. Chem. B **103**, 5320–5327 (1999)
68. S.S. Das, C.G.K. Malek: Mater. Res. Soc. Symp. Proc. **518**, 149–154 (1998)
69. J.R. Smith, G. Bozzolo, A. Banerjea, J. Ferrante: Phys. Rev. Lett. **63**(12), 1269–1272 (1989)
70. U. Dammer, O. Popescu, P. Wagner, D. Anselmetti, H.J. Güntherodt, G.N. Misevic: Science **267**, 1173–1175 (1995)
71. N. Mozes, H.J. Busscher, P.S. Handley, P.G. Rouxhet: *Microbial Cell Surface Analysis* (VCH Publishers, Inc., New York, 1991)
72. S. Kidoaki, T. Matsuda: Langmuir **15**(22), 7639–7646 (1999)
73. H. Egawa, K. Furusawa: Langmuir **15**(5), 1660–1666 (1999)
74. R. Schmidt: *Werkstoffverhalten in biologischen Systemen*, 2nd edn. (Springer-Verlag, Berlin, 1999)
75. Z.B. Ed. Maksić, W.J. Orville-Thomas: *Modern Modelling of the Chemical Bond* (Elsevier, Amsterdam, 1999)
76. M. Morra, C. Cassinelli: J. Biomed. Mater. Res. **31**, 149–155 (1996)
77. S.P. Gorman, D.S. Jones, W.M. Mawhinney, J.G. McGoven, C.G. Adair: J. Mater. Sci. Mater. Med. **8**, 631–635 (1997)
78. F.W. Hyde, M. Aulberg, K. Smith: J. Ind. Microbiol. Biot. **19**, 142–149 (1997)
79. W.R. Bowen, N. Hilal, R.W. Lovitt, C.J. Wright: Colloid Surf. A **136**, 231–234 (1998)
80. C. Rotsch, M. Radmacher: Langmuir **13**(10), 2825–2832 (1997)
81. M. Hermansson: Colloid Surf. B **14**, 105–119 (1999)
82. Z. Borkowska, U. Stimming: "Structure formation in the double-layer at the electrode/aqueous electrolyte interface", in *Structure of Electrified Interfaces*, ed. by J. Lipkowski, P. Ross (VCH Publishers, Inc., New York, 1993), pp. 277–307
83. A.M. Denton, H. Löwen: Thin Solid Films **330**, 7–13 (1998)
84. W.A. Ducker, Z. Xu, J.N. Israelachvili: Langmuir **10**(9), 3279–3289 (1994)
85. S. Sounilhac, E. Barthel, F. Creuzet: J. Appl. Phys. **85**(1), 222–227 (1999)
86. C. Gao: Appl. Phys. Lett. **71**(13), 1801–1803 (1997)
87. J. Crassous, J.L. Loubet, E. Charlaix: Mater. Res. Soc. Symp. Proc. **366**, 33–38 (1995)
88. X. Tian, B. Bhushan: J. Phys. D **29**, 163–178 (1996)

89. L. Xu, A. Lio, J. Hu, D.F. Ogletree, M. Salmeron: J. Phys. Chem. **102**(3), 540–548 (1998)
90. L. Li, H. Bender, G. Zou, P.W. Mertens, M.A. Meuris, M.M. Heyns: J. Electrochem. Soc. **143**(1), 233–237 (1996)
91. J.A. Schaefer: J. Electron Spectrosc. **44**, 163–174 (1987)
92. J.A. Schaefer, W. Göpel: Surf. Sci. **155**, 535–552 (1985)
93. Y. Yasuda, H. Ikeda, S. Zaima: Appl. Surf. Sci. **113/114**, 579–584 (1997)
94. H. Ikeda, K. Hotta, S. Furuta, S. Zaima, Y. Yasuda: Jpn. J. Appl. Phys. **35**, 1069–1072 (1996)
95. G.J. Kluth, R. Maboudian: J. Appl. Phys. **80**(9), 5408–5414 (1996)
96. T. Miura, M. Niwano, D. Shoji, N. Miyamoto: Appl. Surf. Sci. **100/101**, 454–459 (1996)
97. T. Hattori, T. Aiba, E. Iijima, Y. Okube, H. Nohira, N. Tate, M. Katayama: Appl. Surf. Sci. **104/105**, 323–328 (1996)
98. T. Miura, M. Niwano, D. Shoji, N. Miyamoto: J. Appl. Phys. **79**(8), 4373–4380 (1996)
99. M.K. Weldon, B.B. Stefanov, K. Raghavachari, Y.J. Chabal: Phys. Rev. Lett. **79**(15), 2851–2854 (1997)
100. B. Vögli, H. von Känel: "The influence of surface topography on adhesion and its application to micro parts handling", in *ECASIA 97*, ed. by I. Olefjord, L. Nyborg, D. Briggs (John Wiley & Sons, Inc., New York, 1997)
101. S.M. Hsu, X.H. Zhang: "Lubrication: Traditional to nano-lubricating films", in *Micro/Nanotribology and its Applications*, ed. by B. Bhushan (Kluwer Academic Publishers, Netherlands, 1997), pp. 399–420
102. D. Dowson: "Friction and traction in lubricated contacts", in *Fundamentals of Friction: Macroscopic and Microscopic Processes*, ed. by I.L. Singer, H.M. Pollock (Kluwer Academic Publishers, Netherlands, 1992), pp. 325–349
103. S. Ndoni, P. Jannasch, N.B. Larsen, K. Almdal: Langmuir **15**(11), 3859–3865 (1999)
104. C.M. Mate, V.J. Novotny: J. Chem. Phys. **94**(12), 8420–8427 (1991)
105. V.V. Konchits, Y.E. Kirpichenko, N. Myshkin: Tribol. Int. **29**(5), 365–371 (1996)
106. H.P. Evans, R.W. Snidle: Tribol. Int. **29**(8), 659–667 (1996)
107. J.N. Israelachivili, P.M. McGuiggan: J. Mater. Res. **5**(10), 2223–2231 (1990)
108. S.J. O'Shea, M.A. Lantz, M.E. Welland: "AFM at liquid-solid interfaces", in *Micro/Nanotribology and its Applications*, ed. by B. Bhushan (Kluwer Academic Publishers, Netherlands, 1997), pp. 101–120
109. B. Fodi, R. Hentschke: Langmuir **14**(2), 429–437 (1998)
110. K. Koga, X.C. Zeng, H. Tanaka: Phys. Rev. Lett. **79**(26), 5262–5265 (1997)
111. J. Gao, W.D. Luedtke, U. Landmann: J. Phys. Chem. B **101**, 4013–4023 (1997)
112. J. Gao, W.D. Luedtke, U. Landmann: Phys. Rev. Lett. **79**(4), 705–708 (1997)
113. J. Gao, W.d. Luedtke, U. Landmann: J. Phys. Chem. **106**(10), 4309–4318 (1997)
114. P.A. Thompson, G.S. Grest: Phys. Rev. Lett. **68**(23), 3448–3451 (1992)
115. M. Dijkstra: Thin Solid Films **330**, 14–20 (1998)
116. J.N. Israelachvili: J. Vac. Sci. Technol. A **10**(5), 2961–2972 (1992)
117. J.N. van Alsten, S. Granick: Tribol. Trans. **33**(3), 436–446 (1990)
118. P. Pissis, A. Kyritsis, D. Daoukaki, G. Barut, R. Pelster, G. Nimtz: J. Phys. **10**, 6205–6227 (1998)
119. G. Barut, P. Pissis, R. Pelster, G. Nimtz: Phys. Rev. Lett. **80**(16), 3543–3546 (1998)
120. M. Heuberger: J. Phys. Chem. B **102**(26), 5038–5041 (1998)

121. U. Landmann, W.D. Luedtke, J. Gao: Langmuir **12**, 4514–4528 (1996)
122. R. Hentschke: Macromol. Theory Simul. **6**(3), 287–316 (1997)
123. G. Hübner, N. Spencer: Phys. Today pp. 22–27 (1998)
124. B. Capella, G. Dietler: Surf. Sci. Rep. **34**, 1–104 (1999)
125. B. Bhushan, J.N. Israelachvili, U. Landman: Nature **374**, 607–616 (1995)
126. J.M. Carlson, A.A. Batista: Phys. Rev. E **53**, 1–32 (1996)
127. B.N.J. Persson: Phys. Rev. Lett. **71**(8), 1212–1215 (1993)
128. B.N.J. Persson: Comments Condens. Mater. Phys. **17**(5), 281–305 (1995)
129. B.N.J. Persson: Chem. Phys. Lett. **254**, 114–121 (1996)
130. N. Sasaki: Jpn. J. Tribology **39**(3), 291–301 (1994)
131. B.N.J. Persson: *Sliding Friction*, 1st edn. (Springer-Verlag, Berlin, 1998)
132. M. Schoen: Physica A **240**, 328–339 (1997)
133. M.G. Rozman: J. Phys. Chem. **102**(41), 7924–7930 (1998)
134. M.G. Rozman, M. Urbakh, J. Klafter: Phys. Rev. Lett. **77**(4), 683–686 (1996)
135. B.N.J. Persson: Surf. Sci. Rep. **33**, 83–119 (1999)
136. E. Manias, Hadziioannou, G.t. Brinke: J. Chem. Phys. **101**(2), 1721–1724 (1994)
137. E. Manias, G. Hadziioannaou, G.t. Brinke: Langmuir **12**(19), 4587–4593 (1996)
138. E. Manias, G. Hadziioannou, I. Bitsanis, G.t. Brinke: Europhys. Lett. **24**(2), 99–104 (1993)
139. E. Manias, I. Bitsanis, G. Hadziioannou, G.t. Brinke: Europhys. Lett. **33**(5), 371–376 (1996)
140. G. Reiter, A.L. Demirel, S. Granick: Science **263**, 1741–1744 (1994)
141. P. Bordarier, B. Rousseau, A.H. Fuchs: Thin Solid Films **330**, 21–26 (1998)
142. H. Bock, M. Schoen: J. Phys. **12**, 1545–1568 (2000)
143. S. Granick: Phys. Today **52**(7), 26–31 (1999)
144. M. Gutman, E. Nachliel: Solid State Ionics **61**, 229–234 (1993)
145. D.E. Gragson, G.L. Richmond: J. Phys. Chem. B **102**(20), 3847–3861 (1998)
146. J. Slater: J. Chem. Phys. **39**, 3199 (1964)
147. E. Clementi, D. Raimondi, W. Reinhardt: J. Chem. Phys. **38**, 2686 (1963)
148. A. Bondi: J. Chem. Phys. **68**, 441 (1964)
149. G.W. Robinson, S.B. Zhu, S. Singh, M.W. Evans: *Water in Biology, Chemistry and Physics: Experimental Overviews and Computational Methodologies* (World Scientific, Singapore, 1996)
150. S. Woutersen, H.J. Bakker: Phys. Rev. Lett. **83**(10), 2077–2080 (1999)
151. A. Katrusiak: Phys. Rev. Lett. **77**(21), 4366–4369 (1966)
152. O. Mishima, H.E. Stanley: Nature **396**, 329–335 (1998)
153. M. Vedamuthu, S. Singh, G.W. Robinson: J. Phys. Chem. **98**(9), 2222–2230 (1994)
154. H. Bluhm, S.H. Pan, L. Xu, T. Inoue, D.F. Ogletree, M. Salmeron: Rev. Sci. Instrum. **69**(4), 1781–1784 (1998)
155. M. Morgenstern, J. Müller, T. Michely, G. Comsa: Z. Phys. Chem. **198**, 43–72 (1997)
156. H. Bluhm, M. Salmeron: J. Chem. Phys. **111**(15), 6947–6954 (1999)
157. H. Bluhm, T. Inoue, M. Salmeron: Phys. Rev. B **61**(11), 7760–7765 (2000)
158. G. Nimtz: Phys. Scr. **T13**, 172–177 (1986)
159. G. Nimtz, A. Kops, R. Pelster: "Hydrogen bond networks: Water in mesoscopic hydrophobic confinements", in *Hydrogen Bond Networks*, ed. by M.C. Bellisent-Funel, J.C. Dore (Kluwer Academic Publishers, Netherlands, 1994), pp. 419–424
160. E. Stöckelmann, E.M. Aydt, R. Hentschke: J. Mol. Model. **3**, 347–354 (1997)
161. D. Sonntag: *Hygrometrie* (Akademie-Verlag, Berlin, 1966)

162. W.H. Lawnik, U.D. Goepel, A.K. Klauk, G.H. Findenegg: Langmuir **11**(8), 3075–3082 (1995)
163. Q. Dai, J. Hu, M. Salmeron: J. Phys. Chem. B **101**(11), 1994–1998 (1997)
164. M. Morgenstern, T. Michely, G. Comsa: Phys. Rev. Lett. **77**(4), 703–706 (1996)
165. J. Hu, R.W. Carpick, M. Salmeron, X. Xiao: J. Vac. Sci. Technol. B **14**(2), 1341–1343 (1996)
166. C.H. Chung, S.I. Yi, W.H. Weinberg: J. Vac. Sci. Technol. A **16**(3), 1785–1789 (1998)
167. M.C. Flowers, N.B.H. Jonathan, A. Morris, S. Wright: Surf. Sci. **351**, 87–102 (1996)
168. C. Poncey, F. Rochet, G. Dufour, H. Roulet, F. Sirotti, G. Panaccione: Surf. Sci. **338**, 143–156 (1995)
169. R. Konecny, D.J. Doren: J. Chem. Phys. **106**(6), 2426–2435 (1997)
170. J.A. Schaefer: Surf. Sci. **178**, 90–100 (1986)
171. J.A. Schaefer: Appl. Phys. A **51**, 305–316 (1990)
172. J.A. Schaefer, J. Anderson, G.J. Lapeyre: J. Vac. Sci. Technol. A **3**(3), 1443–1447 (1985)
173. J.Q. Broughton, J.A. Schaefer, J.C. Bean, H.H. Farrell: Phys. Rev. B **33**(10), 6841–6845 (1986)
174. J.A. Schaefer, F. Stucki, D.J. Frankel, G. W., G.J. Lapeyre: J. Vac. Sci. Technol. B **2**(3), 359–365 (1984)
175. L. Andersohn, U. Köhler: Surf. Sci. **284**, 77–90 (1993)
176. C. Merkl, T. Pfohl, H. Riegler: Phys. Rev. Lett. **79**(23), 4625–4628 (1997)
177. J. Hu, X.D. Xiao, M. Salmeron: Appl. Phys. Lett. **67**(4), 476–478 (1995)
178. R. Akiyama, F. Hirata: J. Chem. Phys. **108**(12), 4904–4911 (1998)
179. R. Guckenberger, M. Heim, G. Cevec, H. Knapp, W. Wiegräbe, A. Hillebrand: Science **266**, 1538–1540 (1994)
180. P.B. Miranda, L. Xu, Y.R. Shen, M. Salmeron: Phys. Rev. Lett. **81**(26), 5876–5879 (1998)
181. J. Hu, X.D. Xiao, D.F. Ogletree, M. Salmeron: Science **268**, 267–269 (1995)
182. M. Salmeron, L. Xu, J. Hu, Q. Dai: MRS Bull. **XXII**(8), 36–41 (1997)
183. M. Odelius, M. Bernaconi, M. Parrinello: Phys. Rev. Lett. **78**, 2855 (1997)
184. P.B. Miranda, X. L., Y.R. Shen, M. Salmeron: Phys. Rev. Lett. **81**(26), 5876–5879 (1998)
185. M. Heim, R. Steigerwald, R. Guckenberger: J. Struct. Biol. **119**, 212–221 (1997)
186. K.C. Ludema: Jpn. J. Tribology **36**(4), 357–358 (1991)
187. B. Derjaguin, Z. Zorin, N. Churaev, V. Shishin: "Adsorption of water", in *Wetting, Spreading and Adhesion*, ed. by J. Padday (Academic Press, New York, 1978), pp. 201–212
188. M. Binggeli, C.M. Mate: J. Vac. Sci. Technol. B **13**(3), 1312–1315 (1995)
189. M. Binggeli, C.M. Mate: Appl. Phys. Lett. **64**(4), 415–417 (1994)
190. R. Pashley, J. Kitchener: J. Colloid Interface Sci. **71**, 491–500 (1979)
191. D. Ferry, A. Glebov, V. Senz, J. Suzanne, J.P. Toennies, H. Weiss: J. Chem. Phys. **105**(4), 1697–1701 (1996)
192. J. Freund, J. Halbritter, J.K.H. Hörber: Microsc. Res. Techniq. **44**, 327–338 (1999)
193. M.B. McBride, P. Baveye: Soil Sci. Soc. Am. J. **59**(2), 388–394 (1995)
194. G. Peschel, K.H. Adlfinger: Ber. Bunsen **74**(4), 351–357 (1970)
195. A.L. Demirel, S. Granick: Phys. Rev. Lett. **77**(11), 2261–2264 (1996)
196. J. Galle, H.L. Vörtler: Surf. Sci. **421**, 33–43 (1999)
197. J. Halbritter: Appl. Phys. A **68**, 153–162 (1999)

198. I.Y. Sokolov, G.S. Henderson, F.J. Wicks, G.A. Ozin: Appl. Phys. Lett. **70**(7), 844–846 (1997)
199. P.M. McGuiggan, J.N. Israelachivili: J. Mater. Res. **5**(10), 2232–2243 (1990)
200. R. Ho, J.Y. Yuan, Z. Shao: Biophys. J. **75**, 1076–1083 (1998)
201. S.J. Marrink, D.P. Tielemann, A.R. van Buuren: Faraday Discuss. **103**, 191–201 (1996)
202. D.E. Gragson, B.M. McCarty, G.L. Richmond: J. Phys. Chem. **100**, 1472–1475 (1996)
203. M.K. Weldon, Y.J. Chabal, D.R. Hamann, S.B. Christman, E.E. Chaben, L.C. Feldman: J. Vac. Sci. Technol. B **14**(4), 3095–3106 (1996)
204. L.J.M. Schlangen, L.K. Koopal, C. S., M. A., J. Lyklema: Langmuir **11**(5), 1701–1710 (1995)
205. E. Hedborg, F. Winquist, H. Sundgren, I. Lundström: Thin Solid Films **340**, 250–256 (1999)
206. E. Nicollian, J. Brews: *MOS Physics and Technology* (John Wiley & Sons Inc., New York, 1982)
207. B.N.J. Persson: Chem. Phys. Lett. **141**, 366–368 (1987)
208. J.S. Nelson, B.W. Dodson, P.A. Taylor: Phys. Rev. B **45**(8), 4439–4444 (1992)
209. L. Xu, H. Bluhm, M. Salmeron: Surf. Sci. **407**, 251–255 (1998)
210. E. Deulin, A. Gatsenko, B. Loginov: Surf. Sci. **433-435**, 288–292 (1999)
211. A.J. Gellman: J. Vac. Sci. Technol. A **10**(1), 180–187 (1992)
212. V. Derflinger, H. Brändle, H. Zimmermann: Surf. Coat. Technol. **113**, 286–292 (1999)
213. C. Donnet, T. Le Mogne, J.M. Martin: Surf. Coat. Technol. **62**, 406–411 (1993)
214. J.M. Martin, H. Pascal, C. Donnet, T. Le Mogne, J.L. Loubet, T. Epicier: Surf. Coat. Technol. **68/69**, 427–432 (1994)
215. J.M. Martin, C. Donnet, T. Le Mogne: Phys. Rev. B **48**(14), 10 583–10 586 (1993)
216. T. Le Mogne, C. Donnet, J.M. Martin, A. Tonck, N. Millard-Pinard: J. Vac. Sci. Technol. A **12**(4), 1998–2004 (1994)
217. C. Donnet, J.M. Martin, T. Le Mogne, M. Belin: Tribol. Int. **29**(2), 123–128 (1996)
218. V. Rigato, G. Maggioni, D. Boscarino, L. Sangaletti, L. Depero, V.C. Fox, D. Teer, C. Santini: Surf. Coat. Technol. **116-119**, 176–183 (1999)
219. D.Y. Wang, C.L. Chang, W.Y. Ho: Surf. Coat. Technol. **111**, 123–127 (1999)
220. V.C. Fox, N. Renevier, D.G. Teer, J. Hampshire, V. Rigato: Surf. Coat. Technol. **116-119**, 492–497 (1999)
221. T. Jun, X. Qunji: Appl. Surf. Sci. **108**, 471–475 (1997)
222. Y. Tian, G. Liu, Y. Hong, S. Fu, Y. Hu: "The microtribology and self-lubrication microsystems", in *SPIE Conference on Materials and Device Characterization in Micromachining*, ed. by C.R. Friedrich, Y. Vladimirsky (SPIE, Santa Clara, 1998), pp. 236–240
223. A. Berman, S. Steinberg, S. Campbell, A. Ulman, J. Israelachvili: Tribol. Lett. **4**, 43–48 (1998)
224. G. Goodyear, R.E. Larsen, R.M. Stratt: Phys. Rev. Lett. **76**(2), 243–246 (1996)
225. C.L. Bauer, M. Scherge, W.W. Mullins: Mater. Sci. Forum **2**, 207–209 (1996)
226. M. Scherge, C.L. Bauer, W.W. Mullins: Acta Metall. Mater. **43**(9), 3525–3538 (1995)
227. V.A. Muratov, J.E. Olsen, B.M. Gallois, T.E. Fischer, J. Bean: J. Electrochem. Soc. **145**(7), 2465–2470 (1998)
228. S. Mori: Jpn. J. Tribology **37**(11), 1351–1361 (1992)
229. M. Cieplak, E.D. Smith, M.O. Robbins: Science **265**, 1209–1212 (1994)

230. W. Zhong, D. Tománek: Phys. Rev. Lett. **64**(25), 3054–3057 (1990)
231. S.A. Zinenko: Wear **156**, 351–363 (1992)
232. T. Nakano: Jpn. J. Tribology **36**(4), 487–495 (1994)
233. C. Daly, J. Krim: "Sliding friction of compressing Xe monolayers", in *Micro/Nanotribology and its Applications*, ed. by B. Bhushan (Kluver Academic Publishers, Netherlands, 1996), pp. 311–316
234. O. Marti: Phys. Scr. **T49**, 599–604 (1993)
235. D. Tabor: "Friction as a dissipative process", in *Fundamentals of Friction: Macroscopic and Microscopic Processes*, ed. by I.L. Singer, H.M. Pollock (Kluwer Academic Publishers, Netherlands, 1992), pp. 3–24
236. A. Dayo, W. Alnasrallah, J. Krim: Phys. Rev. Lett. **80**(8), 1690–1693 (1998)
237. G.V. Dedkov: Technol. Phys. Lett. **24**(10), 766–768 (1998)
238. P.A. Taylor, J.S. Nelson, B.W. Dodson: Phys. Rev. B **44**(11), 5834–5835 (1991)
239. D. Fuhrmann, C. Wöll: New J. Phys. **1**, 11–19 (1998)
240. J. Krim: Comments Condens. Mater. Phys. **17**(5), 263–280 (1995)
241. G. He, M.H. Müser, M.O. Robbins: Science **284**, 1650–1652 (1999)
242. J. Krim, R. Chiarello: J. Vac. Sci. Technol. B **9**(2), 1343–1346 (1991)
243. A.L. Demirel, S. Granick: Phys. Rev. Lett. **77**(21), 4330–4333 (1996)
244. L. Knopoff, J.A. Landoni, M.S. Abinante: Phys. Rev. A. **46**, 7445–7449 (1992)
245. H.J. Xu, L. Knopoff: Phys. Rev. E **50**, 3577–3581 (1994)
246. I.G. Main, P.R. Sammonds, P.G. Meredith: Geophys. J. Int. **115**, 367–380 (1993)
247. R.A. Ibrahim: Appl. Mech. Rev. **47**(7), 209–253 (1994)
248. R.Y. Siegwart, H. Bleuler: "Micro- and nanorobotics in Switzerland", in *International Conference on Intelligent Robots and Systems*, ed. by M. Mitsuishi, T. Sato (IEEE/RSJ, Pittsburgh, 1995)
249. S.H. Chang, S.S. Li: Rev. Sci. Instrum. **70**(6), 2776–2782 (1999)
250. C.M. Mate, G.M. McClelland, R. Erlandsson, S. Chiang: Phys. Rev. Lett. **59**, 1942–1945 (1987)
251. E. Gnecco, R. Bennewitz, T. Gaylog, C. Loppacher, M. Bammerlin, E. Meyer, H.J. Güntherodt: Phys. Rev. Lett. **84**(6), 1172–1175 (2000)
252. E. Liu, B. Blanpain, J.P. Celis, J.R. Roos: J. Appl. Phys. **84**(9), 4859–4865 (1998)
253. E. Rabinowicz: *Friction and Wear of Materials* (John Wiley & Sons, Inc., New York, 1965)
254. E. Rabinowicz: "Friction fluctuations", in *Fundamentals of Friction: Macroscopic and Microscopic Processes*, ed. by I.L. Singer, H.M. Pollock (Kluwer Academic Publishers, Netherlands, 1992), pp. 25–34
255. H. Czichos, K.H. Habig: *Tribologie Handbuch*, 1st edn. (Friedr. Vieweg & Sohn Verlagsgesellschaft mbH, Braunschweig/Wiesbaden, 1992)
256. D. Cule, T. Hwa: Phys. Rev. Lett. **77**(2), 278–281 (1996)
257. D.E. Kim, N.P. Suh: Wear **149**, 199–208 (1991)
258. R. Bennewitz, T. Gyalog, M. Guggisberg, M. Bammerlin, E. Meyer, H.J. Güntherodt: Phys. Rev. B **60**(16), R11 301–R11 304 (1999)
259. M. Fujihira, Y. Morita: J. Vac. Sci. Technol. B **12**(3), 1609–1613 (1994)
260. S. Fujisawa, E. Kishi, Y. Sugawara, S. Morita: Jpn. J. Appl. Phys. **33**(6), 3752–3755 (1994)
261. S. Morita, S. Fujisawa, E. Kishi, Y. Sugawara: Jpn. J. Tribology **39**(11), 1319–1327 (1994)
262. U.D. Schwarz, O. Zwörner, P. Köster, R. Wiesendanger: Phys. Rev. B **56**(11), 6987–6996 (1997)

263. H. Hölscher, U.D. Schwarz, R. Wiesendanger: Europhys. Lett. **36**(1), 19–24 (1996)
264. H. Hölscher, U.D. Schwarz, R. Wiesendanger: Surf. Sci. **375**, 395–402 (1997)
265. H. Yioshizawa, Y.L. Chen, J. Israelachvili: J. Phys. Chem. **97**, 11 300–11 313 (1993)
266. N.M. Alexeyev, N.N. Kuzmin, G.R. Trankovskaya, E. Shuvalova: Wear **156**, 251–261 (1992)
267. M. Woydt, J. Schwenzien: Tribol. Int. **26**(3), 165 (1993)
268. C.D. Hahm, B. Bhushan: J. Appl. Phys. **81**(8), 5384–5389 (1997)
269. S. Sasaki: Wear **134**, 185 (1989)
270. T. Baumberger: "Dry friction dynamics at low velocities", in *Physics of Sliding Friction*, ed. by B. Persson, E. Tosatti (Kluwer Academic Publishers, Netherlands, 1996), pp. 1–26
271. M. Weiss, F.J. Elmer: Phys. Rev. B **53**, 7539 (1996)
272. M. Weiss, F.J. Elmer: Phys. Rev. B **104**, 55 (1997)
273. M. Weiss, F.J. Elmer: "A simple model for wearless friction: The Frenkel-Kontorova-Tomlinson model", in *Physics of Sliding Friction*, ed. by B.N.J. Persson, E. Tosatti (Kluwer Academic Publishers, Netherlands, 1996), pp. 163–178
274. E. Meyer, R.M. Overney, K. Dransfeld, T. Gyalog: *Nanoscience*, 1st edn. (World Scientific Publishing, P O Box 128, Farrer Road, Singapore 9122805, 1998)
275. A. Dayo, J. Krim: Int. J. Thermophys. **19**(3), 827–834 (1998)
276. C. Daly, J. Krim: Phys. Rev. Lett. **76**(5), 803–806 (1996)
277. A. Liebsch, S. Gonçalves, M. Kiwi: Phys. Rev. B **60**(7), 5034–5043 (1999)
278. C. Caroli: "Dry friction as a hysteretic elastic response", in *Physics of Sliding Friction*, ed. by B. Persson, E. Tosatti (Kluwer Academic Publishers, Netherlands, 1996), pp. 27–49
279. G.G. Adams: J. Tribol. **118**, 819–823 (1996)
280. O. Zwörner, H. Hölscher, U.D. Schwarz, R. Wiesendanger: Appl. Phys. A **66**, 263–267 (1998)
281. J. Takadoum, Z. Zsiga: J. Mater. Sci. Lett. **12**, 791–793 (1993)
282. K. Demizu, R. Wadabayashi, H. Ishigaki: Tribol. Trans. **33**(4), 505–510 (1990)
283. R.S. Gates, S.M. Hsu, E.E. Klaus: Tribol. Trans. **32**(3), 357–363 (1989)
284. K. Komvopoulos, H. Li: J. Tribol. **114**, 131–140 (1992)
285. J.K. Lancaster: Tribol. Int. **23**, 371–389 (1990)
286. R. Maboudian, R.T. Howe: J. Vac. Sci. Technol. B **15**(1), 1–20 (1997)
287. H. Tian, T. Matsudaira: J. Tribol.-T. ASME **115**, 28–35 (1993)
288. A.A. Polycarpou, I. Etsion: Tribol. Trans. **41**(2), 217–224 (1998)
289. A.A. Polycarpou, I. Etsion: J. Tribol. **120**, 296–303 (1998)
290. K. Komvopoulos, W. Yan: J. Tribol. **120**, 808–813 (1998)
291. Y. Ando, J. Ino: Wear **216**, 115–122 (1998)
292. E. Meyer, R. Lüthi et al.: "Friction force spectroscopy", in *Physics of Sliding Friction*, ed. by B. Persson, E. Tosatti (Kluwer Academic Publishers, Netherlands, 1996), pp. 349–367
293. W.W. Tworzydlo, E.B. Becker, J.T. Oden: Appl. Mech. Rev. **47**(7), 255–274 (1994)
294. M.T. Bengisu, A. Akay: "Interaction and stability of friction and vibrations", in *Fundamentals of Friction: Macroscopic and Microscopic Processes*, ed. by I.L. Singer, H.M. Pollock (Kluwer Academic Publishers, Netherlands, 1992), pp. 553–566

295. V. Scherer, W. Arnold, B. Bhushan: "Active friction control using ultrasonic vibration", in *Tribology Issues and Opportunities in MEMS*, ed. by B. Bhushan (Kluwer Academic Publishers, Netherlands, 1998), pp. 463–470
296. J. Gao, W.D. Luedtke, U. Landman: J. Phys. Chem. **102**(26), 5033–5037 (1998)
297. D.P. Hess, A. Soom: "Unsteady friction in presence of vibrations", in *Fundamentals of Friction: Macroscopic and Microscopic Processes*, ed. by I.L. Singer, H.M. Pollock (Kluwer Academic Publishers, Netherlands, 1992), pp. 535–552
298. H.H. Langen, T. Masuzawa, M. Fujino: Ann. CIRP **44**, 173–76 (1995)
299. K.F. Böhringer, V. Bhatt, K. Goldberg: "Sensorless manipulation using transverse vibrations of a plate", in *International Conference on Robotics and Automation (ICRA)* (IEEE, Leuven, Belgium, 1998), pp. 1989–1996
300. R.F. Hariadi, S.C. Langford, J.T. Dickinson: J. Appl. Phys. **86**(9), 4885–4891 (1999)
301. L. Scudiero, S.C. Langford, J. Dickinson: Tribol. Lett. **6**, 41–55 (1999)
302. K. Hokkirigawa, M. Mizumuto: Jpn. J. Tribology **37**(11), 1363–1373 (1992)
303. P.A. Campbell, B. O'Rourke, P. Dawson, R.J. Turner, D.G. Walmsley, P. Spedding, E.P. Watters: Appl. Phys. A **66**, 867–871 (1998)
304. J.T. Dickinson, N. Park, M. Kim, S.C. Langford: Tribol. Lett. **3**, 69–80 (1997)
305. K. Schouterden, B.M. Lairson: Thin Solid Films **340**, 40–44 (1999)
306. T. Sumomogi, T. Endo, K. Kuwahara, R. Kaneko: J. Vac. Sci. Technol. B **13**(3), 1257–1259 (1995)
307. C.J. Lu, Z. Jiang, B.D. B., T. Miyamato: J. Tribol. **117**, 244–249 (1995)
308. V. Radhakrishnan: Proc. Natl. Acad. Sci. USA **95**, 5448–5455 (1998)
309. W. Nachtigall: *Biological mechanisms of attachment* (Springer-Verlag, Berlin, 1974)
310. Y.C. Fung: *Biomechanics: Motion, Flow, Stress, and Growth* (Springer-Verlag, Berlin, 1998)
311. W. Nachtigall: "Biophysik des Schwimmens", in *Biophysik*, ed. by W. Hoppe, W. Lohmann, H. Markl, H. Iegler (Springer-Verlag, Berlin, 1977), pp. 525–536
312. J. Videler: Symp. Soc. Exp. Biol. **49**, 1–20 (1995)
313. W. Fahrenbach, D. Knutson: J. Invest. Dermatol. **65**, 39–44 (1975)
314. W. Nachtigall: *Bionik: Grundlagen und Beispiele für Ingenieure und Naturwissenschaftler* (Springer-Verlag, Berlin, 1998)
315. E. Reif, A. Dinkelacker: Neues Jahrb. Geol. P. **164**, 184–187 (1982)
316. O.M. Bocharova-Messner, A.Z. Dmitriev: "Morphological and functional analysis of the wing venation in odonata according to the data of the scanning electron microscopy", in *IX Congress of All-Union Entomological Society. Abstr. Theses.*, ed. by N. Dumka, Kiev (1984), pp. 65–65
317. R. Kristic: *Human Microscopic Anatomy* (Springer-Verlag, Berlin, 1991)
318. J. Buckwalter: Instr. Course Lect. **32**, 349–370 (1983)
319. R.M. Aspden: Proc. R. Soc. Lond. B **258**, 195–200 (1994)
320. G. Ateshian: J. Biomech. Eng.-T. ASME **119**, 81–86 (1997)
321. J. Clark, A. Norman, M. Kaab, H. Notzli: J. Anat. **195**, 45–56 (1999)
322. J. Smeathers: "Cartilage and joints", in *Biomechanics - Materials: A Practical Approach*, ed. by J.F.V. Vincent (IRL Press, Oxford, 1992), pp. 99–132
323. K. Date: Arch. Histol. Jap. **42**, 517–531 (1979)
324. D. Leach, S. Caldwell, J. Ferguson: Acta Anat. **133**, 234–246 (1988)
325. J. Chappuis, I. Sherman, A. Neumann: Ann. Biomed. Eng. **11**, 435–449 (1983)
326. D. Swann, F. Silver, H. Slayter, W. Stafford, E. Shore: Biochem. J. **225**, 195–201 (1985)
327. G. Jay: Connect. Tissue Res. **28**, 71–88 (1992)

328. D. Swann, R. Hendren, E. Radin, S. Sotman, E. Duda: Arthritis Rheum **24**, 22–30 (1981)
329. B. Hills, B. Butler: Ann. Rheumat. Dis. **43**, 641–648 (1984)
330. B. Hills: Ann. Biomed. Eng. **23**, 112–115 (1995)
331. B. Hills, M. Monds: Brit. J. Rheumatol. **37**, 137–142 (1998)
332. R. Fein: Proc. Inst. Mech. Eng. **181**, 125–128 (1967)
333. M. Kwan, W. Lai, V. Mow: Ann. Biomech. Eng. **12**, 537–558 (1984)
334. V. Mow: J. Lubric. Technol.-T. ASME **91**, 320–329 (1969)
335. J. Hou, V. Mow, W. Lai, M. Holmes: J. Biomech. **25**, 247–259 (1992)
336. Z. Jin, D. Dowson, J. Fisher: Proc. Inst. Mech. Eng. H **206**, 117–124 (1992)
337. M. MacConaill: J. Anat. **66**, 210–227 (1932)
338. A. Ogston, J. Stanier: J. Physiol. **199**, 244–258 (1953)
339. H. Chikama: Nippon Seikeigeka Gakkai Zasshi **59**, 559–572 (1985)
340. R.J. Wootton, D.J.S. Newman: Nature **280**, 402–403 (1979)
341. W. Gronenberg: J. Comp. Physiol. A **178**, 727–734 (1996)
342. W. Gronenberg: J. Exp. Biol. **199**, 2021–2033 (1996)
343. P. Walker, D. Dowson, M. Longfeld, V. Wright: Ann. Rheum. Dis. **27**, 512–520 (1968)
344. A. Maroudas: Proc. Inst. Mech. Eng. H **181**, 122–124 (1967)
345. D. Dowson, M. Longfeld, P. Walker, V. Wright: Proc. Inst. Mech. Eng. H **181**, 45–54 (1968)
346. G. Ateshian, H. Wang: Proc. Inst. Mech. Eng. H **211**, 419–424 (1997)
347. E. Radin, D. Swann, P. Weisser: Nature **228**, 377–377 (1970)
348. D. Guerra, L. Frizziero, M. Losi, B. Bacchelli, G. Mezzadri, I. Pasquali-Ronchetti: J. Submicr. Sytol. Path. **28**, 385–393 (1996)
349. J. Jurvelin, D. Muller, M. Wong, D. Studer, A. Engel, E. Hunziker: J. Struct. Biol. **117**, 45–54 (1996)
350. R. Rowe: Tissue Cell **133**, 681–690 (1981)
351. P.P. Purslow, J.A. Trotter: J. Muscle Res. Cell M. **15**, 299–308 (1994)
352. H.K. Ault, A.H. Hoffman: J. Biomech. Eng.-T. ASME **114**, 142–146 (1992)
353. J. Fredberg, K. Jones, M. Nathan, S. Raboudi, Y. Prakash, S. Shore, J. Butler, G. Sieck: J. Appl. Physiol. **81**, 2703–2713 (1996)
354. B. Hills: Lymphology **23**, 39–47 (1990)
355. S.N. Gorb: "Reibungssysteme bei Insekten", in *Technische Biologie und Bionik. 4. Bionik - Kongress, München 1998*, ed. by W. Nachtigall, A. Wisser (Gustav Fisher Verlag, Stuttgart, 1998), pp. 185–189
356. O.T. Thiaw, X. Mattei: Can. J. Zool. **70**, 1064–1068 (1992)
357. W. Wickler: Zool. Anz. **163**, 90–107 (1959)
358. S.C. Saxena, M. Chandy: J. Zool. (London) **148**, 315–340 (1966)
359. A. Singh, N.K. Agarwal: Acta Ichthyologica et Piscatoria **21**, 29–35 (1991)
360. N. Singh, N.K. Agarwal: "Organs of adhesion in four hillstream fishes, a comparative morphological study", in *Advances in Limnology*, ed. by H.R. Singh (Narendra Publishing House, Delhi, 1993), pp. 311–315
361. N. Singh: J. Anim. Morphol. Physl. **40**, 109–114 (1993)
362. N. Singh, N.K. Agarwal, H.R. Singh: Advances in Fish Biology and Fisheries **1**, 281–291 (1994)
363. G. Zaccone: Archives de Biologie **92**, 317–342 (1981)
364. D.M. Green, D.L. Barber: Can. J. Zool. **66**, 1610–1619 (1988)
365. G. Courtney: Can. J. Zool. **68**, 556–578 (1990)
366. R.M. Alexander: *Animal Mechanics* (Blackwell, Oxford, 1983)
367. M. Braun: Arbeiten aus dem Zoologisch-Zootomischen Institut in Würzburg **4**, 231–237 (1878)
368. W.D. Dellit: Jenaer Zeitschrift für Naturwissenschaften **68**, 613–656 (1934)

369. U. Hiller, R. Blaschke: Naturwissenschaften **54**, 344–345 (1967)
370. U. Hiller: Z. Morphol. Tiere **62**, 307–362 (1968)
371. B. Roll: J. Zool. **235**, 289–300 (1995)
372. R. Ruibal, V. Ernst: J. Morphol. **117**, 271–294 (1965)
373. P.F.A. Maderson: Nature **203**, 780–781 (1964)
374. A.M. Bauer: J. Morphol. **235**, 41–58 (1998)
375. A.P. Russell: Can. J. Zool. **64**, 948–955 (1986)
376. E.E. Williams, J.A. Peterson: Science **215**, 1509–1511 (1982)
377. D.J. Irschick, C.C. Austin, K. Petren, R.N. Fisher, J.B. Losos, O. Ellers: Biol. J. Linn. Soc. **59**, 21–35 (1996)
378. V. Ernst, R. Ruibal: J. Morphol. **120**, 233–266 (1967)
379. A. Cua, K. Wilhelm, H. Maibach: Brit. J. Dermatol. **123**, 473–479 (1990)
380. M. Cartmill: Am. J. Phys. Anthropol. **50**, 497–509 (1979)
381. M. Hamrick: Am. J. Phys. Anthropol. **106**, 113–127 (1998)
382. E. Perkins, D. Ford: Am. J. Phys. Anthropol. **42**, 383–393 (1975)
383. J. Hazel, M. Stone, M. Grace, V. Tsukruk: J. Biomech. **32**, 477–484 (1999)
384. J.H. Waite: Biol. Rev. **58**, 209–231 (1983)
385. M.J. Naldrett: "Adhesives", in *Biomechanics - Materials: A Practical Approach*, ed. by J.F.V. Vincent (IRL Press, Oxford, 1992), pp. 219–240
386. M.S. Steinberg: Dev. Biol. **180**, 377–388 (1996)
387. C.J. Strange: Bioscience **47**, 5–8 (1997)
388. R.E. Baier: "Surface properties influencing biological adhesion", in *Adhesion in Biological Systems*, ed. by R.S. Manly (Academic Press, New York, 1970), pp. 15–48
389. L. Weiss: "A biophysical consideration of cell contact phenomena", in *Adhesion in Biological Systems*, ed. by R.S. Manly (Academic Press, New York and London, 1970), pp. 1–12
390. A.C. Taylor: "Adhesion of cells to surfaces", in *Adhesion in Biological Systems*, ed. by R.S. Manly (Academic Press, New York and London, 1970), pp. 51–70
391. P. Flammang: Echinoderm Studies **5**, 1–60 (1996)
392. S. Tyler: Fortschritte der Zoologie **36**, 331–347 (1988)
393. L.A. Thomas, C.O. Hermans: Biol. Bull.(Woods Hole) **169**, 675–688 (1985)
394. D.P. DeVore, R.J. Gruebel: Biochem. Bioph. Res. Co. **80**, 993–999 (1978)
395. N. Maruyama, H. Etoh, K. Sakata, K. Ina: Agr. Biol. Chem. Tokyo **55**, 2887–2889 (1991)
396. J. Pardo, E. Gutierrez, C. Saez, M. Brito, L.O. Burzio: Protein Expres. Purif. **1**, 147–150 (1990)
397. J.H. Waite, D. Hansen, K.T. Little: J. Comp. Physiol. B **159**, 517–525 (1989)
398. H. Yamamoto, S. Yamauchi, S. Ohara: Biomimetics **1**, 219–238 (1992)
399. M.W. Denny: Dissertation Abstracts International B Sciences and Engineering **40**, 5166–5167 (1980)
400. J.H. Waite: Chem. Ind.-London **1991**, 607–611 (1991)
401. E.C. Bell, J.M. Gosline: J. Exp. Biol. **199**, 1005–1017 (1996)
402. J. Smeathers, J.F.V. Vincent: Journal of Molluscan Studies **45**, 219–230 (1979)
403. H.M. Feder: Ecology **36**, 764–767 (1955)
404. M.E. Lavoie: Biol. Bull.(Woods Hole) **111**, 114–122 (1956)
405. V.L. Paine: Am. Nat. **62**, 517–529 (1929)
406. J.D. McKenzie: Cell Tissue Res. **248**, 187–199 (1987)
407. P. Flammang, M. Jangoux: Zoomorphology **113**, 47–60 (1993)
408. M.W. Denny: J. Exp. Biol. **91**, 195–217 (1981)
409. A.M. Weigl: Trans. Am. Microsc. Soc. **113**, 276–301 (1994)
410. G.A. Young, A.B. Yule, G. Walker: Biofouling **1**, 137–146 (1988)
411. A.B. Yule, G. Walker: Crustacean Issues **5**, 389–402 (1987)

412. A.B. Yule, G. Walker: J. Mar. Biol. Assoc. UK **64**, 147–156 (1984)
413. D. Lacombe, V.R. Liguori: Biol. Bull.(Woods Hole) **137**, 170–180 (1969)
414. J.S. Edwards: Int. J. Insect Morphol. **21**, 369–371 (1992)
415. T.P. Cogley, J.R. Anderson, J. Weintraub: Int. J. Insect Morphol. **10**, 7–18 (1981)
416. E.K. Tillinghast: Naturwissenschaften **68**, 526–527 (1981)
417. T. Strohmenger, W. Nentwig: Zool. Anz. **218**, 9–16 (1987)
418. H. Günzl: Zoomorphologie **90**, 197–204 (1978)
419. E. Braum, N. Peters, M. Stolz: Int. Rev. Ges. Hydrobiol. **81**, 101–108 (1996)
420. H.M. Peters: Zool. Jahrb. Allg. Zool. **71**, 287–300 (1965)
421. H.M. Peters, S. Berns: Zool. Jahrb. Anat. **109**, 59–80 (1983)
422. R. Bennemann, I. Pietzsch-Rohrschneider: Cell Tissue Res. **193**, 491–501 (1978)
423. X. Turon: Cah. Biol. Mar. **32**, 295–309 (1991)
424. M. Gianguzza, G. Dolcemascolo: Eur. Arch. Biol. **105**, 51–62 (1994)
425. V. Ernst: Tissue Cell **5**, 83–96 (1973)
426. V. Ernst: Tissue Cell **5**, 96–104 (1973)
427. U. Welsch, V. Storch, W. Fuchs: Cell Tissue Res. **148**, 407–416 (1974)
428. D.M. Green: Copeia **1981**, 790–796 (1981)
429. G. Hanna, W.J.P. Barnes: J. Exp. Biol. **155**, 103–125 (1990)
430. S.B. Emerson, D. Diehl: Biol. J. Linn. Soc. **13**, 199–216 (1980)
431. D.M. Green, J. Carson: J. Nat. Hist. **22**, 131–135 (1988)
432. A. Ohler: Asiatic Herpetological Research **6**, 85–96 (1995)
433. H. Schliemann, M. Hoeber: Natur Mus. **108**, 44–48 (1978)
434. H. Schliemann, C. Rehn: Z. Säugetierkd. **45**, 29–39 (1980)
435. H. Schliemann: Z. Wiss. Zool. Abt. A **181**, 353–400 (1970)
436. H. Schliemann: Funkt. Biol. Med. **2**, 169–177 (1983)
437. H. Schliemann: Natur Mus. **104**, 15–20 (1974)
438. H. Schliemann: Mitteilungen aus dem Hamburgischen Zoologischen Museum und Institut **72**, 249–259 (1975)
439. J.G.M. Thewissen, S.A. Etnier: J. Mammal. **76**, 925–936 (1995)
440. A.E. Sorensen: Annu. Rev. Ecol. Syst. **17**, 443–463 (1986)
441. M.G. Buonocore: "Bonding to hard dental tissues", in *Adhesion in Biological Systems*, ed. by R.S. Manly (Academic Press, New York, 1970), pp. 225–253
442. J. Sorensen, J. Condon: Journal of Dental Research **79**, 1602–1602 (2000)
443. S. He, Q. Liu, Y. Mai, R. Bryant: J. Mater. Sci. **7**, 611–616 (1996)
444. S.D. Eigenbrode: "Plant surface waxes and insect behaviour", in *Plant Cuticles - An Integral Functional Approach*, ed. by G. Kerstiens (BIOS, Oxford, 1996), pp. 201–222
445. S.D. Eigenbrode, K.E. Espelie: Ann. Rev. Entomol. **40**, 171–194 (1995)
446. W. Barthlott, C. Neinhuis: "Lotusblumen und Autolacke: Ultrastruktur pflanzlicher Grenzflächen und biomimetische unverschmutzbare Werkstoffe", in *4. Bionik - Kongress, München 1998*, ed. by W. Nachtigall, A. Wisser (Gustav Fischer Verlag, Stuttgart, 1998), pp. 281–293
447. W. Barthlott, C. Neinhuis: Planta **202**, 1–8 (1997)
448. P.J. Holloway: J. Sci. Food Agric. **20**, 124–128 (1969)
449. P.J. Holloway: "Plant Cuticles: Physicochemical Characteristics and Biosynthesis", in *Air Pollution and the Leaf Cuticle*, ed. by K.E. Percy, J.N. Cape, R. Jagels, C.J. Simpson (Springer-Verlag, Berlin, 1994), pp. 1–13
450. P.J. Holloway: Ann. Appl. Biol. **63**, 145–153 (1969)
451. T. Wagner, C. Neinhuis, W. Barthlott: Acta Zool.-Stockholm **77**, 213–225 (1996)

452. A.R. Parker, Z. Hegedus, R.A. Watts: Proc. R. Soc. Lond. B **265**, 811–815 (1998)
453. W. Büttiker, H.W. Krenn, J.F. Putterill: Zoomorphology **116**, 77–83 (1996)
454. S.N. Gorb: J. Morphol. **234**, 213–222 (1997)
455. H.R. Hepburn: "Structure of the Integument", in *Comprehensive Insect Physiology Biochemistry and Pharmacology*, ed. by G.A. Kerkut, L.I. Gilbert (Pergamon Press, Oxford, 1985), pp. 1–58
456. M. Locke: Science **147**, 295–298 (1965)
457. M. Locke: "The structure and formation of the integument in insects", in *The Physiology of Insecta*, ed. by M. Rockstein (Academic Press, New York, 1964), pp. 123–213
458. Y. Bouligand: Tissue Cell **18**, 621–643 (1986)
459. Y. Bouligand: Tissue Cell **4**, 189–217 (1972)
460. T. Weis-Fogh: J. Exp. Biol. **37**, 889–907 (1960)
461. T. Weis-Fogh: J. Mol. Biol. **3**, 648–667 (1961)
462. S.N. Gorb: Proc. R. Soc. Lond. B **266**, 525–535 (1999)
463. H. Mittelstaedt: Z. für Vergleichende Physiologie **32**, 442–463 (1950)
464. S.N. Gorb: Odonatologica **25**, 371–376 (1996)
465. S.N. Gorb: Zhurnal Evolutionnoj Biochimii i Physiologii **27**, 46–51 (1991)
466. S.N. Gorb: Vestn. Zool. **1989**, 62–67 (1989)
467. S.N. Gorb: Annual Journal of SGSPO **1**, 3–20 (1993)
468. S.N. Gorb: Zool. Zh. **69**, 148–154 (1990)
469. S.N. Gorb: Zool. Zh. **69**, 37–45 (1990)
470. S.N. Gorb: Vestn. Zool. **1990**, 59–62 (1990)
471. S.N. Gorb: Microsc. Res. Techniq. **37**, 583–591 (1997)
472. S.N. Gorb: J. Morphol. **234**, 1–10 (1997)
473. S.N. Gorb: Vestn. Zool. **1991**, 61–68 (1991)
474. H. Weber: *Biologie der Hemipteren* (Verlag von Julius Springer, Berlin, 1930)
475. A. Schrott: Berichte Naturwiss.-Med. Ver. Innsbruck **73**, 159–168 (1986)
476. F. Haas: Syst. Entomol. **20**, 85–98 (1995)
477. M. Baehr: Entomol. Gen. **6**, 311–333 (1980)
478. S.N. Gorb: Int. J. Insect Morphol. **27**, 205–225 (1998)
479. P.M. Hammond: "Wing-folding mechanism of beetles, with special reference to investigations of adephagan phylogeny (coleoptera)", in *Carabid beetles: Their Evolution, Natural History, and Classification*, ed. by T. Ervin, G.E. Ball, D.R. Whitehead (Junk Publishers: The Hague, Boston, 1989), pp. 113–180
480. G.A. Samuelson: "An elytron to body meshing mechanism of possible significance in the higher classification of chrysomelidae (coleoptera)", in *Proceedings of the Third International Symposium on the Chrysomelidae*, ed. by B. Publishers, Leiden (1994), pp. 136–147
481. G.A. Samuelson: "Binding sites: Elytron-to-body meshing structures of possible significance in the higher classification of chrysomeloidea", in *Chrysomelidae Biology, The Classification, Phylogeny and Genetics*, ed. by P.H.A. Jolivet, M.L. Cox (SPB Academic Publishing, Amsterdam, 1996), pp. 267–290
482. M. Karner: "Bifunctional muscles and interlocking mechanism in cantharis: How not to move the wings", in *Abstract Papers II. Biomechanic Workshop of the Studygroup Morphology*, ed. by R. Blickhan (Innovationskolleg Bewegungssysteme, Jena, 1997), pp. 191–192
483. B.N. Shvanvich: *General entomology* (Sovetskaja nauka, Moscow, Leningrad, 1949)
484. C. Walther: Z. für Vergleichende Physiologie **62**, 421–460 (1969)
485. S.C. Goel: J. Entomol. **46**, 167–173 (1972)

486. S.N. Gorb: J. Morphol. **230**, 219–230 (1996)
487. R. Conde-Boytel, E.H. Erickson, S.D. Carlson: Int. J. Insect Morphol. **18**, 59–69 (1989)
488. J. Heinzeller, P. Seifert, B. Aschauer: Verhandlungen der Deutschen Zoolog. Gesellschaft **82**, 260–260 (1989)
489. P. Seifert, T. Heinzeller: Zoomorphology **109**, 71–78 (1989)
490. K. Sander: Zool. Jahrb. Anat. **75**, 383–388 (1956)
491. S. Heilig, K. Sander: Zool. Jb. Syst. **113**, 307–317 (1986)
492. A.F. Emeljanov: Trudy VEO **69**, 19–109 (1987)
493. S.N. Gorb: "The jumping mechanism of cicada cercopis vulnerata: skeletal structures, musculature, frictional surfaces and kinematic model of leg movements", in *Abstract Papers I. International Conference on Motion Sytems*, ed. by R. Blickhan (Innovationskolleg Bewegungssysteme, Jena, 1997), pp. 121–122
494. T. West: Trans. Linn. Soc. Lond. **23**, 393–421 (1862)
495. H. Dewitz: Archiv für die Gesammte Physiologie des Menschen und der Thiere **33**, 440–481 (1884)
496. J.E. Rombouts: Zool. Anz. **7**, 619–623 (1884)
497. E. Bauchhenss, M. Renner: Int. J. Insect Morphol. **6**, 225–227 (1977)
498. E. Bauchhenss: Zoomorphologie **93**, 99–123 (1979)
499. G. Walker, A.B. Yule, J. Ratcliffe: J. Zool. (London) **205**, 297–307 (1985)
500. S.N. Gorb: Proc. R. Soc. Lond. B **265**, 747–752 (1998)
501. N.E. Stork: J. Exp. Biol. **88**, 91–107 (1980)
502. N.E. Stork: Zoological Journal of the Linnean Society **68**, 173–306 (1980)
503. N.E. Stork: J. Nat. Hist. **17**, 583–597 (1983)
504. A.M. Lees, J. Hardie: J. Exp. Biol. **136**, 209–228 (1988)
505. A.F.G. Dixon, P.C. Croghan, R.P. Gowing: J. Exp. Biol. **152**, 243–253 (1990)
506. J.D. Gillett: Entomologist **65**, 123–123 (1932)
507. M. Carver, G.F. Gross, T.E. Woodward: "Hemiptera", in *The insects of Australia*, ed. by C.U. Press, New York (1991), pp. 429–510
508. E.H. Slifer: Ann. Entomol. Soc. USA **43**, 173–188 (1950)
509. U.D. Kendall: Z. Zellforsch. Mik. Ana. **109**, 112–137 (1970)
510. W. Hennig: "Diptera (Zweiflügler)", in *Handbuch der Zoologie*, ed. by M. Beier (W. de Gruyter, Berlin, 1973), pp. 1–335
511. L.M. Roth, E.R. Willis: J. Exp. Biol. **119**, 483–517 (1952)
512. J.W. Arnold: Int. J. Insect Morphol. **3**, 317–334 (1974)
513. R.E. Snodgrass: *Anatomy of the honey bee* (Comstock Publishing Associates, New York, 1956)
514. W. Federle, K. Rohrseitz, B. Hölldobler: J. Exp. Biol. **203**, 505–512 (2000)
515. B.S. Heming: Can. J. Zool. **49**, 91–108 (1970)
516. B.S. Heming: Can. J. Zool. **50**, 751–766 (1972)
517. R. Beutel, S. Gorb: J. Zool. Syst. Evol. Res. (in press) (2000)
518. O. Breidbach: Mikrokosmos **69**, 200–201 (1980)
519. A.C. Neville: *Biology of the Arthropod Cuticle* (Springer-Verlag, Berlin, 1975)
520. A.G. Richards: Entomol. News **76**, 253–256 (1965)
521. H.R. Hepburn: J. Georgia Entomol. Soc. **4**, 159–167 (1969)
522. J.D. Gillett, V.B. Wigglesworth: Proc. R. Soc. Lond. B **111**, 364–376 (1932)
523. B. Henning: Z. Morphol. Anthropol. **79**, 323–342 (1974)
524. E.L. Brainerd: Am. Zool. **34**, 128–128 (1994)
525. W. Barthlott, N. Ehler: Trop. Subtrop. Pflanzenwelt **19**, 367–467 (1977)
526. W. Barthlott, C. Neinhuis, D. Cutler, F. Ditsch, I. Meusel, I. Theisen, H. Wilhelmi: Botanical Journal of the Linnean Society **126** (1998)

527. C. Neinhuis, M. Wolter, W. Barthlott: Z. Pflanzenk. Pflanzen. **99**, 542–549 (1992)
528. C.E. Kennedy: J. Ecol. Entomol. **11**, 291–300 (1986)
529. W. Federle, U. Maschwitz, B. Fiala, M. Riederer, B. Hölldobler: Oecologia **112**, 217–224 (1997)
530. S. Ishii: Appl. Entomol. Zool. **22**, 222–228 (1987)
531. J.S. Edwards, M. Tarkanian: Proc. R. Soc. Lond. **45**, 1–5 (1970)
532. I. Hasenfuss: Zoomorphology **87**, 51–64 (1977)
533. A. Kosaki, R. Yamaoka: Jpn. J. Appl. Entomol. **40**, 47–53 (1996)
534. J. Wallentin, M. Mondon, H. Stadler, S. Gorb, C. Ziegler: "The secretes of fly secretes: Adhesion properties probed by force-distance curves", in *Scanning-Probe Microscopes and Organic Materials VIII*, ed. by D.J. Müller, H.F. Knapp, Basel (1999), pp. 14–14
535. S.N. Gorb: Journal of Insect Physiology **44**, 1053–1061 (1998)
536. B. Bhushan, V.N. Koinkar: Wear **180**, 9–16 (1995)
537. M. Binggeli, R. Christoph, H.E. Hintermann: Tribol. Lett. **1**, 13–21 (1995)
538. A. Harnisch, D. Hülsenberg: Glass. Sci. Technol. **73**(3), 67–72 (2000)
539. D. Hülsenberg: Microelectron. J. **28**, 419–432 (1997)
540. T.J. Garrison, G.S. Settles, N. Narayanswami, D.D. Knight: AIAA J. **32**, 1234–1241 (1994)
541. R. Larsson, J. Lundberg: Wear **190**, 184–189 (1995)
542. P.M. Cann, H.A. Spikes: Acta Tribologica **2**(1), 45–50 (1993)
543. M.J. Downs, W.R.C. Rowley: Precis. Eng. **15**, 281–286 (1993)
544. O. Mollenhauer, A. Karguth, M. Scherge: PCT/DE 99/00981 **WO99/51965** (1999)
545. C.S. Yust: Tribol. Trans. **37**(1), 201–205 (1994)
546. X. Li, B. Bhushan: Thin Solid Film **340**, 210–217 (1999)
547. N. Ohmae, M. Tagawa: Jpn. J. Tribology **37**(3), 245–254 (1992)
548. H. Deng, T.W. Scharf, J.A. Barnard: J. Appl. Phys. **81**(8), 5396–5399 (1997)
549. F. Iwata, M. Kawaguchi, H. Aoyama, A. Sasaki: Thin Solid Films **302**, 122–126 (1997)
550. H. Kamusewitz, W. Possart, D. Paul: "Measurements of thermodynamic quantities by means of solid/water/vapour and solid/water/hexadecane systems", in *Polymer Surfaces and Interfaces: Characterization, Modification and Application*, ed. by K.L. Mittal, K.W. Lee (VSP, Netherlands, 1997), pp. 125–143
551. A. Kawai, K. Shimada, E. Andoh: Solid State Phenom. **65-66**, 191–194 (1999)
552. J.C. Berg: *Wettability* (Marcel Dekker, Inc., New York, 1993)
553. H.F. Knapp, A. Stemmer: Surf. Interface Anal. **27**, 324–331 (1999)
554. F. Rieutord, M. Salmeron: J. Phys. Chem. B **102**(20), 3941–3944 (1998)
555. D. Tabor, R. Winterton: Proc. R. Soc. Lond. A **312**, 435–450 (1969)
556. J.N. Israelachivili: J. Colloid Interface Sci. **44**(2), 259–273 (1973)
557. J.T. Yates Jr., V.S. Smentkowski, A. Linsebigler: "Surface science and extreme pressure lubrication-CCl_4 chemistry on Fe(110)", in *Fundamentals of Friction: Macroscopic and Microscopic Processes*, ed. by I.L. Singer, H.M. Pollock (Kluwer Academic Publishers, Netherlands, 1992), pp. 313–322
558. P.B. Merrill, S.S. Perry: Surf. Sci. **418**, 342–351 (1998)
559. R. Kaneko: Tribol. Int. **28**(1), 33–37 (1995)
560. B. Bhushan, S. Venkatesan: J. Mater. Res. **8**(7), 1611–1628 (1993)
561. S.C. Moulzolf, R.J. Lad, P.J. Blau: Thin Solid Films **347**, 220–225 (1999)
562. R. Kaneko: Jpn. J. Tribology **37**(11), 1335–1341 (1992)
563. E.M. Stuve, N. Kizhakevariam: J. Vac. Sci. Technol. A **11**(4), 2217–2224 (1993)
564. J.M. Martin, M. Belin: Thin Solid Films **236**, 173–179 (1993)

565. P.A. Gerin, P.B. Dengis, P.G. Rouxhet: J. Chem. Phys. **92**, 1043–1065 (1995)
566. P.B. Dengis, M.J. Genet, P.G. Rouxhet: Surf. Sci. Spectra **4**(1), 21–27 (1997)
567. B.W. Clare, P.J. Jennings, C.P. Lund, J.C.L. Cornish, G.T. Hefter: Thin Solid Films **326**, 160–165 (1998)
568. P.B. Dengis, P.G. Rouxhet: J. Microbiol. Meth. **26**, 171–183 (1996)
569. H. Bishop: *Auger Electron Spectroscopy* (Cambridge University Press, Cambridge, 1989)
570. B.C. Smith: *Fundamentals of Fourier Transform Infrared Spectroscopy* (CRC Press, Boca Raton, 1996)
571. H. Mantsch, D. Chapman: *Infrared Spectroscopy of Biomolecules* (Wiley-Liss, New York, 1996)
572. D.A. Myrzakozha, T. Hasegawa, J. Nishijo, T. Imae, Y. Ozaki: Surf. Sci. **427-428**, 107–110 (1999)
573. S. Ji, T. Jiang, K. Xu, S. Li: Appl. Surf. Sci. **133**, 231–238 (1998)
574. J. Pendry: *Low Energy Electron Diffraction* (Academic Press, New York, 1974)
575. M. van Hove, S. Tong: *Surface Crystallography by LEED* (Springer-Verlag, Berlin, 1979)
576. G. Ertl, J. Küppers: *Low Energy Electrons and Surface Chemistry* (VCH Publishers, New York, 1985)
577. M. Edelwirth, J. Freund, S. Sowerby, W.M. Heckl: Surf. Sci. **417**, 201–209 (1998)
578. E.R.E. van der Hage, T.L. Weeding, J.J. Boon: "Mass spectrometry of natural and synthetic lignin polymers: On the detection of coumaryl units by in-source pyrolysis mass spectrometry", in *Applications of Modern Mass Spectrometry in Plant Science Research*, ed. by R. Newton, T. Walton (Oxford University Press, Oxford, 1996), pp. 117–138
579. B. Bhushan, Y. Cheng: J. Appl. Phys. **81**(8), 5390–5395 (1997)
580. B. Bhushan: *Handbook of Micro/Nano Tribology*, 2nd edn. (CRC Press, Boca Raton, 1999)
581. E. Meyer: Progr. Surf. Sci. **41**, 3–49 (1992)
582. E. Meyer, H. Lüthi, L. Howald, H.J. Güntherodt: "Friction force microscopy", in *Forces in Scanning Probe Methods*, ed. by H.J. Güntherodt, Anselmetti, E. Meyer (Kluwer Academic Publishers, Netherlands, 1995), pp. 285–306
583. C. Rivetti, C. Walker, C. Bustamante: J. Mol. Biol. **280**, 41–59 (1998)
584. C. Rivetti, M. Guthold, C. Bustamante: J. Mol. Biol. **264**, 919–932 (1996)
585. I.T. Dorn, R. Eschrich, E. Seemüller, R. Guckenberger, R. Tampe: J. Mol. Biol. **288**, 1027–1036 (1999)
586. F. Jamitzky, R.W. Stark, G. Morfill, W.M. Heckl: Journal of computer-assisted microscopy ((submitted)) (2000)
587. N.C. Santos, E. Ter-Ovanesyan, J.A. Zasadzinski, M. Prieto, M.A.R.B. Castanho: Biophys. J. **75**, 1869–1873 (1998)
588. S. Magonov, M.H. Whangbo: *Surface Analysis with STM and AFM* (VCH Verlagsgesellschaft, Weinheim, 1996)
589. R. Wiesendanger: *Scanning Probe Microscopy: Analytical Methods* (Springer-Verlag, Berlin, 1998)
590. G.U. Lee, L.A. Chrisey, E. O'Ferral, D. Pilloff, N.H. Turner, R.J. Colton: Israel J. Chem. **36**, 81–87 (1996)
591. A. Noy, C.H. Sanders, D.V. Vezenov, S.S. Wong, C.M. Lieber: Langmuir **14**, 1508–1511 (1998)
592. M. Radmacher, M. Fritz, H.G. Hansma, P.K. Hansma: Science **265**, 1577–1579 (1994)
593. N.H. Thomson, M. Fritz, M. Radmacher, J.P. Cleveland, C.F. Schmidt, P.K. Hansma: Biophys. J. **70**, 2421–2431 (1996)

594. H.G. Hansma: J. Vac. Sci. Technol. B **14**(2), 1390–1394 (1996)
595. C. Rotsch, K. Jacobson, M. Radmacher: Proc. Natl. Acad. Sci. **96**, 921–926 (1999)
596. H. Lüthi, E. Meyer, H. Haefke, H.J. Güntherodt: Science **266**, 1979–1981 (1994)
597. A. Ikai: Surf. Sci. Rep. **26**(8), 261–332 (1996)
598. M. Radmacher, M. Fritz, H.G. Hansma, P.K. Hansma: Science **265**, 1577–1579 (1994)
599. C. Bustamante, C. Rivetti, D.J. Keller: Curr. Opin. Struct. Biol. **7**, 709–716 (1997)
600. K.A. Barbee, P.F. Davies, R. Lal: Circ. Res. **74**(1), 163–171 (1994)
601. Z. Shao: News Physiol. Sci. **14**, 142–149 (1999)
602. Z. Shao, J. Mou, D.M. Czajkowsky, J. Yang, J.Y. Yuan: Adv. Phys. **45**(1), 1–86 (1996)
603. G. Binnig, C. Rohrer, C. Gerber, E. Weibel: Phys. Rev. Lett. **49**, 57–61 (1982)
604. R. Wiesendanger, H.J. Güntherod: *Scanning Tunneling Microscopy*, 2nd edn. (Springer-Verlag, Berlin, 1995)
605. O. Marti, M. Amrein: *STM and SFM in Biology* (Academic Press, San Diego, 1993)
606. G.A.D. Briggs, A.J. Fisher: Surf. Sci. Rep. **33**(1/2), 1–82 (1999)
607. A.A. Bukharaev, E.A. Samarsky, V.M. Janduganov, N.V. Berdunov, P.G. Antonov: J. Vac. Sci. Technol. B **13**(3), 1274–1279 (1995)
608. J. Halbritter: Appl. Phys. A **66**, S181–S186 (1998)
609. A. Ohi, W. Mizutani, H. Tokumoto: J. Vac. Sci. Technol. B **13**(3), 1252–1255 (1995)
610. Z.L. Ma, N. Liu, W.B. Zhao, Q.J. Gu, X. Ge, Z.Q. Xue, S.J. Pang: J. Vac. Sci. Technol. B **13**(3), 1212–1215 (1995)
611. J.T. Woodward, J.A. Zasadzinski: Biophys. J. **72**, 964–976 (1997)
612. C. Bustamante, D. Keller: Phys. Today **12**, 32–38 (1995)
613. G. Binnig, C. Quate, C. Gerber: Phys. Rev. Lett. **56**, 930–933 (1986)
614. J.W. Carlson, A. Jonas, S.G. Sligar: Biophys. J. **73**, 1184–1189 (1997)
615. M. Radmacher, R.W. Tillmann, M. Fritz, H.E. Gaub: Science **257**, 1900–1905 (1992)
616. M. Radmacher: IEEE Eng. Med. Biol. pp. 47–57 (1997)
617. U. Schwarz, H. Haefke: Appl. Phys. A **59**, 33–40 (1994)
618. H. Hölscher, W. Raberg, U.D. Schwarz, A. Hasbach, K. Wandelt, R. Wiesendanger: Phys. Rev. B **59**(3), 1661–1664 (1999)
619. B. Bhushan, G.S. Blackman: J. Tribol. **113**, 452–457 (1991)
620. M. Farina, A. Schemmel, G. Weissmüller, R. Cruz, B. Kachar, P. Bisch: J. Struct. Biol. **125**, 39–49 (1999)
621. L. Howald, H. Haefke, H. Lüthi, E. Meyer, G. Gerth, H. Rudin, H.J. Güntherodt: Phys. Rev. B **49**, 5651–5656 (1994)
622. K. Fujii, Y. Fujihara, S. Sugisawa, K. Fukada, T. Kato, T. Seimiya: Jpn. J. Appl. Phys. **33**(6B), 3768–3770 (1994)
623. M. Ohta, Y. Sugawara, F. Osaka, S. Ohkouchi, M. Suzuki, S. Mishima, T. Okada, S. Morita: J. Vac. Sci. Technol. B **13**(3), 1265–1267 (1995)
624. A.L. Shluger, A.L. Rohl, R.M. Wilson, R. Williams: J. Vac. Sci. Technol. B **13**(3), 1155–1161 (1995)
625. A.A. Baker, W. Helbert, J. Sugiyama, M.J. Miles: J. Struct. Biol. **119**, 129–138 (1997)
626. H. Bluhm, U.D. Schwarz, R. Wiesendanger: Phys. Rev. B **57**(1), 161–169 (1998)

627. H. Bluhm, U.D. Schwarz, K.P. Meyer, R. Wiesendanger: Appl. Phys. A **61**, 525–533 (1995)
628. U.D. Schwarz, H. Hölscher: Phys. Bl. **54**(12), 1127–1130 (1998)
629. J. Ruan, B. Bhushan: J. Appl. Phys. **76**(1), 5022–5035 (1994)
630. J. Ruan, B. Bhushan: J. Tribol. **116**, 378–387 (1994)
631. A. Arnoldi, C.M. Kacher, E. Bäuerlein, M. Radmacher, M. Fritz: Appl. Phys. A **66**, 1–5 (1998)
632. S. Sheng, D.M. Czajkowsky, Z. Shao: J. Microsc. **196**(1), 1–5 (1999)
633. S.S. Wong: Nature **394**, 52–55 (1998)
634. S.J. Tans, M.H. Devoret, H. Dai, A. Thess, R.E. Smalley, L.J. Geerligs, C. Dekker: Nature **386**, 474–477 (1997)
635. S.S. Wong, J.D. Harper, P.T. Lansbury, C.M. Lieber: J. Am. Chem. Soc. **120**(3), 603–604 (1998)
636. G. Neubauer, S. Cohen, G. McClelland, D. Horne, C. Mate: Rev. Sci. Instrum. **61**, 2296–2308 (1990)
637. G. Meyer, N. Amer: Appl. Phys. Lett. **53**, 1045–1047 (1988)
638. O. Marti: "Instrumentation for scanning force microscopy and friction force microscopy", in *Micro/Nanotribology and its Applications*, ed. by B. Bhushan (Kluwer Academic Publishers, Netherlands, 1997), pp. 17–34
639. S. Grafström, J. Ackermann, T. Hagen, R. Neumann, O. Probst: J. Vac. Sci. Technol. B **12**(3), 1559–1565 (1994)
640. C. Ascoli, F. Dinelli, C. Frediani, D. Petracchi, M. Salerno: J. Vac. Sci. Technol. B **12**(3), 1642–1645 (1994)
641. C.J. Lu, Z. Jiang, D.B. Bogy, Y. Miyamoto: J. Tribol. **117**, 334–343 (1995)
642. R. Linnemann, T. Gotszalk, I.W. Rangelow, P. Dumania, E. Oesterschulze: J. Vac. Sci. Technol. B **14**(2), 856–860 (1996)
643. R. Kassing, E. Oesterschulze: "Sensors for scanning probe microscopy", in *Micro/Nanotribology and its Applications*, ed. by B. Bhushan (Kluwer Academic Publishers, Netherlands, 1997), pp. 35–54
644. U. Schwarz, H. Haefke, P. Reiman, H.J. Güntherodt: J. Microscopy **173**, 183–197 (1994)
645. P. Markiewicz, M.C. Goh: J. Vac. Sci. Technol. B **13**(3), 1115–1119 (1995)
646. U.D. Schwarz, O. Zwörner, P. Köster, R. Wiesendanger: J. Vac. Sci. Technol. B **15**(4), 1527–1530 (1997)
647. U.D. Schwarz, O. Zwörner, P. Köster, R. Wiesendanger: Phys. Rev. B **56**(11), 6997–7000 (1997)
648. P.E. Sheehan, C.M. Lieber: Science **272**, 1158–1161 (1996)
649. B. Bhushan, J. Ruan: J. Tribol. **116**, 389–397 (1994)
650. J.P. Aime, G. Couturier, R. Boisgard, L. Nony: Appl. Surf. Sci. **140**, 333–338 (1999)
651. B. Anczykowski, B. Gotsmann, H. Fuchs, J.P. Cleveland, V.B. Elings: Appl. Surf. Sci. **140**, 376–382 (1999)
652. G. Bar, R. Brandsch: Langmuir **14**, 7343–7347 (1998)
653. G. Bar, R. Brandsch, M.H. Whangbo: Surf. Sci. **422**, L192–L199 (1999)
654. R.W. Stark, T. Drobek, W.M. Heckl: Appl. Phys. Lett. **24**(22), 3296–3298 (1999)
655. T. Drobek, R.W. Stark, M. Gräber, W.M. Heckl: New J. Phys. **1**(22), 1–11 (1999)
656. G.C. Ratcliff, D.A. Erie: Appl. Phys. Lett. **72**(15), 1911–1913 (1998)
657. M.Q. Li: Appl. Phys. A **68**, 255–258 (1999)
658. D.M. Czajkowsky, M.J. Allen, V. Elings, Z. Shao: Ultramicroscopy **74**, 1–5 (1998)

659. A. Manivannan, C.R. Cabrera, A. Fujishima: J. Vac. Sci. Technol. B **13**(3), 1172–1177 (1995)
660. H. Suzuki, S. Mashiko: Thin Solid Films **331**, 176–180 (1998)
661. M. Radmacher, M. Fritz, P.K. Hansma: Biophys. J. **69**, 264–270 (1995)
662. J.T. Dickinson, K.H. Siek, K.W. Hipps: Tribol. Lett. **1**, 159–175 (1995)
663. S. Morita, Y. Sugawara: Appl. Surf. Sci. **140**, 406–410 (1999)
664. F.J. Giessibl: Science **267**, 68–71 (1995)
665. I.Y. Sokolov, G.S. Henderson, F.J. Wicks: Appl. Surf. Sci. **140**, 362–365 (1999)
666. D.M. Czajkowsky, Z. Shao: FEBS Lett. **430**, 51–54 (1998)
667. J. Yang, J. Mou, J.Y. Yuan, Z. Shao: J. Microsc. **182**(2), 106–113 (1996)
668. J.B.D. Green, A. Novoradovsky, G.U. Lee: Langmuir **15**(1), 238–243 (1999)
669. M. Luna, F. Rieutord, Q. Melman N. A. Dai, M. Salmeron: J. Phys. Chem. A **102**(34), 6793–6800 (1998)
670. M. Radmacher, R.W. Tillmann, H.E. Gaub: Biophys. J. **64**, 735–742 (1993)
671. C.D. Frisbie, L.F. Rozsnyai, A. Noy, M.S. Wrighton, C.M. Lieber: Science **265**, 2071–2074 (1994)
672. M. Rief, F. Oesterhelt, B. Heymann, H.E. Gaub: Science **275**, 1295–1297 (1997)
673. N.H. Thomson, S. Kasas, B. Smith, H.G. Hansma, P.K. Hansma: Langmuir **12**, 5905–5908 (1996)
674. F. Oesterhelt, M. Rief, H.E. Gaub: New J. Phys. **1**, 6.1–6.11 (1999)
675. H.F. Knapp, R. Guckenberger, A. Stemmer: Probe Microscopy **1**, 247–257 (1998)
676. S. Umemura, M. Igarashi, Y. Andoh, R. Kaneko, S. Aizawa, K. Noguchi, T. Dekura, A. Toda: J. Vac. Sci. Technol. B **16**(1), 38–42 (1998)
677. L. Howald, R. Lüthi, P. Guethner, H.J. Güntherodt: Z. Phys. B **93**, 267 (1993)
678. J.H. Hoh, A. Engel: Langmuir **9**, 3310–3312 (1993)
679. E.L. Florin, V.T. Moy, H.E. Gaub: Science **264**, 415–417 (1994)
680. G.U. Lee, L.A. Chrisey, R.J. Colton: Science **266**(4), 771–773 (1994)
681. G. Sagvolden, I. Giaever, J. Feder: Langmuir **14**, 5984–5987 (1998)
682. M. Rief, M. Gautel, F. Oesterhelt, J.M. Fernandez, H.E. Gaub: Science **276**, 1109–1112 (1997)
683. M. Guthold, X. Zhu, C. Rivetti, G. Yang, N.H. Thomson, S. Kasas, H.G. Hansma, B. Smith, P. Hansma, C. Bustamante: Biophys. J. **77**, 2284–2294 (1999)
684. S.S. Wong, H. Takano, M.D. Porter: Anal. Chem. **70**, 5209–5212 (1998)
685. W.M. Heckl: BIOforum Int. **22**(7-8), 425–429 (1999)
686. S. Thalhammer, R.W. Stark, S. Müller, J. Wienberg, W.M. Heckl: J. Struct. Biol. **119**, 232–237 (1997)
687. J.A. Hammerschmidt, W.L. Gladfelter, G. Haugstad: Macromolecules **32**, 3360 – 3367 (1999)
688. D.M. Czajkowsky, H. Iwamoto, T.L. Cover, Z. Shao: Proc. Natl. Acad. Sci. USA **96**, 2001–2006 (1999)
689. C. McHargue: "Surface mechanical properties using nanoindentation", in *Micro/Nanotribology and its Applications*, ed. by B. Bhushan (Kluwer Academic Publishers, Netherlands, 1997), pp. 467–492
690. J. Ruan, B. Bhushan: J. Mater. Res. **8**(12), 3019–3023 (1993)
691. U.G. Hofmann, C. Rotsch, W.J. Parak, M. Radmacher: J. Struct. Biol. **119**, 84–91 (1997)
692. W. Oliver, G. Pharr: J. Mater. Res. **7**, 1564 (1992)
693. J. Bozzola, L. Russell: *Electron Microscopy* (Jones and Bartlett Publishers, Boston, 1992)

694. A.G.E. Pearse: *Histochemistry: Theoretical and Applied* (Churchill Livingstone, Edinburgh, 1980)
695. A.C. Neville: *Biology of Fibrous Composites* (Cambridge University Press, Cambridge, 1993)
696. S.O. Andersen, T. Weis-Fogh: Adv. Insect. Physiol. **2**, 1–65 (1964)
697. S.N. Gorb: Naturwissenschaften **86**, 552–555 (1999)
698. S.F. Frazier, G.S. Larsen, D. Neff, L. Quimby, M. Carney, R.A. DiCaprio, S.N. Zill: J. Comp. Physiol. A **185**, 157–172 (1999)
699. B. Romeis: *Mikroskopische Technik* (Urban und Schwarzenberg, München, 1989)
700. D.H. Meissner, H. Schwarz: J. Electron Micr. Techn. **14**, 348–356 (1990)
701. H. Schwarz, B.M. Humbel: Inst. Phys. Conf. Ser. **93**, 543–544 (1988)
702. W. Bloom, D. Fawcett: *A Textbook of Histology* (W.B. Saunders Company, Philadelphia, London and Toronto, 1972)
703. D.P. Abrol: J. Anim. Morphol. Physl. **3**, 107–112 (1986)
704. S.N. Gorb, S. Anton, F.G. Barth: J. Morphol. **217**, 129–136 (1993)
705. A.R. Spurr: J. Ultra. Struct. R. **26**, 31–43 (1969)
706. K.C. Richardson, L. Jarett, E.H. Finke: Stain Technol. **35**, 313–325 (1960)
707. S.L. Tuxen: *Taxonomist's Glossary of Genitalia of Insects* (Munksgaard, Copenhagen, 1970)
708. A.G. Richards, P.A. Richards: Int. J. Insect Morphol. **8**, 143–157 (1979)
709. S.N. Gorb: J. Zool. (London) (in press) (2000)
710. S.N. Gorb: J. Morphol. **240**, 101–113 (1999)
711. M.H. Maxwell: J. Microsc. **112**, 253–255 (1978)
712. D.M. Martin, A.S. Hallsworth, T. Buckley: J. Microsc. **112**, 345–352 (1978)
713. A. Boyde, S.J. Jones: Microsc. Res. Techniq. **33**, 92–120 (1996)
714. J.F.V. Vincent: "Composites", in *Biomechanics - Materials: A Practical Approach*, ed. by J.F.V. Vincent (IRL Press, Oxford, New York, Tokyo, 1992), pp. 57–74
715. M.M. Giraud-Guille: Calcified Tissue Int. **42**, 167–180 (1988)
716. M.M. Giraud-Guille: Microsc. Res. Techniq. **27**, 420–428 (1994)
717. B. Satiat-Jeunemaître: Bull. Soc. Bot. France **136**, 87–98 (1989)
718. A.C. Neville, T. M.G., B. Zelazny: Tissue Cell **1**, 183–200 (1969)
719. R. Ker: "Tensile fibres: Strings and straps", in *Biomechanics - Materials: A Practical Approach*, ed. by J.F.V. Vincent (IRL Press, Oxford, 1992), pp. 75–98
720. J. Eylers: "Liquids", in *Biomechanics - Materials: A practical Approach*, ed. by J.F.V. Vincent (IRL Press, Oxford, 1992), pp. 9–32
721. J.F.V. Vincent: "Plants", in *Biomechanics - materials: A practical approach*, ed. by J.F.V. Vincent (IRL Press, Oxford, New York, Tokyo, 1992), pp. 165–192
722. S.N. Gorb, M. Scherge: Proc. R. Soc. Lond. B **267**(1449), 1239–1244 (2000)
723. Y. Jiao, M. Scherge, S.N. Gorb: Journ. Exp. Biol. **203**(12), 1887–1895 (2000)
724. M. Scherge, S.N. Gorb: J. Micromech. Microeng. **10**(1449), 359–364 (2000)
725. M. Scherge, S.N. Gorb: Tribol. Lett. **8**, 1–7 (2000)
726. R. Carpick, D. Ogletree, M. Salmeron: J. Colloid Interface Sci. **211**, 395–400 (1999)

Index

Location: http://www.springer.de/phys/

You are one **click** *away from a* **world of physics** *information!*

Come and visit **Springer's**

Physics Online Library

Books

- Search the Springer website catalogue
- Subscribe to our free alerting service for new books
- Look through the book series profiles

You want to order? Email to: orders@springer.de

Journals

- Get abstracts, ToC's free of charge to everyone
- Use our powerful search engine LINK Search
- Subscribe to our free alerting service LINK *Alert*
- Read full-text articles (available only to subscribers of the paper version of a journal)

You want to subscribe? Email to: subscriptions@springer.de

Electronic Media

- Get more information on our software and CD-ROMs

You have a question on an electronic product? Email to: helpdesk-em@springer.de

Bookmark now:

http://www.springer.de/phys/

Springer · Customer Service
Haberstr. 7 · 69126 Heidelberg, Germany
Tel: +49 (0) 6221 - 345 - 217/8
Fax: +49 (0) 6221 - 345 - 229 · e-mail: orders@springer.de
d&p · 6437.MNT/SFb

Springer

Druck: Strauss Offsetdruck, Mörlenbach
Verarbeitung: Schäffer, Grünstadt

Lightning Source UK Ltd.
Milton Keynes UK
UKOW030616270912

199719UK00001B/23/P